Marxian Economics

"The contribution of Japanese writers to the study of Marxist political economy remains underestimated in the Anglophone world . . . Among the important achievements of Japanese writers is the capacity to shed original light not only on the 'pure economics' of Marx's thinking but on his political economy in the broadest sense. Hiroshi Onishi's book is an invaluable addition to the canon which the late Makoto Itoh helped establish. It serves both as a textbook and as a historical introduction to the entire development of modern capitalist society from its pre-capitalist origins to the present day from a historical materialist perspective, which all scholars and researchers will welcome. An indispensable read."

—Alan Freeman, University of Manitoba

Marxian Economics
A New Japanese Tradition

Hiroshi Onishi

English-language edition first published 2023 by Pluto Press
New Wing, Somerset House, Strand, London WC2R 1LA
and Pluto Press Inc.
1930 Village Center Circle, Ste. 3-834, Las Vegas, NV 89134

www.plutobooks.com

Copyright © Hiroshi Onishi 2023

The right of Hiroshi Onishi to be identified as the author of this work has been asserted in accordance with the Copyright, Designs and Patents Act 1988.

British Library Cataloguing in Publication Data
A catalogue record for this book is available from the British Library

ISBN 978 0 7453 4718 9   Paperback
ISBN 978 0 7453 4721 9   PDF
ISBN 978 0 7453 4720 2   EPUB

This book is printed on paper suitable for recycling and made from fully managed and sustained forest sources. Logging, pulping and manufacturing processes are expected to conform to the environmental standards of the country of origin.

Typeset by Riverside Publishing Solutions, Salisbury, England

Printed in the United Kingdom

Contents

List of Tables and Figures   vii
Preface and Acknowledgments   xi

1   Human Beings in Marxist Materialism: Humans, Nature, and the Relations of Production   1
    1.1   Production as the Base   1
    1.2   A Materialistic Understanding of the Superstructure   18
    1.3   How We Can Understand the Determination of Superstructure by the Base   31

2   Capitalism as a Commodity Producing Society: The Quantitative Character of Capitalistic Production, and Capital as a Self-Valorizing Value   47
    2.1   Productivity to Generalize Commodity Production   47
    2.2   Product and Commodity: Commodity Exchange in a Concrete and Materialistic Concept of the Human Being   53
    2.3   Money as a Commodity: Production for Earning Money as a Concrete and Materialistic Concept of the Human Being   59
    2.4   Capital as Self-Valorizing Money: Production for Producing Surplus Value as Concrete and Materialistic Concepts of Enterprises   66

3   Capitalism in Industrial Societies: The Qualitative Character of Capitalistic Production and Capital as the Command Over Labor   71
    3.1   The Command Over Labor   71
    3.2   Changes in the Magnitude of Surplus Value   90
    3.3   Industrial Revolutions and Capitalistization in Non-Manufacturing Sectors   99

4   The Growth and Death of Capitalism: Accumulation Theory: A New Quality Created by Quantity   105
    4.1   The Birth, Growth, and Death of Capitalism: The Marxian Optimal Growth Model   105
    4.2   Conditions for Reproduction Without Accumulation   112
    4.3   Conversion of Surplus Value Into Capital   115
    4.4   General Law of Capitalist Accumulation: The End of Capitalist Accumulation   126
    4.5   Primitive Accumulation and State Capitalism   155

5	The Distribution of the Surplus Value Among Industries and to Non-productive Sectors	171
	5.1 The Subjects and Structure of *Capital*	171
	5.2 The Circulation, Turnover, and Social Reproduction of Capital: The Circulation of Capital	172
	5.3 The Conversion of Surplus Value into Profit, Interest, and Rent: The Process of Capitalist Production as a Whole	177
6	Precapitalistic Economic Formations	199
	6.1 Agriculture as Roundabout Production	199
	6.2 Slavery and Serfdom in Craft Production, Animal Husbandry, and Fishing	217
	6.3 The Part Played by Hunting in the Transition From Ape To Man	225

ADDENDA

I	The Fundamental Marxian Theorem (FMT) with Joint Production and Fixed Capital	231
II	Decentralized Market Model of the Marxian Optimal Growth Theory	235
III	Price Term of Reproduction Scheme	245
IV	Three-Sector Reproduction Scheme Including the Commercial Sector	251
V	Incorporating Class Dynamics in the Marxian Optimal Growth Model	255
VI	Converting the Analytical Marxist Model to a Labor Hire Model to Express Historical Trends of Disparities in Firm Sizes	263
Mathematical Appendix: How to Solve the Dynamic Optimization Problem		269

References   279
Index   287

Tables and Figures

TABLES

1.1	Situations in which the ruled class members cannot unite (Prisoner's Dilemma)	32
1.2	Situations in which ruled class members divide into revolutionaries and free riders (the game of chicken)	33
1.3	Situations in which all ruled class members consolidate and revolt (non-problematic)	33
1.4	Cases in which members of ruled class are satisfied with the present situation and do not revolt (non-problematic)	34
1.5	Payoff structure that determines the unity/free ride problem of the ruled class members	34
1.6	Transformed model to explain the changes in hegemonic countries in the world system	37
3.1	Capital lending and exploitation in Analytical Marxism	78
3.2	Trends in the proportion of union members by job category of the Fushimi branch of the All Kyoto Construction Workers' Union (%)	100
4.1	Decomposition of labor input in the optimal state of the Marxian optimal growth model	114
4.2	Decomposition of labor input in the growth process of the Marxian optimal growth model	129
4.3	Substitution of K^* and s^* for K and s in Table 4.2	131
4.4	Payoff structure of the two classes by their chosen attitudes	144
4.5	Two stages of capitalism and their ruling political parties	166
5.1	Relationship between the amount of invested capital by input factors and the annual turnover	176
5.2	Reproduction scheme in the level of price of production	180
6.1	Historical development of nomadic societies	218
A.1	Reproduction scheme of the price of production under the equalized return on invested capital	246
A.2	Reproduction scheme of the Marxian optimal growth model (simplified version) at the price level	247
A.3	Reproduction scheme of the Marxian optimal growth model (complete version with capital input in both sectors) at the price level	247

A.4	Conversion of capital lending in Table 3.1 to labor hire	264
A.5	Generalized case of Table A.4	265
A.6	Images of historical changes in disparity expansion and contraction	267

FIGURES

1.1	Productive activity as the relationship between humans and nature	3
1.2	Production of the means of production and final products	12
1.3	Use of the means of production as a roundabout production system	15
1.4	Social classes and their advocates	28
1.5	Various social conditions that determine the frequency of altruism	43
2.1	Commodity-producing society as a society maintained by product exchange	54
2.2	Commodity economy with and without money	60
3.1	Symbols for the fundamental Marxian theorem	73
3.2	A case of diminishing marginal productivity	76
3.3	A case of constant marginal productivity	76
3.4	Contestable exchange of wage and workers' effort	81
3.5	Mechanism of capitalist development	83
3.6	Logical structure of the analysis of capital and capitalism	89
3.7	Union participation ratio of the construction industry in Kyoto Prefecture in 1990	102
4.1	Redefinition of the Marxian optimal growth model	107
4.2	Long-term decline in net domestic investment ratio in Japan	117
4.3	Dynamics of capital accumulation toward a steady state	121
4.4	Improvement in input labor per unit in the Japanese economy (in terms of hour/standardized million yen in 2000)	128
4.5	Capital accumulation over time since the industrial revolution	140
4.6	The newer the Japanese expressway, the lower its efficiency	143
4.7	Catch-up process and income disparity between early starters and latecomer countries when both have the same subjective discount rate	147
4.8	Catch-up process and income disparity between early starters that need to reach a higher target and latecomer countries	149
4.9	Catch-up process and income disparity between early starters with a lower target to reach and latecomer countries	150

4.10	Growth path and income disparity between the rich and poor within one country	151
4.11	Income paths of the three classes during and after primitive accumulation	158
4.12	Cost structure of small farmers	161
4.13	Investment rate increases rapidly immediately following the industrial revolution	165
5.1	Infinitely repeated circulation of capital	173
5.2	Differential rent	189
5.3	Differential rent and absolute rent	191
5.4	Multiple transitional dynamics including bubble process in the Marxian optimal growth model	195
6.1	Ratio of total labor used for land accumulation and capital accumulation	205
6.2	Changed target of accumulation by replacing non-intensive agriculture with intensive agriculture	214

Preface and Acknowledgments

This book is an English translation of a textbook on Marxian economics designed for undergraduate students who have a background in modern economics. For some time I have been using this textbook at Keio University, which can be considered representative in Japan. Chinese and Korean editions were published in the 2010s, testifying to the importance that these countries still attach to Marxian economics. The text hence gives Western readers access to an East Asian standard for Marxian economics.

Mainstream economics divides the subject into more subdivisions than Marxian economics, such as "intermediate macroeconomics" and "intermediate microeconomics," which is one reason why I think we also need more advanced textbooks of Marxian economics. It uses mathematical tools selected from the basic framework of mainstream economics, which readers can be assumed to know, in line with my aim of competing with mainstream economics in the capitalist world of Japan. However, many of the unavoidable mathematical notations in this book are in most cases followed by relatively plain explanations, so readers who lack these mathematical tools can skip the mathematical parts.

Thus, in order to present a Marxian economics textbook that can be understood at the intermediate level and contains mathematical explanations, this book has summarized the contents of all three volumes of Marx's *Capital*, and also devoted a considerable amount of space to explaining the ideas of historical materialism. As you can see from the contents, Chapters 2 to 4 correspond to the contents of the first volume of *Capital* and Chapter 5 corresponds to the contents of the second and third volumes of *Capital*. Chapter 1 discusses the whole economic formation of society in historical materialist terms and Chapter 6 discusses pre-capitalistic economic formations. Unlike comparable textbooks, this one places considerable weight on explaining historical materialism and of the theory of surplus value that correspond with Marx's thought in *Capital*.

As Engels noted in his *Socialism: Utopian and Scientific*, historical materialism, together with the theory of surplus value, constitutes Marxism. Moreover, the theory of surplus value itself, which elucidates the essence of capitalism, is part of historical materialism. What makes capitalism as a class society different from other class societies is that there is an appearance or illusion of equality between capitalists and workers and the absence of exploitation. Therefore to reveal the appearance to be the illusion, is precisely the task that clarifying the historical position of capitalism.

MARXIAN ECONOMICS

Marx's *Capital* is a part of a project to depict the entire history of humankind and this project can be complete only when the whole history of humankind has been explained. This view is supported by the fact that Marx began to concentrate on historical research after he had almost finished the entire draft of *Capital*, and Engels did the same. This is the reason why we, who live after the publication of the three volumes of *Capital*, must focus on this field of research. Incidentally, I explain not only in Chapters 1 and 6, but in every chapter of this book, that any social institution is no more than a unique historical entity that arises only under specific conditions.

Another novel characteristic of this book is that it is not based on Keynesian economics but on neoclassical mainstream economics. This is different from the approach favored by other Marxian economists, who perceive neoclassical economics to be the main enemy, from the standpoint of anti-market economics. For this reason, I would like to make it clear that for Marx's *Capital* (the project of the theory of surplus value), it does not matter whether or not the market mechanism adjusts supply and demand smoothly. According to the fundamental Marxian theorem proved by Nobuo Okishio,[1] the greatest Marxist economist to be born in Japan, as long as profit exists, exploitation exists, no matter what mechanism is used to adjust prices. Therefore, we do not need any other mechanism to generate surplus, for example, price rigidity, externalities, information incompleteness, or "violence," which prevents the normal operation of the market mechanism. It was important to show in *Capital* that exploitation exists even in a pure market equilibrium with no extraordinary circumstances. This is why this book excludes all such particular factors and accepts the general equilibrium theory. Incidentally the world-renowned Japanese school of Marxian economics called "Uno Economics" also assumes that *Capital* must be conceptualized as a "theory of pure capitalism."

However, as long as we use a model that is usually considered to be non-Marxian, we must at least explain the reason for doing so. Thus, I must reiterate that the choice of each economic agent is explicitly stated in mainstream economics and that the concepts of the maximization of utility by individuals and the maximization of profit by capitalistic firms are materialistic in nature. Historical materialism assumes that humans are not driven by justice; instead, they are driven by interest, and the fact that the working class and the capitalist class have contradictory interests drives the history of capitalist societies.

As a matter of fact, this issue is also closely related to the importance of not overlooking the physical dimension in economics. Traditional Marxian economics tends to discuss various issues only in terms of their value dimension, but what workers want is not the quantity of labor that is involved in the consumption goods they receive, but the quantity of goods for consumption they receive in physical terms. For this reason, the utility function must not be written as $U = f(V)$: it should be $U = f(Q)$ where V refers to the value consumed

---

[1] He was also a Communist Party member and activist.

PREFACE AND ACKNOWLEDGMENTS

and Q to the quantity of goods consumed, and the production function that determines the productivity of labor must be written not in value terms (terms of embodied labor) but in physical terms.

As for the analytical framework of this book, I would like to add one more point on the Marxian optimal growth theory developed in Chapter 4 of this book. Although this model is also described as a decentralized market model in Addendum II at the end of the book, it is originally introduced in Chapter 4 as the social planner model, which shows the optimal path for a society as a whole, rather than the natural path of the actual economy. This is because this model aims to explain the core thesis of historical materialism: why capitalism arose, how it developed, and why it must die out in the end, although this thesis is often misunderstood. We think that only in this way can we prove the inevitability and the necessity of the death of capitalism.

As mentioned above, since this book discusses Marx's propositions using the language of mainstream economics, we have had many discussions with mainstream economists in the vicinity, and I am indebted to them for many valuable comments for correcting errors and model development. In particular, I express my sincere gratitude to my colleague at Keio University, Professor Satoshi Ohira, and my appreciation to Professor Yuuho Yamashita and my graduate students from Kyoto and Keio universities, with whom I have developed the Marxian optimal growth model.

Furthermore, I need to express my sincere gratitude to Professor Alan Freeman, one of my best friends in the field of mathematical Marxism, and to Pluto Books. Alan made many useful comments on my draft via Pluto Books and, guided by those comments, I have revised many of my descriptions. This book could not have been published without all their help.

Last but not least, I offer my sincere gratitude to the Keio Economic Society for its generous financial support for publishing this book.

Tokyo, Japan
July 2022

1
Human Beings in Marxist Materialism: Humans, Nature, and the Relations of Production

This book offers a fresh explanation of Marxist political economy as a part of Marx and Engels' theory of historical materialism. People often understand—and rightly so—that this is what differentiates Marxism from mainstream economics. The Marxist understanding of the characteristics of humans differs in crucial aspects from that of mainstream economics. For example, Marxist materialism also discusses human characteristics or mentality. We know that these can change at each historical stage. However, we must understand that materialism is at the core of Marx's understanding of human nature, and therefore, that the aim of humans is to make a better life for ourselves. We humans pursue our material interests; or our utility (called use value in Marxist economics). Therefore, the first objective of this book is to demonstrate that both Marxist economics and mainstream economics have the same basis: the notion of the materialistic human being. This is the only objective understanding we can have of human nature.

This book starts with the general understanding that a human being is a materialist entity in Marx's basic view of society. However, my explanation is necessarily not only materialistic but also dialectic. In other words, the point of this explanation is to explore the relationships between many categories. For example, we know labor, nature, production, and so on as individual categories with individual characteristics. However, what we do not know so well is which of these categories is the most fundamental, and which is the next most fundamental, and broadly, the logical relationship between each category. Therefore, readers need to read this book carefully and assess whether it may neglect any essential category and whether the relationships between the categories put forward here are correctly captured or not. Based on this methodology this chapter first explains historical and non-historical categories that are not specific to capitalism.

1.1 PRODUCTION AS THE BASE

1.1.1 Humans, Nature and Production: Labor as the Primal Factor of Production

The starting point of this book is the abstract human being as a species being. The existence of the species being presumes the act of living; a very abstract category that implies that the species being not only eats but also gathers or produces (that is, works).[1] What is important here is that "production" itself is not the purpose of humans but the means to live throughout history. In this way, Marxist materialism is based on the need to eat, or in other words, on material interest and utility. This is the essence and purpose of production and can be translated into the terminology of mainstream economics thus: the objective function of humans is utility maximization (the principle of utility).

The next point is the object of labor. Because it is lies beyond the subject of labor, it is part of nature. "Nature" includes all the objects that are processed during production, and in this sense, can sometimes include humans themselves. For example, the object of labor in the personal service industry—such as education, hairdressing, and entertainment—is not a non-human object, but the humans themselves. Furthermore, we can work for ourselves directly. For example, we can cut our own hair or study by ourselves. All these objects make up nature, which exists as the object of labor outside the subject of labor.

The point is that productive activity (labor) needs an object, which I call nature here; thus, the principal activity of the humans, namely, labor, is taking conscious steps to connect themselves with the nature that lies outside themselves. In other words, the relation between humans, nature, and production is the first primitive and fundamental relation that needs to be explained. Marxist economics calls this tripartite relationship the metabolisation between humans and nature.

Furthermore, if we observe the productive activity of humans, we can add one more very important factor: acquisition from nature. This is the direct result of productive activity, and by acquiring things from nature humans can live and derive utility. **Figure 1.1** shows the relation between humans and nature in terms of production and acquisition.

---

[1] Strictly speaking, the activity of living also includes defending, reproducing, and child-rearing. Here, while defense (and attack) can be regarded as a part of the widest meaning of productive activity, reproducing and child-rearing mean the "reproduction of the human being." Without this activity, our species would not continue to exist socially or historically. Marx distinguishes between productive and reproductive activities as the "(re)production of matter" and the "reproduction of life." Engels (1884) discusses the causal relationship between them in *The Origin of the Family, Private Property and the State*. He says that, at least so far as monogamous society, the "(re)production of matter" determined the "reproduction of life." That is, the social necessity to provide an heir for private property determined the form of marriage. Furthermore, the !Kung hunting society in the Kalahari Desert is important because in it male domination is based on their dominant role in hunting animals. This is also the expression of a materialistic determination based on the specific form of the family in this group.

HUMAN BEINGS IN MARXIST MATERIALISM

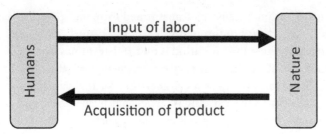

Figure 1.1 Productive activity as the relationship between humans and nature

Figure 1.1 shows that humans input their labor in nature and take something from nature in return. The human's purpose in engaging in labor is to acquire something from nature. If we express this relation quantitatively, we can say that the amount of work required in this activity depends on how much we can expect to acquire from nature. This is the marginal utility or productivity principle of mainstream economics. This can be explained in the mathematical language of mainstream economics. The non-mathematical reader may skip the next subsection.

1.1.2 The Marginal Utility Principle of Mainstream Economics

Let us assume that the disutility of marginal labor input is less than the utility derived from the marginal product. Where an additional marginal product gives us a net increase in "utility minus disutility" we produce more. However, if this additional product increases the disutility of the marginal labor input (increasing disutility) or diminishes the utility of the marginal product (the diminishing return of product), we reach an equilibrium in which the disutility of the marginal labor input would be equal to the utility derived from the marginal product.

In this case, if we continue to increase production, the disutility of the marginal labor input becomes larger than the utility gained by the marginal product. Hence, to increase production beyond this equilibrium point is irrational because it will decrease the net utility, defined as the utility gained by additional production minus the disutility from an additional labor input. Therefore, the humans should stop production when the disutility of the marginal labor input = the utility derived from the marginal product. In this case, the loss of utility by the expenditure of labor would be equal to the gain in utility by the goods produced and the net utility is maximized. If we now indicate the labor input as l, the disutility of the expense of labor as D, the produced goods as y, and the utility of consumption as U, we can express the relation above as

$$\frac{dD}{dl} = \frac{dU}{dy} \cdot \frac{dy}{dl} \qquad (1.1)$$

This formula is important because its transformed formula, $dD = dU$, means that the increase in utility and disutility through this labor are equal. This is the condition to maximize utility, and where this condition is met, it determines labor input (to nature).

Therefore, human metabolism with nature = production by labor, which is a proactive human activity that obtains utility as a consideration for the hardship undergone in the activity. In this case, humans make decisions on how much labor to expend. In other words, they decide the amount of labor they will expend by comparing the utility and disutility of that labor. The labor theory of value (LTV) is a way of thinking of value by measuring it in terms of the amount of embodied labor which is the total expense of human activity into nature. Although its founder, Adam Smith, did not have a marginalist theory, he regarded labor as "toil and troubles," or "disutility," and its total expense, as "the true price of everything" or value.

Here, we can express the value of one unit of product by the formula (l^*/y^*); that is, the ratio of optimized input labor (l^*) to the amount of goods produced (y^*).

What variables affect this unit value? We can answer this question mathematically by formulating the determination of the amount of labor input (= its value). Therefore, let us specify the relation that determines the amount of labor input. We can translate the equilibrium $dD = dU$, introduced above, into a maximization problem using the following utility function:

$$U = U(y, l) \quad \frac{\partial U}{\partial y} > 0, \frac{\partial U}{\partial l} < 0 \tag{1.2}$$

where y is the amount of expenditure on goods for consumption and l is the input of labor time. However, to translate this function into the most popular Cobb–Douglas form, we replace labor expense with free time as a variable explaining this utility function, as follows:

$$U = Cy^\alpha (H-l)^\beta \quad \frac{\partial U}{\partial y} > 0, \frac{\partial U}{\partial (H-l)} > 0 \tag{1.3}$$

where H is the total time available to the human (e.g., 24 hours in one day). Therefore $(H - l)$ is the free time, and α, β have the standard meaning for the Cobb–Douglas function. We assume they are positive, in order to set this function as an increasing function of consumption (y) and a decreasing function of l. Furthermore, to maintain consistency with the above explanation, we also assume $0 < \alpha, \beta < 1$, which means diminishing marginal utility of both factors.

On the other hand, because the amount of goods produced for consumption is also a function of the expense of labor, we can also specify the following production function:

$$y = Al^\gamma \tag{1.4}$$

where A and γ express different kinds of labor productivity and we assume $A > 0$ under $0 < \gamma \leq 1$. In this case, the above-mentioned maximization problem becomes the problem of choosing the optimal labor input (l) to maximize $U = Cy^{\alpha}(H-l)^{\beta}$.

Therefore, partially differentiating U with respect to l must yield zero. That is,

$$\frac{\partial U}{\partial l} = CA^{\alpha}\gamma\alpha l^{\gamma\alpha-1}(H-l)^{\beta} - CA^{\alpha}\beta l^{\gamma\alpha}(H-l)^{\beta-1} = 0 \qquad (1.5)$$
$$\Leftrightarrow CA^{\alpha}l^{\gamma\alpha-1}(H-l)^{\beta-1}\{\gamma\alpha(H-l) - \beta l\} = 0$$

We can simplify this expression as $\gamma\alpha H = (\gamma\alpha + \beta)l$. Therefore, the optimal labor input becomes

$$l^* = \frac{\gamma\alpha}{\gamma\alpha + \beta}H^2 \qquad (1.6)$$

Furthermore, we can calculate the value for one unit of product as follows:

$$\frac{l^*}{y^*} = \frac{\frac{\gamma\alpha H}{\gamma\alpha + \beta}}{A\left(\frac{\gamma\alpha H}{\gamma\alpha + \beta}\right)^{\gamma}} = \frac{1}{A}\left(\frac{\gamma\alpha H}{\gamma\alpha + \beta}\right)^{1-\gamma} \qquad (1.7)$$

which indicates that this value is affected by A, γ, H, and $\gamma\alpha/(\gamma\alpha + \beta) = 1/(1 + \beta/\gamma\alpha)$. In concrete terms:

(1) An increase in labor productivity A leads to a decrease in value for one unit of product.
(2) Although γ reflects another kind of labor productivity, its effects are not clear.[3]
(3) Although H is fixed in nature, if necessary, domestic labor can be cut by using household electric appliances, thus increasing the actual sense of free time, which is the

---

[2] Although it not the main point of this discussion, this equation also explains the "industrious revolution" in economic history that occurred (1) when new consumer goods such as tobacco, tea, and coffee appeared (2) in countries where it was challenging to create arable land. This is because condition (2) implies that factors other than labor are invariant in the production function, and condition (1) implies that the effect of direct consumption has increased in the utility function, which can be mathematically understood as a jump in α while A is invariant. At this time, the optimal labor expenditure per capita, l^*, rises. In other words, it becomes more "industrious". The effect of (1) was emphasized by de Vries (1975) for the pre-modern European period.

[3] Here, differentiating $\{\gamma\alpha H/(\gamma\alpha + \beta)\}^{1-\gamma}$ with respect to γ yields $\{\gamma\alpha H/(\gamma\alpha + \beta)\}^{1-\gamma}[-\log\{\gamma\alpha H/(\gamma\alpha + \beta)\} + (1-\gamma)\{\beta/(\gamma\alpha + \beta)\}]$. It shows, by a complex calculation, that a larger γ increases l^*/y^* if γ is close to zero, and decreases l^*/y^* if γ is close to one in the range $0 < \gamma \leq 1$.

same as an increase in H. In this case, the disutility of labor decreases and labor supply increases, and the value of one unit of product becomes larger.

(4) Because an increase in $\gamma\alpha/(\gamma\alpha+\beta)$ makes leisure time less important, the disutility of the expense of labor decreases and the value of one unit of product increases, as in the case of H.

These results show that value is determined by technical conditions (the parameters of the production function), such as (1) and (2), and the various conditions in the preferences, such as (3) and (4) (the parameters of the utility function). Additionally, since Marx did not conduct a detailed analysis on the disutility of labor, he discussed only (1) and (2) and did not explain the effects of (3) and (4). However, further examination shows that the case of $\gamma=1$ is the most Marxian situation in the following sense: the production function becomes

$$y = Al \tag{1.8}$$

which is a typical LTV situation since the input labor per unit product (l/y) is as constant as the inverse of labor productivity $1/A$. Furthermore, this situation can be a solution to the optimization problem when $\gamma=1$, independent of the various properties of the utility function. We can explain why Marx ignored the utility side, such as (3) and (4) in this way.

Here, because the condition $\gamma=1$ expresses constant labor productivity (in other words, the proportionality of the amount of product to the amount of input labor), we do not need a marginal productivity principle to explain how input labor (= value) is determined. The marginal utility principle is a sufficient explanation.

1.1.3 The Concept of Value

Though the concept of value did not arise before the spread of the market mechanism, the utility or disutility of labor has been the criterion of human activities throughout human history. As Marx said, "In all states of society, the labor time that it costs to produce the means of subsistence, must necessarily be an object of interest to mankind."[4] However, this criterion is historically specific.

Here, we are discussing society before it achieved a significantly high level of productivity. Marx calls such a society the "realm of freedom" in *Capital*, Volume 3, and distinguishes it from the "realm of necessity" in which "labor ... is determined by necessity and mundane considerations." This depiction is very interesting because he says here that the ultimate aim in the "realm of physical necessity" is "achieving this" (i.e., rationally regulating our interchange with nature) "with the least expenditure of energy and under conditions most

---

[4] Marx, 1887, 47.

favorable to, and worthy of, their human nature."[5] This is the most efficient expenditure of labor, which I describe above. It shows that Marx also understood that humans aim to minimize their labor expenditure compared to the utility, or "use value" in the "realm of necessity." In this description, Marx also says that the fundamental condition of the "realm of freedom" beyond the "realm of necessity" is a shortened working-day. Thus, it is also the minimization of the expenditure of labor.

Incidentally, it is true that establishing the category of value needed the generalized commodity production of the capitalist system. However, my explanation above implies that the disutility of labor input has been the measure of value from the beginning of human society, long before the spread of commodity production. In less complex societies, in which products did not appear as commodities, the disutility associated with any given labor input did not necessarily determine value, but just the amount of production. However, at all stages of history, human societies have always compared the utility and disutility to be gained by and needed for production.

In conclusion, the point here is that there is only one ultimate factor of input: labor. This is because humans are innately separate from nature and must rely upon their labor to consume; and that all of nature is just the object that must be processed. Nature has its own law of movement, as illustrated by the example of crop yields. William Petty, the first advocate of the LTV, expressed this point saying, "Labor is the Father and active principle of Wealth, as Lands are the Mother." Here, Petty's "land" is our "nature."[6] In this way, it is true that nature itself also has positive utility. However, we humans are the first to have some awareness of the laws of nature and to conduct our actions according to such laws. We generally acquire more from nature by the additional expense of labor in this way.

For example, when we plant crops we calculate and compare the expected amount of harvest and labor input to determine the volume of production, referring to the natural law that sunshine grows crops. Therefore, the law of nature applies to the side of nature, not to the human side, and humans put in some quantity of labor and take certain products from nature under this law. Thus, for humans as the subjects of production, labor is the only factor of input, and the ultimate purpose of labor is the additional utility gained by this productive activity.

Therefore, the value, which is the object to be measured by the activity as a proactive human activity, is nothing but labor value. Anti-Marxist critiques oppose this with various theories of value, such as the "sun power theory of value," the "oil theory of value", the "water theory of value," and the "land theory of value," and want to set them on par with the LTV. However, sun power, oil, water, and land are not criteria for measuring human

---

[5] Marx, 1907, 593.
[6] Marx (1887) quotes this sentence positively in Chapter 1, section 2, 31.

productive activities. If there were another type of subjective activity conducted by the sun, earth or clouds, we could assign to them a corresponding theory of value, but there is none. Subjective activities to produce goods are conducted only by humans by their nature.[7]

In addition, Marxian economics usually does not say that the purpose of production is utility. Marx's terminology is "use value," which is needed for all products and is therefore the purpose of production. However, this "use value" theory also means that producers should make customers derive utility from their products,[8] and therefore, "use value" and "utility" have the same meaning, at least in this context. Although Marxian economists thought that accepting the category "utility" necessarily denies the LTV, they missed the point that productive activity is the selection of the amount of labor input to balance the utility gained by production. Here, labor is the only input factor into nature, and therefore value can be measured only by the amount of that labor. This is because the amount of the input labor determined aims to balance the utility acquired by this exchange with nature. In this way, the LTV is a very natural understanding of value.

1.1.4 Humans, the Means of Production, Production: Mental and Physical Labor

However, the analyses above are still abstract, even if we have identified the basic triangle consisting of humans, nature, and production. This is because nature in the productive process includes not only the object of labor, but also the means of labor, such as tools and machines, and the means of production consist of both the object of labor and the means of labor. Furthermore, this fact tells us that the many so-called "natures" involved in productive activity are not pure nature, but processed nature, and this is the very characteristic of human labor that makes it different from the labor of other creatures. This distinction is very clear in the case of the means of labor.

---

[7] I add the words "by their nature" because the narrower sense of the purpose of productive activities included profit-maximizing in the capitalist era and surplus-product-maximizing in the feudal era. However, these optimization behaviors are also only special cases of balancing the input labor with the acquired utility. Matsuo (2007) argues that the methods to measure the amount of oil, water, or land used can be chosen by the "standpoints." However, it is not a matter of the standpoint but an issue of objective fact, because humans engage in productive activities subjectively. In this case, only the amount of their input labor can be the measure of value.

[8] Strictly speaking, use value is heterogeneous whereas utility is not. Use values are certainly quantified and Marx (1887) says this in its chapter 1, e.g., "tons of iron", "yards of linen" etc. But they are also differentiated, which utility is not. Therefore (for cardinal utility) we can add up the utility from wearing a yard of linen to that from driving around in a ton of iron. For ordinal utility, this is expressed as the fact that utility is a total ordering, but use value is partial ordering.

To show this point, I cite the following sentence by Marx. He emphasized the means of labor are an indicator of the degree of human development and how products are made, rather than of what products are made. That is to say,

> [H]ow they are made, and by what instruments, that enables us to distinguish different economic epochs. Instruments of labor not only supply a standard of the degree of development to which human labor has attained, but they are also indicators of the social conditions under which that labor is carried on.[9]

This shows that the means of labor divide human history because they indicate the relative development of human labor over time. This understanding comes from another recognition; namely, that the crucial aspect of human productive activity, as opposed to that of other creatures, is the human capacity for imagination and creativity. A bit before the above-quoted sentence, Marx said:

> A spider conducts operations that resemble those of a weaver, and a bee puts to shame many an architect in the construction of her cells. But what distinguishes the worst architect from the best of bees is this, that the architect raises his structure in imagination before he erects it in reality. At the end of every labor-process, we get a result that already existed in the imagination of the laborer at its commencement.[10]

This quotation does not contain the phrase, "the means of labor" explicitly, but nevertheless stresses the fact that laborers exercise their own imagination before engaging in labor and engage in production guided by this imagination. Furthermore, his insight is that such imaginative products are formed prior to creating the means of labor. The act of making a hammer assumes that it will be used to strike something. Making a bow and arrow implies that these tools will be used to shoot something. Thus, human tool-making relies on the profound imaginative ability of humans well before the stage in which the productive activity takes place. Furthermore, the earlier it is conceived and more profound the act of imagination, the more intelligent the means of labor. It is for this reason that Marx stated that the development of the means of labor determines the development of the human being.

Incidentally, some other primates use very primitive tools. For example, the Brazilian longtail monkey of the Brazilian highlands uses big stones to break walnuts, and some non-primates such as sea otters and Egyptian vultures together with some invertebrates such as octopuses use stones to break shells and ostrich eggs. These activities show that

---

[9] Marx, 1887, 128.
[10] Marx, 1887, 127.

they must also imagine the results of using these tools, but this ability is the mere ability to *use* tools and not to *make* tools. However, some exceptional special species can indeed *make* very simple tools. The first one is chimpanzees that can bend twigs, insert these implements into the holes of white ants, and earn a tasty treat. Another example is the New Caledonian crow, which folds twigs to take insects from the insides of the trees. These creatures' abilities are exceptional because they make tools. However, the level of complexity of these tools is completely different from that of humans. Thus, we can say that even the most inventive animals have crucially different levels of imaginative ability. Citing the words of Benjamin Franklin positively, Marx defined humans as tool-making animals.[11] Marx perhaps did not know of the above exceptions, but this is not crucial because we cannot compare the tools of chimpanzees and the New Caledonian crow with those of humans. Because the "[i]nstruments of labor ... supply a standard of the degree of development," it is enough that we focus on level of sophistication of the tool itself.

Furthermore, we again note that almost all human means of labor are artificial, unlike the tools of the Brazilian longtail monkey, sea otters, and Egyptian vultures. We have developed our tools artificially for an unimaginably long time, ever since the beginning of human history.

Let us imagine the world of seven million years ago, when we humans branched out from our common ancestors with chimpanzees. At that time, *Sahelanthropus tchadensis* had just started hunting, consuming carrion like the modern chimpanzee, and utilizing bipedal locomotion—and perhaps it is possible they used weapons for hunting and defense, because humans had lost their sharp nails and fangs[12] (as well as limbs and tails designed for climbing and living in trees). For this evolutionary reason, we could not live in safety without making weapons. We must remember that humans at the time did not know how to light fires and could not harvest and eat nuts, and grains. In this case, the acquisition of uncooked meat, including marrow and brains, was a very important means of sourcing protein. We can understand this importance when we watch the behaviors of gorillas in the jungle and the long digestive organs of herbivores in the savanna. They eat continually all day from morning to evening.

---

[11] Marx, 1887, 117. In the post-war period, Oakley (1959) developed this academic hypothesis. Although it lost popularity due to the observations of exceptional creatures mentioned above, Stout (2016) restores this hypothesis by developing a new neurological experimental archeology, suggesting that humans developed their brains and linguistic capability by inventing and developing tools.

[12] In fact, having smaller canine teeth is one of the crucial characteristics of Hominidae in paleoanthropology.

However, a much more important fact is that the tools that were made by humans, including their cooking technology, evolved very rapidly and at a much faster rate than their biological evolution. The development of tools accelerated the development of the human brain. Other animals, who do not and cannot have tools are unable to change their ways of life without the physical evolution of their own bodies. By contrast, we humans can achieve new lifestyles by transforming our tools, which have come an extension of our own bodies. This is the essence of Marx's words cited above. That is, humans have become creatures who transform tools as extensions of our own being and *evolve socially*. Therefore, from this point onwards, this creature started a history of undergoing social evolution without biological evolution. In this way, tools led to a decisive revolution.

Besides this revolutionary characteristic, tools have contributed another crucial change in our history. That is, we can distinguish between the makers and the users of tools—and identify a division of labor between tool makers and tool users. This division is much more basic than a horizontal division between hunters and gatherers or agriculture and manufacturing because the former is a vertical division between planning and practice in the course of production. It is also a division between mental and physical labor. On the other hand, compared with this vertical division, the horizontal division is simpler and may be compensated by reciprocity or exchange.

Let us consider that the existence of the means of production also deepens our understanding of the LTV by widening the category of nature. Here, the means of production are also a part of nature, even if only after created by humans. This relation can be shown as in **Figure 1.2**.

Figure 1.2 shows that the means of production are also part of nature, acquired from the nature produced by human labor. To demonstrate this relationship, I draw two black arrows to show how the entire flow of human labor goes into nature in the broad sense. Both black arrows represent inputs into nature, and the ultimate gains (acquisitions) from nature are only the "final products;" that is, the purpose of production. Thus, even if the production system has become a dual one, the essence of production is the relationship (exchange) between nature and humans.

This understanding is decisive because mainstream economics does not understand this point and thus denies the LTV. Mainstream economists simply consider two factors of production: labor and the means of production, and they miss that the ultimate factor of production of humans is labor alone. The LTV embraces the conclusion that the value of one unit of the final product consists both of the consumed part of the value of the means of production and of the amount of labor directly used to produce this final product. Marx called the former "dead labor" or "past labor" and the latter "living labor." Summing up, together they make up the total value of the product.

MARXIAN ECONOMICS

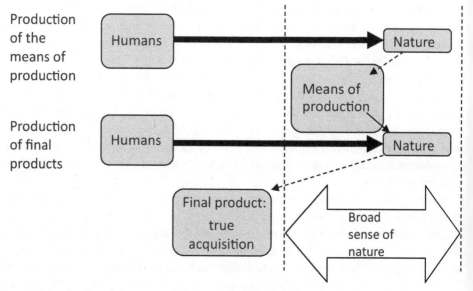

Figure 1.2 Production of the means of production and final products

On the other hand, counting dead labor as part of embodied labor has another implication for the LTV. This is because the proportionality between input (embodied) labor and the amount of substance seen as valuable is more visible in machine-based production systems where the means of production has become more important. I explain why below, but first, we need to confirm that "the substance that is seen as valuable" is usually called "value" and appears directly as its price. Thus, this proportional relation can be between the input (embodied) labor and "value" or the quantity of products, and this is the basis of the narrower sense of the LTV: the embodied labor theory of value. If we call this the LTV situation, we cannot find it using a simple production function without a means of production, as in $y = Al^\gamma$ in the former subsection 1.1.2. For this production function, the proportionality can be realized only when $\gamma = 1$, an unrealistic situation, and this situation is not consistent with the original realistic assumption in the former subsection 1.1.2.

However, if we introduce the means of production into the production function, this difficulty disappears. Before offering a full-fledged explanation of this new production function in Chapter 4, I want to introduce an easy formula for the production function that includes a means of production as "k" and provide an explanation:

$$y = Al_y^{\gamma_1} k_y^{\gamma_2} \tag{1.9}$$

Here, k is assumed to be a kind of flow variable, and we also assume $0 < \gamma_1 < 1$ and $0 < \gamma_2 < 1$, which indicate a diminishing return to both factors. Furthermore, we need to introduce one more production function for producing the means of production:

$$k = B l_k^{\gamma_3} k_k^{\gamma_4} \tag{1.10}$$

In this case, assuming $k = k_y + k_k = \varphi k + (1-\varphi)k$ $(0 \leq \varphi \leq 1)$, we can rewrite this equation as

$$k = B^{\frac{1}{1-\gamma_4}} l_k^{\frac{\gamma_3}{1-\gamma_4}} (1-\varphi)^{\frac{\gamma_4}{1-\gamma_4}} \tag{1.11}$$

and by inserting it into the former production function, we have

$$y = A l_y^{\gamma_1} \left(\varphi B^{\frac{1}{1-\gamma_4}} l_k^{\frac{\gamma_3}{1-\gamma_4}} (1-\varphi)^{\frac{\gamma_4}{1-\gamma_4}} \right)^{\gamma_2} = \varphi^{\gamma_2}(1-\varphi)^{\frac{\gamma_4 \gamma_2}{1-\gamma_4}} A B^{\frac{\gamma_2}{1-\gamma_4}} l_y^{\gamma_1} l_k^{\frac{\gamma_2 \gamma_3}{1-\gamma_4}} \tag{1.12}$$

In addition, if we express the total labor necessary to produce y as $l_y + l_k = l$, and set $l_y = sl$, $l_k = (1-s)l$ $(0 < s < 1)$, the above equation (1.12) becomes

$$y = \varphi^{\gamma_2}(1-\varphi)^{\frac{\gamma_4 \gamma_2}{1-\gamma_4}} A B^{\frac{\gamma_2}{1-\gamma_4}} s^{\gamma_1} l^{\gamma_1} (1-s)^{\frac{\gamma_2 \gamma_3}{1-\gamma_4}} l^{\frac{\gamma_2 \gamma_3}{1-\gamma_4}}$$

$$= \varphi^{\gamma_2}(1-\varphi)^{\frac{\gamma_4 \gamma_2}{1-\gamma_4}} A B^{\frac{\gamma_2}{1-\gamma_4}} s^{\gamma_1} (1-s)^{\frac{\gamma_2 \gamma_3}{1-\gamma_4}} l^{\gamma_1 + \frac{\gamma_2 \gamma_3}{1-\gamma_4}} \tag{1.13}$$[13]

Therefore, if both $\gamma_3/(1-\gamma_4)$ and $\gamma_1 + \gamma_2\gamma_3/(1-\gamma_4)$ can be one in both production function (1.11) and (1.13), then we can call it an LTV situation. In fact, this condition can be realized in the case of an constant return to scale, which is shown as $\gamma_1 + \gamma_2 = 1$ and $\gamma_3 + \gamma_4 = 1$[14],

---

[13] Strictly speaking, k is a stock variable, and the production function of the means of production should be written as

$$\dot{k} = B\{(1-s)l\}^{\gamma_3}\{(1-\phi)k\}^{\gamma_4} - \delta k$$

However, the production necessary to produce this good is only to produce y amounts of final goods, and therefore, we can set $\dot{k} = 0$, and the above production function ca be transformed into

$$k = \frac{B}{\delta}\{(1-s)l\}^{\gamma_3}\{(1-\phi)k\}^{\gamma_4}$$

In this case, this equation does not have any essential difference from the original production function in which k is treated as a flow variable.

[14] Contrary to this condition, it is possible that Marx assumed increasing returns to scale, as shown in Negishi (1985), Chapter 4. If so, this may differ from our assumption.

even if all γ_1, γ_2, γ_3 and γ_4 are larger than zero and smaller than one (that is, a diminishing return to each factor of production).[15]

1.1.5 A Special Stage of Production: The Roundabout Production System

I show above that the proportionality between the physical quantity of products, which is taken as their value, and the input (embodied) labor could become a reality, just as in capitalism, where the means of production plays a more important role in the production process than it did in pre-market societies. As I wrote in the previous subsection, there was no concept of commodity value in pre-market societies before capitalism; and the spread of the market system was the crucial condition for the modern concept of value to emerge. We now need to identify one condition for the proportionality between the physical quantity of a product and the input (embodied) labor introduced by the machine-based production system after the Industrial Revolution.[16]

The dual structure in Figure 1.2 can be shown differently, as shown in **Figure 1.3**, which simplifies the relationship between the two direct factors of production, the means of production and input labor. Here, the two factors of production serve as inputs in the production of a final product, and one of these factors is also the product of human labor. Therefore, we can understand that our total labor is divided into two sectors: one for producing the means of production and one for producing the means of consumption. This is a kind of roundabout production system in which a part of the total labor is consumed to prepare for future production, and the degree of circularity became increasingly deeper through history, first with the use of simple tools, then by using sophisticated tools, and finally by using machines. However, the main point is that humans have been tool-making

---

[15] Based on the concept of the minimum necessary investment, Onishi (2019) introduces each firm's horizontal long-term cost curve. Similar curves can be found in textbooks of microeconomics in terms of price, but my cost curve is based on input labor. Thus, Onishi (2019) proves there is a constant return to labor in the long run; that is, the proportionality between the physical amount of product and its embodied labor.

[16] In fact, the machine-based production system, which was decisive for capitalism, introduced another reason to create the concept of commodity value because it destroyed workers' skills and standardized human labor into simple labor and substantiated the concept of abstract human labor. It also became the material basis of the concept of commodity value. Thus, the concept of value is historical.

Additionally, this standpoint is contrary to Engels' supplement to *Capital*, Volume 3, where he states that the labor theory of value matches pre-capitalistic production only by labor without the means of production. Our standpoint is based on the assumption of diminishing returns to labor, which made the amount of goods produced by labor only non-proportional to the amount of labor input in the pre-capitalistic era.

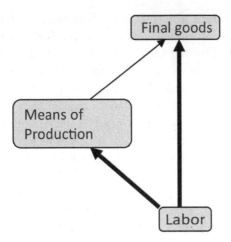

Figure 1.3 Use of the means of production as a roundabout production system

primates from the beginning and have produced goods by a roundabout production system. That is an essential characteristic of humans.

1.1.6 Humans, Means of Production Owned by Others, Production: Ownership, Class, and Relations of Production

As we have seen, the system of production is a process of exchange between humans and nature. The crucial element is the means of production, which should logically be produced before the final product can be produced. Therefore, as shown in Figure 1.2, the production of the means of production and the production of final products are separate: the producers of the final goods use tools or machines that were made by others. For example, carpenters must use saws and chisels purchased from others and fishers usually must have fishing tools made by others. However, they could buy such tools inexpensively.

However, what happens if the means of production become expensive—unlike saws, chisels, and fishing tools—and the producers are poor? For example, small-scale fishers generally lack the resources to own dragnets, and dragnets cannot be used individually. Their ownership is often shared or they are owned by a strong single fisher. This pattern of ownership became typical for machines after the Industrial Revolution. Machines used by many workers were often owned only by capitalists, apart from the few exceptions in which they were owned in common. In this way, as the means of production have increased in importance, so has the property system.

Furthermore, as this happens, we would expect to see a crucial gap between the interests between the owners of the means of production and the users of these means, or even outright conflicts between them. This is the *relations of production* Marx identifies, which typically involves the categories of capitalists and workers. The former are the rulers in the production process and the latter are the ruled. Of course, such class divisions have a much longer history. In the era of serfdom the most important factor of production—land—was "owned" dually by peasants and landlords,[17] and in the era of slavery,[18] the most important factor of production—the slaves—were owned by the masters. Thus, compared with the system of slavery in which living humans were owned by others, capitalism is better, because workers are independent. However, the status of the capitalist worker is devalued because they have lost control of the means of production, whereas the serfs retained this control.

Strictly speaking, serfs had lower levels of property rights, termed "occupancy," than their landlords, who retained upper property rights over the land. In this system peasants could independently control their own labor and land in the production process but could not be freed from the land because the land was the only and necessary condition for them to make their livelihood. This sense of indivisibility from the land is another crucial characteristic, which is called "bondage to the land." In this case, the peasant's bondage to the land was also the land's bondage to the peasants.

Now, since the means of productions have come to be owned by capitalists and the laboring class has become unable to live without using others' means of production, the situation changed completely. Now, the power of capitalists over labor has become absolute, based on property rights, and they can impose heavy rental charges on rented machines and set lower wages and longer labor hours as the owners of the means of production. The workers' bargaining power against the capitalists has become weak because capitalists can fire workers readily if their demands are not met. On the other hand, if workers are fired, they lose the means to live. These new situations were brought about by the new condition in which no one can produce without machines, and modern machines have become the fundamental basis of the production process. Marx termed this compulsive power of capitalists over labor the command over labor and the command over unpaid labor. This is the definition of capital that Marx provided in the conclusion of *Capital*, Volume 1, Chapter 18,[19] and it shows that capitalistic relations of production or class are, in fact, relations between the commander and the commanded.

It is useful to remember that in pre-capitalist societies (1) the means of production were less important in the production process and (2) they were owned by the workers (who were direct producers). In these societies, the most important factor of production was direct labor itself, not the means of production, and while craftspeople's skills were crucial, who owned the craftsmen's tools was not important. The only important thing was who was

---

[17] This view of the serfdom system is based on Nakamura (1977).

[18] At that time, tools were not so crucial and were usually very small.

[19] Another definition of capital by Marx is "self-valorizing value." See subsection 2.4.3.

skilled and could improve their skills further; it did not matter who controlled the means of production. In other words, the crucial factor of production was direct labor, indivisible from humans. Thus, humans themselves are the crucial factor of production when the kind of competition unique to a capitalist system does not exist. Furthermore, we can imagine another "human-oriented productivity society" in the next mode of production, in which humans are much more important than the means of production. New business trends demand distinctly human traits and abilities, such as creativity or individuality. This stress on human ability is completely different from that in the feudal mode of production, because the latter did not require creativity or individuality, but is also identical in that human ability is central to the production process in both cases. This is the condition of a post-capitalist society. The new society needs a new technological base that distinguishes it from the societies of the past. Thus, the present effort of our society to secure the right of education for all children can be understood not as a task of distributive equalization, but as a task to improve social productivity.

The second condition that I mentioned above—namely, that workers once owned the means of production—is also very important. Command over labor loosened under systems of state capitalism, such as in the former Soviet Union, Eastern Europe, and Mao's China,[20] where factory managers did not own the means of production. That is, "state ownership" was in reality a very ambiguous system of ownership and therefore factory managers could not be effective as decision-makers and managers. Thus, communist party members had to command labor in place of the managers and ensure that they were working hard. In other words, in this party-state system, command over labor was too weak. The autonomous management enterprises in Yugoslavia were similar in this respect because the ownership of the means of production was unclear. For these reasons, the "socialization of the means of production" cannot be done via state ownership or the autonomous management system in Yugoslavia. Instead, a type of the "socialization of the means of production" needs to be imagined that is radically different from all the past experiments in these state capitalist systems. We could suppose that a more realistic and positive path will include reforms of shareholding and stock listing systems to bring about a much more open and transparent system with strict compliance measures.[21]

Therefore, comparing the above-mentioned societies with ambiguous ownership property rights, we can understand clearly that only after the means of production became decisive in the production process and became owned by non-producers, did the means of production become the power to enforce the formers' command over labor. To describe these new relations, frictions, and conflicts, Marx presented a new conceptual category: the *alienation of labor*. This category captures the idea that although the means of production

---

[20] Later, we discuss why these systems should be called state capitalism.
[21] This point is discussed in subsection 4.4.6.

are made by human labor (the first sector in Figure 1.3), they stand outside humans and against them. As stated in subsection 1.1.4, Marx called the value transferred to the means of production dead or past labor, and the direct labor used to produce the final products living labor. Therefore, we can say that the alienation of labor consists of the rule of dead over living labor.[22] Furthermore, we must express this relation differently, because only the owners of the means of production can plan and command labor; while the direct producers, under this command and in a rote manner, merely put into practice the commander's plans. In this way, the *possibility* of a division between mental and physical labor was elevated to the *inevitability* of that division of labor. Here, we must remember that the imaginative ability before productive activity is an essential capacity of humans. When this aspect of our nature is stolen by dead labor, humans are reduced to being mere tools of those with command over labor.

1.2 A MATERIALISTIC UNDERSTANDING OF THE SUPERSTRUCTURE

1.2.1 Superstructure for Nature: The Productive Roles of the State and Ideology

The previous section focused exclusively on the productive activities of humans, but humans are active along other dimensions, such as through ideology, religion, culture, and politics. All these activities comprise a sort of superstructure, and are also very important, in addition to their strictly productive activities necessary to human life. This section discusses these activities. The first is the most primitive religion: the worship of nature before the formation of states.

The worship of nature consists of veneration of and awe for the sea and mountains, heavenly bodies such as the sun and the moon, and living things such as animals and trees. These natural elements were considered both blessings and threats, depending on the circumstance. This was because human's had little ability to control nature and humans themselves were basically ruled by nature. In this state, because human livelihoods depended on gathering and hunting, they asked for blessings and aimed to suppress nature's anger.[23] Therefore, we should understand that worship was viewed as indispensable work for their lives, and we can therefore regard it as a kind of productive activity. Of course, we now

---

[22] "The past labour, which dominates living labour" (Marx, 1907, 30).

[23] Copper bells and copper swords were used in ancient Japan. Copper bells were used to praise the gods and copper swords were used to expel demons. Zoroastrianism and Brahmanism also refer to gods of right and wrong. Therefore, the sense of right and wrong came about from the fact that nature provides humans with both good and bad things. Apart from this, the ancients were concerned with the reproduction of life and left cultural artifacts such as genital organs and clay figures with the image of the moon. The sun is of decisive importance in agricultural societies, but primitive societies emphasized the moon in relation to the reproduction of life. See Neumann (1963) for this point.

know that sacrifices were not effective in changing the ways of nature, but in the ancient days, we thought them effective and useful for creating a better life. Thus, this type of activity should be understood as both religious and productive. Religion and production were not differentiated. Even if someone was killed as a sacrifice, it was deemed a kind of necessary cultural and productive activity for the people to live.

As communities grew in size, religions became differentiated from productive activities and became ideologies for controlling societies. Increasingly, religions relied on special expert shamans who could communicate with the gods and therefore control nature, and they had the special task to hold sway over the whole society, which needed a more advanced and larger social formation for expanded social exchange and distribution, such as during the chiefdom period. In this case, the worship of the gods became translated into worship for shamans, who then became the leaders of their communities. If shamans demanded that war should be undertaken against neighboring communities, then the people went to war; if they required greater sacrifices, then people made more sacrifices. For example, the wars fought by the ancient Yin dynasty in China and the ancient Inca civilization were often waged to seize prisoners for human sacrifice from other communities. Therefore, in this historical stage, religions were the most important tools for controlling a growing society, and for this purpose various types of ritual ceremonies were brought into being.

However, the emergence of social leaders should not be confused with the formation of states. In my opinion, in order to recognize the formation of states we need another condition: the ability to trace the lines of descent of the social leaders. Without the concept of descent, social leaders could not be properly termed royal families. Archeological excavations sometimes find the bones of carefully buried infants, together with various treasures. This may indicate that the infants' status did not come from their talents, but from their parents' power; namely, via descent. Over the course of history, humans gradually differentiated between political orders and religious ones.

In fact, this new class of social leaders not only called for wars on religious grounds, but also commanded public works (as the general affairs of society) as purely productive activities (e.g., such as constructing levees or managing forests to counteract floods). These activities were not religious but productive, and therefore, social leaders sometimes had to hold community meetings and thus ordered the people to build spaces for such meetings. All these commands and the decisions made were both political and productive. In this way, political activities became clearly differentiated from religious ones. Here, we see the appearance of pure politics for a productive purpose.

This process advanced dramatically with the advent and development of agriculture.[24] Agricultural practices required scientific knowledge about the natural laws of rainfall,

---

[24] This characteristic was also true in the livestock industry because their productivity both depended on the growing cycles of plants. See subsection 6.2.2.

temperature and seasonality, which simple religions could not determine correctly. Without such knowledge, farmers could not work out when and how to plant and harvest agricultural products. This new and fundamental situation changed the objects of religions from nature to human relations. This shift marks the independence of natural science from religion. After this divorce, religions basically focused on how moral issues such as what kind of society was needed, or how people should behave. In this way, religions became representative ideologies that reflect various human social relations.[25]

A typical example is Confucianism, which originated to fulfill several feudal social needs. As we discuss later in subsection 3.1.6, because the feudal system of production depended on the use of human skills, it required a special type of human relationship, which gave rise to the concept of apprenticeship, for honing the skills of craftspeople. A central component of this process of skill formation was that the skill could not be mastered by scientific knowledge alone, but required the unswerving obedience of the apprentices to their older masters. Thus, Confucianism, which prioritizes seniority, respecting one's elders, and the value of loyalty to one's master, could support this special mode of production directly. This philosophy characterizes the Chu-tzu doctrine, which spread in Japan and strongly promoted the development of the feudal system of production in Japan.

The productive effects of religion can also be seen in the Protestant religions. Max Weber noticed the very individualistic and entrepreneurial characteristics of Protestant thought and argued that Protestantism was crucial to the formulation of capitalism in the Western world. Marx makes a similar point in his *Capital*:

> for a society based upon the production of commodities, in which the producers in general enter into social relations with one another by treating their products as commodities and values, whereby they reduce their individual private labour to the standard of homogeneous human labour—for such a society, Christianity with its *cultus* of abstract man, more especially in its bourgeois developments, Protestantism, Deism, &c., is the most fitting form of religion (Marx, 1887, 51).

Marx states here that Protestantism is suited to a society based on the production of commodities. Of course, we find fundamental differences between Weber and Marx, in that Marx states that societies formulate thoughts while Weber states that thoughts formulate societies. However, Marx's understanding also implies that these thoughts contribute to social stability. This is the meaning of "suited." Therefore, our discussion must address the question of how Protestantism expanded its influence in the early stages of capitalism

---

[25] Delayed by the establishment of the natural sciences, in the nineteenth century, Marxism presented another type of science of societies—social science—and reduced the space in which religion could work. Thus, many religions became antagonistic to Marxism.

In other words, had Protestantism appeared much earlier, it might not have expanded because it did not suit the modes of serfdom, slavery, or primitive communism. Only when there is a fit between a belief and the society at large can the former take hold and gain popular currency. Thus, we should understand that Protestantism did not make capitalism, but instead was itself made by the needs of capitalism.

In addition, we Japanese also had a sort of Reformation in the spread of Zen Buddhism in the Edo era. According to Yamamoto (1971), a new line of Buddhist thought held that all productive activities such as agriculture, manufacturing, and commerce are holy in the sense of being moral. However, because this doctrine did not praise profit-seeking behavior and admired only productive labor, we could regard it as a special ideology that strengthened the bases of both feudalism and capitalism.

We also need to discuss the relation between productivity and culture. Culture, distinct from religion and other types of thought, also plays a productive role in addition to its role in justifying class rule. For example, nomad culture is very well suited to the nomadic way of living.

Nomad peoples can breed herbivores only on large expanses of grassland, and therefore their houses need to be small and movable, and their furniture should be limited and easy to move. In light of these practicalities, nomads came to love frugally and hate the farmers' way of living with their more elaborate furniture and decorations. I can understand this feeling well because I have had a long and close connection with Chinese minorities in the Xinjiang Uygur Autonomous Region. Furthermore, this preference for having few belongings naturally breeds a love of multipurpose goods. For example, short knives were prized possessions. Nomad peoples used such knives for cooking, eating, and fighting. This culture was created by the nature of their needs and mode of productivity, and we must therefore understand that every culture has been created in a similar way. Each special characteristic of the mode of production determines the way of life of each different ethnic group; and in turn, their way of life determines people's tastes. Therefore, cultures are also productive in this sense. However, this also means that if we change the mode of production by, say, an agricultural or industrial revolution, we also need cultural revolutions to accompany these changes.

In addition to this explanation of how the quality or characteristics of productivity determine cultures, we need to explain how the quantity of productivity likewise determines a culture. For example, almost all modern people now wear the same type of "modern" clothing. This cultural revolution is also a result of the quantitative development of the productivity of the textile industry, which now is able to give us various sizes of clothing without extra expense according to our body type and enables us to move quickly. Japanese women fold the left front fold over the right when they wear a traditional kimono, and the extent of this fold of course depends on their own girth. From the perspective of production, Japanese traditional clothes like the kimono could be worn flexibly by people of different body sizes. The kimono sash is a clever device that adjusts the fit to all heights. The loincloth and turban were also contrived with the same flexibility: they are easily adjusted to bodies of all shapes and sizes.

Furthermore, the headhunting culture of tropical ethnic groups was also a result of their very low productivity. When I visited an Iban village in northern Borneo, which is famous for headhunting, they taught me of the old customary rule they enforce before a marriage could take place. The prospective groom had to hunt the head of a man in another village. Of course, they already discarded this rule by the time of my visit, but previously, they needed it to prevent overpopulation in their jungle. Because a hunting and gathering society cannot sustain a great population in a certain area, this tribe was forced to limit the number of married men who could have children. Therefore, they tested each adult man's ability to survive as a hunter to fight crocodiles in the river with bare feet, for instance. It was a good screening system.

However, what we must note here is that no human right to live had been articulated in these headhunting societies. Victims were killed without committing any crimes; but they did not raise any opposition to the practice because it was a custom. Only the strong were accepted in this society. Therefore, the absence of a human right to live was a necessary and inevitable social condition based on the low productivity of this society, showing that the notion of human rights is the historical gift of high productivity in modern society.

1.2.2 The Superstructure in Class Societies: The Class State and Ideology

Above examples could show us how the form of productivity determines the superstructure directly. However, this determination has become much more significant in the capitalist era, since ownership of the means of production has become crucial to production and the means of production are not owned by the direct producers (i.e., workers or farmers). This was a class relation forged by the new technology and this specific class relation has formulated its own special superstructure of culture, ideology, and politics.

Of course, it is not only social classes that are concerned with politics. Generally, different industries, regions, ethnic groups, and genders have different interests, and to secure their interests, they are all eagerly trying to access the political sphere. This is what happens in real and concrete societies, although modern economics identifies only the abstract differences in preferences. Differing interests are clear in the field of politics.

For example, the impact of the introduction of election systems on the different family groups in China is very interesting because many social groups have come to assert their special interests through this system (Onishi, 2011). In many cases, big clans are able to field their own candidates and mostly win against smaller clans. Thus, big clans can realize their individual interest by saying, "This is democracy," even if their interest is very special and damages the interests of others. Therefore, we can say that the essence of politics is that it is a different way of realizing special interests that cannot be realized in civil society. It is the same when the working class says, "Let's send our representative to Parliament." Politics is a special vehicle whereby each social group can realize its interests.

The question now arises: what kind of social groups can realize their interests through politics? Although every social group, such as industries, regions, ethnic groups, and genders seek to influence politics, they all have their own particular and different relations to the available means of production. For example, the most important difference in the interests among industries frequently lies between manufacturing and agriculture, and this sometimes leads to severe conflicts between rural and urban regions. Political struggles over free trade agreements, the Trans-Pacific Partnership, or agricultural protectionism are good examples of such conflicts. What we need to know here is that farmers constitute a social class that is different from the working class. They have their own small plots of land, at least in East Asia, while workers do not have any means of production. Therefore, farmers belong neither to the working class nor to the capitalist class that employs the working class. Thus, the essence of existing regional and industrial conflicts should be understood as a class struggle.

This is same in the case of ethnic conflicts. Above, I explained that nomad culture originated from its particular industrial characteristics. Therefore, ethnic conflicts between nomads and other peoples are essentially industrial conflicts. For example, on the plains of the Chinese Inner Mongolia Autonomous Region, Mongolians have their own land where they rear livestock to sell to the Han Chinese merchants. Here, Mongolians have an interest in selling their products at higher prices and Han Chinese have an interest in buying at lower prices. Therefore, if there is a conflict between them, it is an industrial conflict between nomads and merchants. Another type of ethnic conflict in China—such as those being played out in the Tibetan, Uyghur, and Inner Mongolia regions—is the conflict between capitalists and workers. In these areas almost all capitalists are Han Chinese and many or most of the workers are ethnic minorities. Therefore, class conflicts between capitalists and workers take the appearance of being ethnic conflicts between Han Chinese and minorities, although in essence they are class conflicts.[26]

Finally, we need to understand that gender issues are also closely related to class relations or working conditions, particularly in Japan. The strong public demand for support for day nurseries, elder care, and shortened working hours for those in full-time employment is, in reality, a demand for women's right to work. In fact, the most basic type of gender discrimination in Japan is the discrimination between full-time workers and non-regular workers.[27] Furthermore, Aoyagi (2010) claimed that throughout the history of gender relations and family systems, women have been oppressed to provide much population as slaves, serfs, and wage workers, and exploited by the ruling classes.

---

[26] See Onishi (2008, 2012).
[27] See the Institute for Fundamental Political Economy (1995).

MARXIAN ECONOMICS

Therefore, we take "classes" here to be the most representative social groups, each bearing different interests. By doing this we can see clearly that every class is seeking its interest in the political field by the use of political means. It is a struggle between the rich and the poor, because the propertied classes are generally rich. Struggles for or against progressive taxation, or for or against consumption tax, income tax, and property tax are examples of these struggles. Because each class formulates its own political party or supports an existing party to realize its interest, such class struggles must include power struggles among political parties or political groups.

The superstructure consists not only of politics and the state but also culture, thought, and ideology. That is, each class forms its own representatives in the areas of both politics and ideological struggle. When Fukushima nuclear power plant had exploded, Japanese people learnt that the capitalist government and electric power companies had employed several "scientists" as their mouthpieces to realize their interests by legitimizing their pro-nuclear policies. However, this is not only true in the case of nuclear issues: there are many other examples of conflicts of interest in a society, for example, trade policies, labor policies, taxation policies, education polices and so on. No theory that addresses social issues can be neutral in terms of social classes, for every social proposal affects the interests in one way or another of all the constituent groups of society.[28]

To explain this situation, I use the example of a small village that is to have a railroad built through it, connecting it with another neighboring town. Of course, we agree that the construction of a railroad is an example of historical progress, but the connectivity afforded by the railroad may destroy small retail merchants in a rural area by robbing them of their former customers. In this way, the interests of the majority of villagers can harm the interests of the minority, namely, the small retail merchants. Because human societies do not consist of homogeneous human beings, every social change and social proposal can represent the interests of only a part of the society—not all its social constituents. This means that each social group must act in their own interest by forming their own political parties and employing academic ideologues to speak for them. It is also possible to use the social sciences as a possible advocate for a social cause. No social science is neutral in the sense that it encompasses all social interests. Therefore, Marxism explains that ideological struggles, together with economic and political struggles, are all part of class struggles.

However, Marxism does not always support the working class. This is a very important principle that is usually misunderstood, and the point of contention is in the difference between "neutrality" and "objectivity." Let us consider the following passage in Marx:

---

[28] Strictly speaking, policies that realize net increases in production or utility can meet the interests of all social constituents by some form of perfect redistribution. This is the so-called win–win situation. In other words, if the situation is zero-sum, then confrontation among social constituents must become more and more severe. The present situation in advanced countries is the best example of this process.

In France and England, the bourgeoisie conquered political power. Thenceforth, the class struggle, practically as well as theoretically, took on more and more outspoken and threatening forms. It sounded the knell of scientific bourgeois economy. It was thenceforth no longer a question, whether this theorem or that was true, but whether it was useful to capital or harmful, expedient or inexpedient, politically dangerous or not. In place of disinterested inquirers, there were hired prize fighters; in place of genuine scientific research, the bad conscience and the evil intent of apologetic.[29]

This citation contains many implications. The first is that bourgeois economics was also scientific before a certain period, and the disinterested and genuine attitude that bourgeois economists assumed could make them representatives of the capitalist class. This was the state of bourgeois economics before a certain period. In other words, "disinterested" and "genuine" science can sometimes support a special social interest in certain periods and under certain conditions, but this attitude comes not from the *interested* attitudes of the scientists but from their *disinterested* attitudes. And in reality, after the bourgeoisie took political power, modern economics became an interested pseudo-science when it was coopted by the bourgeois.[30] Therefore, a disinterested attitude needed for objectivity in science.

From this viewpoint, I note that Marx did not oppose capitalism. He thought that all countries need capitalism at certain periods, and within these limitations, we must support the capitalist class. Similarly, the bourgeois economics that existed before capitalists assumed political power was revolutionary because it supported the capitalist class from a disinterested point of view. This is also true of Marxist economics. Even in Marxist economics, and even after capitalists have taken political power, it is not always necessary to oppose capitalism. Whether we should oppose it or not depends on whether capitalism is still progressive or has become reactionary. The Marxist assessment of a social system centers on whether or not it is still effective or useful for the growth of productivity. If capitalism is still good for growth, then Marxists must support and maintain this system. Nevertheless, even in this case Marxists must judge capitalism disinterestedly and the reasons why they might support capitalism might be quite different from those of bourgeois

---

[29] Marx, 1887, 11.

[30] Related to this point is the present position of Chinese Marxian economists. "Western economics" has become the mainstream economics taught in the Chinese academy, and therefore Marxist economists have no special treatment, at least in university economics departments. In my opinion, this guarantees their academic sincerity. However, having no advantage does not mean they face disadvantage. Unlike in Western countries, Marxists are not oppressed so greatly in China, and both Marxist and "Western economics," are properly balanced and coexist.

MARXIAN ECONOMICS

economists employed by the capitalist class. For example, Marxist political parties in China and Vietnam support capitalism. This is easy to understand because capitalism has been developing the economies of China and Vietnam effectively. However, we need to check whether their approach is disinterested or not; in other words, whether or not they lend too much support for capitalists.[31]

However, we must again note that to take such a disinterested attitude is very difficult. We could call it essentially impossible because the Marxist view of humans is materialistic. We cannot rightly say that scientists are exceptions to this general rule, while other people generally seek out their own interests. Instead, we need to understand objectively that scientists also seek their own interests and therefore look for clients who want to employ them. Only this way of understanding things is objective and disinterested.

This is the real situation of living scientists. For example, when various ideologues intentionally or unintentionally represent the interest of certain social groups, they cannot say that they are just mouthpieces. However, the Marxian understanding that scholars are essentially the representatives of certain interests relativizes all social theories, and therefore every interested ideologue must deny such a Marxian understanding. By all means, scholars need to say, "my conclusion is the result of neutral thinking." If they did not, then their ideological activity would not be useful to their employers and no one would employ them. Therefore, all scholars must surely pretend to be neutral. To do so, they create beautiful methods of analysis that look strictly neutral, and they formulate "objective" academic evaluation mechanisms such as peer reviews and award systems.[32] Needless to say, Marxists regard all these as deceptive.

Moreover, a very important point that we must not miss is that there are such representatives that act not only on behalf of the ruling class, but also the ruled class; the representatives of the ruled class must be generally warmhearted. The ruled and oppressed are generally weak and poor, and therefore basically it is only those with extremely generous spirits who want to represent their interests. Of course, professors can also recognize their own interests as wage earners, find that their interests are basically the same as other ruled people, and want to express their solidarity with them. However, the point here is that none of these activities is disinterested. As in the former example of the impact of

---

[31] From this viewpoint, the attitude of Chinese political leaders can be summarized in the phrase: "Jiang Zeming supported capitalists too much while Hu Jintao adjusted this bias and Xi Jinping changed to support the working class."

[32] Here, we discuss mainly academic ideologies, but the same is true for their religious counterparts. They need to feign neutrality by discussing morals, ethics, customs, and conventions. For example, in Tibetan Buddhism clockwise worshipping is holy while in Bon Buddhism counterclockwise worshipping is holy. Hinduism and Islam prohibit some foods and enforce fasts. All religions engage in ritual ceremonies. They are apparatuses to pretend neutrality.

a railroad construction in a small village, we cannot oppose such activities only because they harm someone's individual interest. If we need to oppose something only because someone's interest is injured, then we must oppose every policy and all societal change. From the beginning, Marxian ideology did not always support the weaker party, because Marxists supported (and still support) capitalism in its early stages. Onishi (2001) clearly asserts that Marxism is different from Leftism. Needless to say, Leftism defends the weak[33] but Marxism is an objective standpoint that must analyze society independently from the interest of the weak and the strong. **Figure 1.4** provides a simple diagram illustrating this relation.

Although this figure does not seem very special, it basically shows the Marxian understanding of society and the fact that Marxists analyze society objectively. Marxism is neither Advocate A nor Advocate B in Figure 1.4. These two advocates are the objects to be explained by Marxism. Marxism is not the object to be explained but the subject to explain. However, we should note here that Marxist professors also have their own interests as workers employed by universities, and they therefore organize trade unions to realize their interests. I served as the president of the Kyoto University Labor Union for a year and the Faculty and Staff Union of Japanese Universities for two years, and at that time, I functioned as the representative of the professors, not as a neutral social scientist. Therefore, we must recognize the difference between our special interest and the truth. As scientists, we should tell the truth, while as living persons, we have a right to look after our special interests. This relation typifies the very sensitive nature of the activities of Marxist intellectuals and should not be forgotten. Even if they are sincere about their science, sometimes, they need to act only to achieve their own special interests.

---

[33] The opposite perspective is Rightism; which would promote the railroad construction, neglecting the interest of the small retailers in the case mentioned previously. Rightists generally stand up for capitalism and give preference to capitalists asserting their leadership in terms of economic development. For social progress, this perspective is important and sometimes consistent with the Marxian view of society. This is the same way that Marxism sometimes agree in opinion and position with Leftism. Therefore, our problem is when Marxism is consistent with Rightism and when it is with Leftism, and the answer depends upon which standpoint is advantageous for economic development. Racial chauvinism such as Zionism, Yamato racism in Japan, and Han racism in China are examples of Rightism that assert the superiority of the ruling race, while nationalism against racism or imperialism is an example of Leftism. The former is the standpoint of the strong and the latter is that of the weak. Facing these ethnic conflicts, Marxists focus on how each ethnic group or individual develops productivity, and explains that if these groups resist productive development, they might disappear or be assimilated by stronger groups in certain instances. For example, the ancient Japanese Jomon people, who initially had only very primitive agriculture, were assimilated by the Yayoi people who brought rice cultivation from the continent. However, Marxism supports industrial protectionism when it works well for productive development.

MARXIAN ECONOMICS

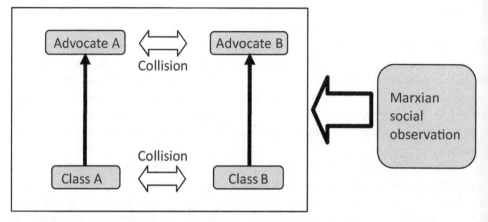

Figure 1.4 Social classes and their advocates

Furthermore, as mentioned above, the essential difficulty is that advocates cannot say that their assertions are not objective if they want to represent certain interests successfully. I myself did not say so when facing negotiating partners as the president of a trade union. Therefore, we Marxists have a dilemma: we need to claim we are disinterested while saying that all ideologies are interested. In other words, in this way, the Marxist understanding of the relation between interests and ideologues is very useful for criticizing other ideologues as frauds, but not for authorizing certain interests, because we understand that all ideologies (including our own) are interested. Hence, Leftists—as the protectors of the interests of the weak—sometimes consciously and sometimes unconsciously want to keep a distance from Marxism. Leftists essentially think that Marxism is outside their control, independent from their interests and bears the potential danger of becoming an enemy.

Therefore, true and objective social science has no supporters. Social groups do not need objective theories: they need only advocates. Marxism is useful only to relativize a counter-ideology, and true and objective scientists lack supporters and cannot maintain their scientific standpoint without aloof and proud souls. Japanese professors and universities are now under strong pressure from outside society to become "useful." This pressure is very damaging to the aims of objectivity because usefulness means being an advocate for someone. Social science is a science that analyzes certain objects objectively in the same manner as the natural sciences, and thus its usefulness or uselessness should not constitute a question. The only concerns should be truth and falsity or right and wrong—independent from the intrusion of the interests of society. Thus, Marxism is a scientism that value science as a science itself without being more than one itself.

In addition to professional scientists and sciences, we should also discuss religions here because they also function as advocates of certain social classes. Unlike the animist religions of less complex society, religions in class societies cannot be divorced from class issues. We must recognize that Buddhism, Christianity, and Islam have spread the principle of egalitarianism under their respective gods, because such thoughts advantaged the oppressed classes. For example, Muhammad was welcomed by the royal family for preaching polytheism but was later oppressed for preaching monotheism. However, this development was better for him in terms of garnering the support of the poor and slaves; with their support, Muhammad earned a great empire of believers. Similarly, Doi (1966) has described the liberatory characteristics of primitive Christianity. Furthermore, since the 1960s, a part of Christianity became central to liberation theology in Africa and Latin America. In seventh-century China Empress Wu Zetian used Buddhism to combat Confucian male chauvinism.

However, it is true that all three religions also functioned as the tools of the ruling classes. For example, the Japanese royal family introduced Buddhism from China in the sixth century to back strong state rule with a strong ideology. After the Roman Empire accepted Christianity, the faith was used to validate the rule of emperors, and then the Christian church itself became a ruling class in the Middle Ages. In Tibet, the notion of *samsara* transmigration in Tibetan Buddhism was used to justify the rule of the lords, claiming that incarnation is the reason why one person is a serf and another is a lord. In this way, various religions functioned to defend the interests of each social group and social class. This sentiment underlies Marx's famous statement, "religion is the opium of the masses."

1.2.3 The Productive Character of Class State and Ideology

The above discussion explains that state, politics, culture, thought, and ideology are determined directly from the mode of productivity and indirectly from class relations. However, the mode of productivity and class relations are themselves related. Although they seem to oppose each other, class relations in fact derive from the mode of productivity.

To understand this we need to see first, that a social class forms as the result of a new characteristic of productivity, in which the means of production has become crucial in the production process. Moreover, each relation of production, including class relations, in general matches each stage of history. As stated, if capitalism is appropriate, then we should maintain capitalism and support the capitalist class; when serfdom was appropriate, it was necessary to support serfdom and the feudal lords. Thus, the politics, culture, thought, and religions of the ruling classes also play useful roles in productivity. For example, a capitalist system cannot develop well without a someone to command labor in the production process. That is, advocates of the ruling classes also contribute indirectly to productive development as well as representing the interest of these classes.[34] Thus, sometimes sincere and disinterested scholars also support such policies.

However, this does not always mean that advocates of the ruled class are opposed to productivity simply because in revolutionary periods the ruled classes pushed history ahead. For example, the ruled class needed to improve their social status, and this perspective has been the engine that has driven the establishment of a more independent human personality: from a slavery-based system to serfdom and from serfdom to capitalism. In other words, if we humans want to secure a central position again in the production process, replacing the mere means of production, we should appeal to our creativity and individuality—traits that are becoming increasingly important in today's society and that cannot be provided by machines. The point here is that to develop this human ability, capitalists need to pay higher wages to laborers and shorten their working hours. I think that in some cases capitalists themselves do so of their own accord. However, in many cases, we need pressure from the working classes and their representatives. Thus, Leftists also have a historical role to play in service of productive development.

Therefore, in general, both the interests of the ruling classes and the ruled classes must be represented and promoted for the sake of historical progress. These two driving forces look like two vectors at 90-degree angles from one another; and whose interaction brings about what is known as historical progress.

This relation between productivity and class issues at the level of the superstructure essentially has the same characteristics as those revealed by the debate on whether the state has a "public" or merely a "class" function. In this argument, while some support public state theory by pointing to the state's public function, others support class state theory[35] by focusing on its function as a tool for the ruling class to oppress the people.

However, the fact that it has a class character does not mean that the state does not also serve productive development. When we need class societies, their formulation is necessary, inevitable, and productive. For example, we humans accumulated capital after the Industrial Revolution by supporting capitalists, and in so doing, our societies

---

[34] Marx expressed this as follows: "Indeed, it is only by dint of the most extravagant waste of individual development that the development of the human race is at all safeguarded and maintained in the epoch of history immediately preceding the conscious reorganisation of society" (Marx, 1907, 57).

[35] This is the standpoint taken by Engels (1880). He says that when specific classes hold state power this is called a "class dictatorship." For example, at the battle in Sekigahara in 1600 in Japan, the dictatorship was transferred from the Toyotomi to the Tokugawa family, but this was only a shift of power within the Samurai class. Although it changed from a one family dictatorship to another, it did not change from being a Samurai class dictatorship. The concept of class dictatorship enables us to understand the nation's characteristics in these dimensions.

developed productivity. In this case, the state's class and public characteristics do not contradict each other. This is clear when states undertake public construction for industries because this type of public function supports the capitalist class by promoting the accumulation of public capital. On the other hand, social welfare policies are also productive because they secure the reproduction of the workforce. Thus, proponents of public state theory must discuss whether the state is productive or not, instead of whether it is public or not. The point here is that the class characteristics and productive characteristics of the state are consistent with each other.

1.3 HOW WE CAN UNDERSTAND DETERMINATION OF THE SUPERSTRUCTURE BY THE BASE

1.3.1 Reaction of the Superstructure and Relative Autonomy I: Relative Autonomy in Political Movements

Although the previous sections point to the ways in which the superstructure is determined by the base, the superstructure can sometimes be autonomous and react to the base. This will be explained in this section, after we show how the class state has relative autonomy from its economic base in human history, using a type of game theory that was developed in mainstream economics.

We start by dispelling the idea held by many readers that game theory can be used only in mainstream economics. Contrary to the misunderstanding of this and other types of harmonistic equilibrium theory, game theory is useful for expressing the discordant results of individually rational behavior, such as the Prisoner's Dilemma. In this case, the rational choices made by two prisoners lead to the worst case for both. There are similar instances in political movements in which social actors face with two choices: whether unite or not.

In the game theory, the choice not to unite is regarded as an instance of free-riding, by which social members want to take the benefits won by other fighters without fighting themselves. Therefore, Marxian political revolution theory needs to look for the conditions in which many social members either want to unite or to free ride, and for this purpose, now we introduce a model based on game theory to explain why it is not easy to unite the majority for a political revolution and why most of the population has difficulty in uniting.[36]

---

[36] Binmore (2007) is a good textbook of general game theory for beginners, and this subsection's model uses the model of Muto (2015). See Onishi (2018) which extends the two-person model in this section into an N-person model. This shows that an increase in the number of members makes unity more difficult. Furthermore, Onishi (2020b) explains that this difficulty is increased if you start with an originally unitary political system and therefore Marxian political science introduces theories such as minority revolution theory, vanguard party theory, and ideological inculcation theory.

MARXIAN ECONOMICS

Table 1.1 Situations in which the ruled class members cannot unite (Prisoner's Dilemma)

		Choice of member B in the ruled class	
		Unite	Free ride
Choice of member A in the ruled class	Unite	68, 68	54, 81
	Free ride	81, 54	60, 60

For this purpose, we can first suppose for simplicity there are two ruled class members. These members face two choices: either to fight against the ruling class or to free ride (take benefits of social improvements won through the struggles of other members by no fighting). The typical case is shown in **Table 1.1**. Here, we consider the case of two players, which is the basic form of game theory. The revolution will succeed if both persons unite, gain only a partial improvement if only one person contributes to the movement, and gain no improvement if no-one participates in it.

In Table 1.1, the first of the two figures in each cell shows the payoff of member A in the ruled class in each case, and the second one shows the payoff of member B in that class. If both members choose to unite, then it is possible that both will benefit more than they would in the case of free riding, but under this payoff structure, freeriding is always more beneficial for each person when the other member selects any choice (to free ride even if the other member tries to unite, or to free ride if the other member tries to free ride). Therefore, here, the case of <free ride, free ride> out of the four combinations ([unite × free ride] × [unite × free ride]) might be selected socially as a "Nash equilibrium,"[37] and both members gain a payoff of 60. This is less than the gain of 68 that both sides can acquire by uniting, and in this way, this case becomes a Prisoner's Dilemma form of a social dilemma. In other words, in this situation, the ruled class cannot unite and cannot get out of that condition, despite their disadvantage.

However, in fact, there is an even worse situation, which is shown in **Table 1.2**. In this case, if the other member (player) free rides, the loss via non-participation in the movement becomes more severe than the loss in the <unite, free ride> or <free ride, unite> combination. In this case, the member should participate to protect their class interest even if there are many difficulties. It is also better for the member to support their class interest (in this case, the member can gain 72 points) than to free ride and obtain much lower gains (60). This represents the game of chicken, in which each member wants to make the other cooperate (unite) unilaterally. However, if the first member fails to make the other participate, and this second member is therefore forced to cooperate (unite)

---

[37] The Nash equilibrium is a category in game theory that refers to a combination of strategies. At this equilibrium, given the strategies of the other players, no player can gain more by changing their strategy.

HUMAN BEINGS IN MARXIST MATERIALISM

Table 1.2 Situations in which ruled class members divide into revolutionaries and free riders (chicken game)

		Choice of member B in the ruled class	
		Unite	Free ride
Choice of member A in the ruled class	Unite	104, 104	72, 108
	Free ride	108, 72	60, 60

unilaterally, resentment and envy will accumulate. In this case, benefiting without fighting (cooperating) becomes the true "freeriding position." In fact, most current Japanese trade unions face this situation.

While the above situation exists in reality, it is also true that the ruled classes have engaged in successful revolutionary revolts and overthrown old social systems in the past. However, this type of situation needs another kind of payoff structure, which is shown in **Table 1.3**. Here, the gain that can be acquired by uniting becomes very large (160 points each). Additionally, members who would otherwise attempt to free ride do not because the outcomes from uniting in this case are better. In other words, the fact that there have been many successful revolutions by the oppressed classes means that there were large-scale gains to achieve through a revolution. Situations where systemic changes are needed should be understood as coming under this payoff structure. If this situation emerges, then the members of the ruled class, which was not united until that time, will unite. In this case, it is also reasonable for the whole society to select the same choice (here, both players select unity), and take the maximum gain of 160 + 160. Thus, this situation is beyond the social dilemma situation and is called non-problematic.

In the other case, each member makes the same choice without hesitation. As shown in **Table 1.4**, the case in which both members unite (gains of 56, 56 in this case) is worse than when they cooperate with the ruling class without showing unity. That is, the combination of both freeriding becomes better for both members because both gains increase to 60, 60. In this case, the current social system basically works well, and if it is overthrown by a revolution, then it results in a much worse situation for all the ruled class members. Strictly speaking, even in this case, it is possible to obtain the maximum gain (72) personally by

Table 1.3 Situations in which all ruled class members consolidate and revolt (non-problematic)

		Choice of member B in the ruled class	
		Unite	Free ride
Choice of member A in the ruled class	Unite	160, 160	100, 150
	Free ride	150, 100	60, 60

MARXIAN ECONOMICS

Table 1.4 Cases in which members of ruled class are satisfied with the present situation and do not revolt (non-problematic)

		Choice of member B in the ruled class	
		Unite	Free ride
Choice of member A in the ruled class	Unite	56, 56	48, 72
	Free ride	72, 48	60, 60

freeriding if the other member chooses "unity." However, we can say this for both members. Consequently, both members select freeriding, and realize the <free ride, free ride> combination, which is better than the <unite, unite> combination. This is a difference from the case in Table 1.1, and this state in this sense is also classified as a "non-problematic situation."

In this way, the unity or free ride problem of members of the ruled class is defined in the payoff structure, but to make it clearer, let us show the payoff structure, not as a numerical example, but as a general form using some parameters. In doing so, we need to express both the cost of participating in the movement and the gain from making the ruling class concede. We set the payoff structure as shown in **Table 1.5**. Here, we express the original gain of each ruled class member before the revolution as "S (status quo)," and the additional gain to each ruled class member taken by additional one participant to the movement as "F (fruit)." This additional gain is assumed to increase in proportion to the number of participants in the movement; that is, both members' participation gives a greater benefit to each member, but the participation of only one member gives only a small benefit. The former case can be understood as a revolution, and the latter as reform. Furthermore, we express the cost of participating in the revolutionary movement as h, by assuming $0 < h < 1$ to reduce the individual gain. We use "h" here because the basic cost of such participation is the loss of the working time by consuming time for the movement. I obtained the four tables above by substituting $F = 21, 48, 90, 12$, respectively for $S = 60$ and $h = 2/3$. This expresses that the degree of development of the movement is determined by the balance between the cost and the benefit to each participant. In other words, the magnitude relationship between

Table 1.5 Payoff structure that determines the unity/free ride problem of the ruled class members

		Choice of member B in the ruled class	
		Unite	Free ride
Choice of member A in the ruled class	Unite	$h(S+2F), h(S+2F)$	$h(S+F), S+F$
	Free ride	$S+F, h(S+F)$	S, S

the present gains represented by (S, S) and the gains after the revolution represented by (h(S + F), h(S + 2F)), both of which depend on these three parameters, determines the success or failure of the revolution.

The difference in the above four situations can be understood as the difference in the following four situations determined by S, F, and h. In particular,

Situation (1) in which $\frac{2h}{1-h}F < S$ is the non-problematic situation shown in Table 1.4, all ruled class members are satisfied.

Situation (2) in which $\frac{h}{1-h}F < S < \frac{2h}{1-h}F$ is the Prisoner's Dilemma case shown in Table 1.1, no one cooperates.

Situation (3) in which $\frac{2h-1}{1-h}F < S < \frac{h}{1-h}F$ is the chicken game case shown in Table 1.2, the ruled class members divide into revolutionaries and free riders.

Situation (4) in which $S < \frac{2h-1}{1-h}F$ is the non-problematic situation shown in Table 1.3, all ruled class members voluntarily participate the movement (this case does not exist when h is 1/2 or less).

This result is very interesting for historical materialists, because the economic foundation and superstructure respond accurately in situations (1) and (4), in that the desired state for the whole society is acquired by the voluntary selection of all the members of the society. However, in situations (2) and (3) the response is inaccurate. In other words, under certain conditions, the superstructure cannot correspond to the economic foundation, showing the relative autonomy of the superstructure from the economic foundation. This is an expression that the subjective conditions are not mature in this case, even if, objectively, social change is required. This model shows that the improvement of the situation expected after the revolution is still not sufficiently higher than the cost of a revolution. In other words, compared to the current situation, extremely bad conditions are necessary for a revolution. In Marxist terminology, this requires the considerable deepening of social contradictions.

Of course, ruling classes do not keep silent and do nothing. For example, they can raise the cost to participate in the movement by discriminating against union members using wages or employment as tools. When this happens, conversion to the new system becomes more difficult if the ruled class members make decisions individually, and therefore, there might be social pressure to prevent individual decision-making among the ruled class and to justify the absolute dominance of the revolutionary parties in controlling the whole social movement of the ruled class. In such cases, the form of revolution becomes violent, like the French Revolution, the October Revolution, and the Meiji Restoration. In this way, both the ruling and ruled classes may become violent and create a social imbroglio. We can see that it is easy to demonstrate the relative autonomy of the superstructure from the economic foundation using a model, but at the same time, we have to know the real difficulty of solving such confusions in reality.

This explains how social movements respond to the base in the theory of historical materialism. The process by which the state, as a part of the superstructure, shifts from policies that support the existing ruling class to those that support the new ruling class can also be analyzed by setting up a somewhat similar payoff structure. For example, when we take "land policy" as Policy A and "tax policy" as Policy B, even if both policies initially supported the existing ruling class, we can imagine a subsequent stage where only one of the policies would be changed to support the new ruling class, and another stage where both policies would be changed to support it. Using the same symbols as in this section, this payoff to the state at each stage would be

- when both policies support the existing ruling class: S
- when only one of the policies supports the existing ruling class: $h(S+F)$
- when both policies support the new ruling class: $h^2(S+F)$.

The h^2 in the last payoffs indicates an assumption that the policy change has twofold a negative impact on the payoff. Here, both situations can exist, $h(S+F) < h^2(S+2F)$ and $h(S+F) > h^2(S+2F)$, but in any case:

(1) If $S < h^2(S+2F)$, i.e., $S < \dfrac{2F}{1-h^2}$, the state changes both policies to support the new ruling class.

(2) If $h^2(S+2F) < h(S+2F)$ and $h^2(S+2F) < S < h(S+2F)$, i.e., $\dfrac{2F}{1-h^2} < S < \dfrac{F}{1-h}$, the state changes only one policy to support the new ruling class.

(3) If in neither of these cases is $\max\{h(S+F), h^2(S+2F)\} < S$, i.e., $\max\left\{\dfrac{F}{1-h}, \dfrac{2F}{1-h^2}\right\} < S$, the state supports the existing ruling class for both policies.

In the same way as described in this subsection, the change in policy accompanies the relative deterioration of the current situation S. In this case, the class character of the state is clear in stages (1) and (3), but the class character in the transitional process (2) becomes ambiguous. This situation must be autonomous from the class relations in the base. Weber's (1924) description of the Roman state at the end of the Empire as a liturgical state that was autonomous from society was a typical example of the process of transformation from slavery to serfdom. Note that the policy that is changed first in the transitional process in (2) depends on the values of F and h, which should differ for each policy. Similarly with the former part of this subsection, we could confirm that the autonomy of the superstructure also arises in this way.

Table 1.6 Transformed model to explain the changes in hegemonic countries in the world system

		Selection of non-hegemonic country B	
		Subordinate to new hegemon country C	Subordinate to old hegemonic country U
Selection of non-hegemonic country A	Subordinate to new hegemonic country C	$h(S+2F), h(S+2F)$	$h(S+F), S+F$
	Subordinate to old hegemonic country U	$S+F, h(S+F)$	S, S

In addition, this model can be applied to analyze the determination of the political power balance by the economic power balance in the world system. For example, let us imagine two non-hegemonic countries that must defer to both of two hegemonic countries. We call the old hegemon U and the new, emerging hegemon C in **Table 1.6**, which is a transformation of Table 1.5. In this case, the non-hegemonic countries consider whether to change their suzerain or not by measuring the benefits (such as to eliminate some of the burdens imposed by hegemon U) and demerits (such as being attacked by hegemon U) of the change. It is especially important that the anticipated benefit depends on the number of countries subordinated to the emerging hegemon C in this model. This is an assumption that the emerging hegemon C can bring its benefits to non-hegemonic countries by its internationally influential power which is strengthened by the number of subordinating non-hegemonic countries. This is a hegemonic system, and as the result of this formulation, this model can explain the possible periods of instability, like the situations (2) or (3) listed above in the course of hegemonic change. Additionally, the painful periods of hegemonic change could be explained by situation (2) (loss by sticking to the old hegemon U) and situation (3) (when the world is divided into two major power spheres and conflicts). This case is very realistic.[38]

1.3.2 Reaction of the Superstructure and Relative Autonomy II: Neutral Appearance of Ideologies

The relative autonomy of this superstructure exists not only in the political movement described above but also at the level of thought and ideology. For example, we know that each mode of life—determined by the mode of production—is recognized as its culture; but why is this so? In the case of nomadic peoples, their special attitudes valuing simplicity might be obtained by each social constituent without this value having been passed down the generations. In other words, to hand it down to the next generation may be a kind of

---

[38] See Onishi (2020a).

compulsion in way of living. In a society in which everything changes rapidly this compulsion can be thwarted and a generation gap is a type of conflict that arises.

However, if we assume that social change is not particularly rapid, then enforcing social values from one generation to the next may be avoidable because everyone may adopt these social values by themselves without undergoing the painful process of trial and error, and this fits their ways of life. For example, this holds true for feudal societies that do not undergo rapid change, but if such a society starts to transform itself into to a capitalist society full of changes, then its culture becomes old-fashioned and some type of cultural revolution generally arises. Conversely, a culture has the tendency to maintain the old society against any change in the social base. The conservative function of culture implies that the superstructure is relative autonomous.

A slightly different example of the relative autonomy of the superstructure is the neutrality of ideologues. We have already discussed in the previous section about scientists and ideologies why ideologues need to appear neutral to function as advocates of certain social interests. Therefore, these individuals claim neutrality to the extent beyond which they cannot represent their clients' interests. This implies that ideologies and ideologues also have a certain autonomy even if they are ultimately determined by the base.

The same issue is also present in the relationship between the public and the class characteristics of states. This is because they cannot work properly as a tool of the ruling class if they do not appear to serve the entire public at large. In this way, a state's public characteristics make them relatively autonomous.

Summing up these relations, we can discuss "justice" as a criterion of human activity that differs from "interests," and we may use this criterion to assert something instead of stating our interests directly. This is because justice is the very opposite of interest and we need to say to our opposition, "Our claim is not based on our interest but on justice." For example, when we oppose a war or corruption, these demands in fact arise from our interest because this war is not good for us, and corrupt politicians always oppose our interests. However, we usually say that our overriding concern is justice. Therefore, the category of justice is a typical cover for hiding the true reason why we support or oppose something. This confirms the materialistic understanding of human behavior and the superstructure, and the point here is that the role of the concept of justice is that it appears to be independent of any interest.

Therefore, the reason why we developed this category is because we need to use objective, fair, or neutral reasoning to make a strong case for our interests. In other words, this is the notion that when humans achieve a somewhat advanced society, objective, fair, and neutral judgments have become part of common sense. This notion is stronger in Western societies, in which monotheistic religions such as Judaism, Christianity, and Islam hold sway. However, in Eastern societies, such as in China, Japan, and Korea, the influence of this notion is relatively weak; people in these places can understand the materialistic truth of human behavior easily but sometimes display a kind of barbarianism in behaving selfishly and not taking justice into account at all. For example, while President Bush justified the Iraq War as a just act, a Chinese newspaper discussed whether that war was profitable or not. While Muslims pray at mosques asking, "What should I do for God?," Chinese people pray at temples asking, "Buddha, what do you do for me?"

The latter type of attitude may strike social scientists as incongruous because our attitudes are expected to be independent of our individual interests. In other words, scientists obtain this necessary attitude only by engaging in the kind of strict mental practice required, not by polytheism, but by monotheism, which makes much of justice. I think that this mentality is similar to that of Marxism-Leninism, and therefore only a monotheistic world could have created Marxism-Leninism; an ideology in which the vanguard should strictly disregard their own personal interests.

However, we have to remember is that the category of justice itself was created in order to justify some interests. That is, justice is the category that justifies someone's interests. At the beginning and in nature, nothing but interest is justified. However, by this justification takes us beyond interest and paints interest as just or unjust.

Nowadays many Marxists discuss the theory of justice (or "the theory of norms"). This is because merely asserting that something is "in the interest of the workers" is weak, and such an assertion must be logically structured in a more objective manner. I half sympathize with this trend, but half disagree. This is because they do not understand that justice in nature is just a notion employed to justify social interests or that society is a battlefield of the interests of various social groups. I think that if they want to criticize bourgeois ideology, criticizing it from the inside does it less damage than *disclosing* its nature, which is to be a mouthpiece for the interests of the bourgeoisie. This is the Marxian understanding of society.

However, I simultaneously sympathize with the theoretical penchant of followers of the theory of justice to seek objectivity and agree with their common-sense recognition that we need to aim for justice to achieve anything (rather than merely to assert our interests directly). The new academic fashion for focusing on justice reflects the people's common-sense approach to justice, in other words, objectivity. In this way, we can see that this new academic fashion reflects a kind of social progress.

1.3.3 Capitalistic and Post-Capitalistic Personalities

The workplace provides the setting for a concrete confrontation between interests and justice. While these people are working to earn a livelihood (i.e., working for money, or their interests), some appear to be devoted to their work apart from their personal interest. A good example is the passion of the true craftsperson, which can be translated as a disinterested passion and in this way it can be understood as a craftsperson's type of justice because it is independent of their personal interest. These individuals do not need to respond to questions of profitability. They believe that they must work with their full effort and achieve the best results. This is justice for craftsperson.

However, it is also true that people work to make a livelihood and thus working is in their interest. Even if we do not respect this way of life, we know that, just as firms maximize their profits, workers maximize their net utility: that is, the net utility they gain through their wage minus the disutility of their labor input. This is a basic Marxist understanding, which is grounded in materialism. However, I want to develop a new horizon here; namely, the personality or human type formulated by the mode of production. The feudal mode of production bred a special type of human; the capitalistic mode of production produce a different type of human; and the communist mode of production will create the future type of human. Thus, there is a way of transmuting interests into justice via historical changes in the stage of production. Our materialism must include this dimension, and only in this way we can understand the relation between justice and interest correctly.

In his *Communist Manifesto*, Marx said that the essence of capitalism is ceaseless change. Therefore, capitalism requires a very flexible type of human who can adjust flexibly to the ceaseless changes of the production system—and initiate such changes successfully. In other words, these new entrepreneurs would identify a chance to build new industries and invest in them; and workers would move quickly to work in the new industries. Needless to say, in these circumstances, such a personality does not care about the value of loyalty to previous bosses, benevolence, or virtue, for these are useless in such a competitive society.[39] Capitalism needs a certain type of person and therefore creates them.

However, this is the story of capitalism, and needs were completely different in a feudal society, whose essence was not change, but stability. Far-sighted talent was not useful in this changeless society; therefore, people tended to be devoted to their work as a vocation without concern for the external conditions of their society. Furthermore, they needed to restrict competition to maintain an industrial order in which crafts were generally small and could be produced in small spaces. The skills that had to be formed in the overall production process needed very intimate human relations to create them and so could be maintained only by small crafts. Confucianism is emblematic of this kind of guiding conservative ideology.

Looking ahead, communist society as a post-capitalist society will also have a different type of stability (which I discuss later in subsection 4.4.4). In this case, I think a similar type of artisanship will return in this society. Let me imagine this future society. Although we needed smart humans in a capitalist society, if the future society is stable, we will not need

---

[39] The overly aggressive Han Chinese business minds formed economic or social disparities with the surrounding people and created antipathy to the Han Chinese (see Onishi 2012). They are not suited to post-capitalism and need to be post-modernized.

such personalities. Instead, we will need and respect the kinds of artisans who concentrate only on improving the qualities of their products. This spirit can be crystallized as a view of life according to which "we will be rewarded only if we produce good products." In this case, we do not have to care for anything but labor as an end of in itself. In fact, the Japanese post-war economic success has been the gift of such craftsmanship to our country. Over the course of its period of high growth, the Japanese economy experienced various types of change, for example in the financial sector and final goods sectors, such as the automobile and home electronic industries. However, in the stable industries, such as agriculture[40] or the components parts industries, producers concentrated on quality products. Thus, there is no apple more delicious than the Japanese apple and no mandarin orange is more delicious than the Japanese mandarin orange. Japan earned worldwide respect for its craftsmanship and became rich. In this case, it is natural for people to believe that "we will be rewarded *someday in the far future* if we produce good products now," that is, in deferred gratification. In sectors, societies, and periods that are more subject to change, the human ability for foresight and adaptation is the most important, but the human talent necessary differs completely in stable sectors, societies, and periods. Thus, the human types needed differ by industry and period.

I want to explain the differences in personalities according to historical periods rather than individual industries. We can express the historical appearance and disappearance of the bright, intelligent personality. Just as capitalism developed on a large scale only by creating the type of human the system requires, its termination and the return to a more stable society will create the need for a new and different type of human being. If so, we must now review and modify the sharp wage curve introduced in the capitalist system, based on the assumption that humans have sensitive reactions to economic incentives. This modification leads to an equal society with equal wages—the communist society! That is, while humans are "interested" materialistically, it is also materialistic that the different needs of different periods formulate different humans.

We can confirm from *The Civil War in France* that Marx recognized that it would take several centuries for such a change in personality, and that this will occur during the true

---

[40] This expression may lead to the misunderstanding that agriculture generally cannot be changed, which is not true. While Japanese agriculture developed without any radical changes after the post-war reform, Chinese agriculture is now under strong pressure to modernize. When I visited a model farm near Yinchuan City of the Ningxia Autonomous Region in China around 2010, collective farmers under a leader were good at identifying profitable agricultural products such as flowers or grapes and they changed their products every year. Before the period of high growth, because the impoverished masses needed only wheat, rice, corn, or kaoliang, there was no point in producing flowers or grapes. However, times have changed and they require entrepreneurs with the smart vision to match. This is why modern China needs capitalism. This will continue until China reaches the point of being a stable society.

and real post-capitalist era in human history. This stage corresponds to what Marx called the realm of freedom I discussed in subsection 1.1.3 in this book. However, Marx also developed a theory of post-capitalism, which is realizable as a short-term goal, distinct from his ultra-long-term vision of post-capitalism. This was discussed in Marx's *Critique of the Gotha Program*, which defined two stages of communism: the first stage being socialism and the second communism. Regarding this point, see Chapter 5 of Kikunami (2018).

1.3.4 From a Collectivist Society to a "Community of Free Individuals"

Note that the issue of personality includes the issue of the rise and fall of the selfish and altruistic personality. Using an evolutionary game model Bowles (2004, Chapter 13) explains that this mechanism results from material conditions. According to Bowles, altruism is not beneficial for every individual in a group in general, but it is useful for the group during between-group/community conflicts. Therefore, groups armed with altruism tend to proliferate, and this tendency can be amplified by cultural conformism, which makes other group members follow the same culture. Furthermore, if some of the community/group members punish selfish free riders, the possibility of altruists increasing becomes higher. To explain this possibility, Bowles built a multi-agent random matching model; the simulation results are shown in **Figure 1.5**.

These results clearly show that factors such as group size, migration rate, and frequency of group conflicts determine whether people are selfish. Furthermore, whether group members share resources or whether groups have different altruism ratios also affect the distribution of human personalities. We can summarize Bowles' research results as follows:

(1) The larger the group, the harder it is for altruism to thrive (and vice versa).
(2) The higher the migration rate, the harder it is for altruism to thrive (and vice versa).
(3) The more frequent the between-group conflicts, the more the numbers of altruists increase (and vice versa).
(4) The more equal the distribution of resources in groups, the more the altruists increase (equal distribution compensates for the loss of altruism).
(5) The more diverse the groups, the more the altruists increase (it is easier for altruists to survive in groups with higher altruism ratios).

Thus, whether humans are selfish or altruistic is determined not exogenously, but endogenously based on the societal conditions in place. This is a materialistic understanding of the formation of human personalities, similar to my discussion on the capitalistic personality and post-capitalistic personality in the previous subsection. However, these conclusions differ from what Bowles initially expected to find. Bowles's initial research aim was to oppose the hypothesis of Homo economicus, and he therefore probably wanted to theorize about the birth and growth of altruism. In fact, he was able to show the birth

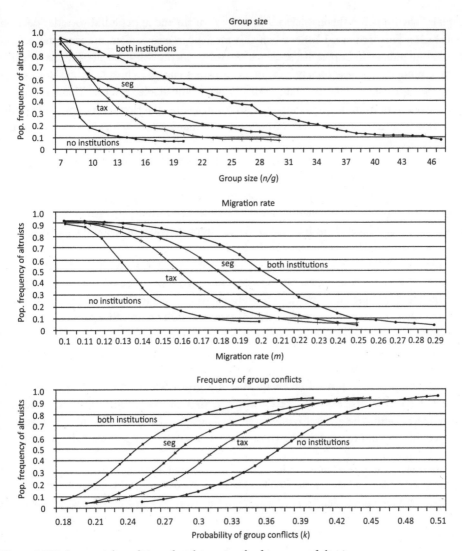

Figure 1.5 Various social conditions that determine the frequency of altruism
Source: Bowles (2004), 464. Note: "seg" in each graph indicates that each group has different ratios of altruists and "tax" indicates that each group member shares all resources to mitigate internal conflicts.

and growth of altruism under the conditions where (1) communities (groups) are small, (2) the migration rate is low, (3) between-group conflicts are frequent, (4) resources are equally distributed, and (5) the group characteristics are diverse. Such conditions can be satisfied typically only in hunting-gathering societies. Therefore, Bowles used this model to explain why hunter-gatherer societies develop altruism and agricultural societies develop selfishness. Furthermore, I believe that this model can also explain the xenophobic and egalitarian policies in imperialist societies and the cooperative industrial relations typically seen in modern Japanese society, in which companies engage in hostile fights. However, if this is so, altruism cannot maintain itself in normal and stable societies. This implication may be different from Bowles's initial research aim against the Homo economicus hypothesis. This is because Bowles's result implies that humans will be much more selfish in the stable future societies.

I believe that Bowles failed to provide evidence against the Homo economicus theory because his research program intended to defend altruism by introducing collectivism in game theory (a collectivist approach to defend altruism inside groups). What we conceive of as a post-capitalistic personality may not represent collectivist altruism, but a "community of free individuals."[41] His framework might support xenophobia and the reconciliation of labor and capital but it is not a Marxist idea.

Therefore, future mature human relations should be based not on collectivism against foreigners or other companies, but on cooperative relations that can go along with the self-interest of people under modern conditions, where people (1) belong to large groups/communities, (2) can move freely, and (3) live peacefully with other groups. This relation is better expressed in an infinitely repeated game theory setting based on the Homo economicus hypothesis.[42]

In the previous subsection, I introduced the idea that being able to accept deferred gratification will be the main characteristic of a post-capitalistic personality. The point is that this view also seeks self-interest in the sense that "we will be rewarded someday." Such a view cannot be generalized without self-interest. Therefore, all generalized values, views of life, or culture need their own individual rationality in a materialistic base. Marx's

---

[41] Marx expressed this free human relationship as *Assoziation* in German. This has been translated variously into Japanese, but it is now expressed not by the old Japanese word but by the original German word itself (see Otani, 2011; Tabata, 2015). Furthermore, Tabata (2015), page 57, states that "*socialisiert*" is derived from *assoziiert*. This is understandable because the meaning of "as" in the prefix of the latter word is "toward" in English (as he explains on page 44 of this book). Thus, this "community of free individuals" should be "a socialized man (*der sozialisierte Menschen*)." I believe that societies consisting of such people should be called "socialized society"; the original meaning of the word socialism. See also page 155 in this book.

[42] Chapter 12 in Bowles (2004) provides a model that explains the collective action to fight against other groups and secure the interest of their own group. Here, a typical collective action is a class struggle. However, even if this is connected with the model in his Chapter 13 as a model of between-group conflicts, we cannot regard it as a model of altruism, which should oppose the Homo economicus hypothesis. This is because the aim of such collective action is collective economic interest, and they act rationally to realize their own interest with a long perspective.

materialism also started from the recognition that the aim of humans is to make a better life for ourselves.

However, we should note that if deferred gratification becomes part of general culture, the latter part of the sentence that "we will be rewarded someday" will not be important for people. This is because the time span of "someday" may be very long for those who hold a long view of their future lives. In this case, people will not care about having an instantaneous reward or the ambiguous relation between effort and reward, but become autotelic (i.e., perceive the work itself as its own reward). Furthermore, in a society in which their material needs are met, people will be concerned about their reputations, that is their social honor. Finally, in a mature and stable society, unlike a capitalist one, it may not be important for people to respond rapidly to social change by using foresight. In this case, people will respect an autotelic effort that does not require external reward and will increase the possibility that reward can be based, not on one's contribution but on one's needs. This is the principle of a society in which labor will be the principal need, and work itself will be its own reward. In his document *Critique of the Gotha Program*, Marx said that this is the second stage of the communist principle of distribution.

In this section we have discussed the relative autonomy of a superstructure from the base, especially with regard to personality and culture. This autonomy can be explained by the Homo economicus hypothesis and as the development of a communist culture that is not concerned with individual interests. Therefore, both personality and culture can be explained through materialism.

2

Capitalism as a Commodity-Producing Society: The Quantitative Characteristics of Capitalistic Production and Capital as Self-Valorizing Value

The last chapter provided an epistemological framework as a precondition for our discussion of capitalism. The next four chapters explain the special characteristic of capitalism as a specific stage of human history. These four chapters correspond to the first volume of Marx's *Capital*, and Chapters 2, 3 and 4 discuss the characteristics of capitalism as a commodity-producing society and an industrial society, and the laws of its birth, growth, and death, respectively. The discussion in Chapter 5 corresponds to the second and third volumes of Marx's *Capital*. This chapter discusses the character of capitalism as a commodity-producing society.

2.1 PRODUCTIVITY TO GENERALIZE COMMODITY PRODUCTION

2.1.1 From a Self-Sufficient Economy to a Commodity Economy

First, we discuss a commodity economy, which is the most basic characteristic of capitalism. Marx starts *Capital* with the statement: "The wealth of those societies in which the capitalist mode of production prevails, presents itself as 'an immense accumulation of commodities.'" We discuss how this characteristic is the result of the higher productivity under capitalism compared to that in slave and serfdom modes of production.

For this purpose, now we imagine a self-sufficient economy with no social division of labor. Its unit might be a household or a village community that is one integrated entity in which there is no exchange. However, this self-sufficiency economy must be transformed gradually into a commodity-producing economy if (1) specialization by the social division of labor increases productivity (economy of specialization) and (2) productivity of the distribution system rises, and the degree of commodity production depends on the degree of the above two conditions. This is explained mathematically on what follows, but readers unfamiliar with mathematics may skip this part. The conclusion itself is simple.

Our mathematical formalization of the above-mentioned relation is as follows. First, for this purpose, we now assume:

(1) Originally, there is a self-sufficient economy with n households, each of whom has k amounts of factors of production and produces m kinds of products. To simplify,

let us assume that amounts of the factors of production necessary to produce n kinds of products are the same. That is, this economy inputs k/m amounts of factors of production for each kind of product. Now, if we set the production function as $f(k/m)$, $f' > 0$, the amount of production for each kind of product must be $n \cdot f(k/m)$.

(2) However, our society is being transformed into a commodity exchange economy, and we can imagine the first step of this transformation. That is, the ith product among m was produced exclusively by x_i agents, and this product was distributed to all other $n - x_i$ agents. In this exchange, x_i agents gave their products to other $n - x_i$ agents, and each x_i agent took all the other kinds of products produced by $n - x_i$ agents as an equal exchange. In this way, most of the x_i agents' products became commodities, and a small part of the $n - x_i$ agents' products became commodities, and all the necessities were exchanged by both type of agents. In this case, the total production of the non-ith products by $n - x_i$ agents becomes

$$(n - x_i) f\left(\frac{k}{m-1}\right) \tag{2.1}$$

under the technological condition

$$y_{j \neq i} = f_{j \neq i}\left(\frac{k}{m-1}\right) \tag{2.2}$$

Therefore, our problem now is to determine the conditions that lead to a commodity economy, and we find the following three conditions do so:

(1) The amount of production of the ith good does not decrease from the original; that is,

$$x_i \cdot f_i(k) \geq n \cdot f\left(\frac{k}{m}\right) \tag{2.3}$$

(2) The amount of production of the $j \neq i$th good does not decrease from the original; that is,

$$(n - x_i) \cdot f\left(\frac{k}{m-1}\right) \geq n \cdot f\left(\frac{k}{m}\right) \tag{2.4}$$

(3) The new condition can exert an additional social cost to make the producers of the ith good provide the ith good to the producers of the $j \neq i$th good and make the latter provide the $j \neq i$th good to the former (thus leading to mutual exchange between the ith goods and the $j \neq i$th goods).

We specify a production function as

$$f(k) = Ak^\alpha \quad (\alpha > 0) \tag{2.5}$$

to identify the above two conditions clearly. In this case, the first two conditions (1) and (2) above become

$$x_i A k^\alpha \geq nA\left(\frac{k}{m}\right)^\alpha \tag{2.6}$$

$$(n - x_i)A\left(\frac{k}{m-1}\right)^\alpha \geq nA\left(\frac{k}{m}\right)^\alpha \tag{2.7}$$

These two inequalities can be summed up into

$$\frac{n}{m^\alpha} \leq x_i \leq n\left[1 - \left(\frac{m-1}{m}\right)^\alpha\right] \Leftrightarrow 1 \leq m^\alpha - (m-1)^\alpha \tag{2.8}$$

Thus, when $\alpha = 1$

$$1 = m^\alpha - (m-1)^\alpha \tag{2.9}$$

In other words, if we need a gap between one and $m^\alpha - (m-1)^\alpha$ for condition (3), α should be larger than one; that is, increasing return to k.

Furthermore, to cover condition (3), we have to measure how much of each factor of production can be saved by conditions (1) and (2). This surplus factor of production must be used to produce the "means of exchange," which has become necessary for the exchange between the ith goods and the $j \neq i$th goods. Therefore, we first identify the surplus factor of production to produce the ith good as $(n/m^\alpha)k$ and that to produce the $j \neq i$th good as $n\left(\frac{m-1}{m}\right)^\alpha k$. Therefore, the total surplus factor of production becomes

$$\left[n - \frac{n}{m^\alpha} - n\left(\frac{m-1}{m}\right)^\alpha\right]k = \left\{1 - \left(\frac{m-1}{m}\right)^\alpha - \frac{1}{m^\alpha}\right\}nk = \left\{1 - \frac{(m-1)^\alpha + 1}{m^\alpha}\right\}nk \tag{2.10}$$

Therefore, if we need $(n-1)T(>0)$ amounts of the means of exchange (typically gold) for this exchange, and if we assume that the production function of this means of exchange is g ($g' > 0$), we should have the condition

$$g\left[\left\{1 - \frac{(m-1)^\alpha + 1}{m^\alpha}\right\}nk\right] > (n-1)T \tag{2.11}$$

This means that a higher productivity of the means of exchange promotes commodification under the given T, and under this condition, the total production of the means of exchange increases.[1] Furthermore, if we assume function g shows an increasing return to the factor of production, a larger number of n and a larger amount of k also promote commodification. Therefore, the expansion of the size of society or cities (urbanization) and the increase in productive wealth are also important for commodification.[2]

In conclusion, (1) and (2) show the importance of the merits of specialization and (3) shows the importance of the wider sense of productivity in the production of the means of exchange for commodification.[3]

2.1.2 The Merit of Specialization and the Rise in the Productivity of the Distribution Sector

If this is true, in what cases do these two conditions hold? First, let us remember Adam Smith, who first proposed the merit of specialization. He imagined a pin factory in which each worker is not engaged in the full process of production, but in only one process, and insisted that such specialization, in other words, the division of labor, increases productivity.

It is true that *manufacture* developed successfully based on the merit of this division of labor. However, because the present process of production is not in this type of *manufacture*, we should explain this merit of specialization by the great mechanized industries. In fact, because such mechanized industries with fixed capital cost a huge initial amount that cannot be owned by every member of society, only a few members of society are specialized to own and use such machinery. Furthermore, in the competitions between capitalists who have a great deal of capital and those who do not have much capital, the former win and become even larger and more specialized. This phenomenon is called "the merits of scale" or the "increasing return to capital."

A small amount of manipulation can reveal how this mechanism works. For this purpose, let κ represent the cost of invested fixed capital and κ_0 denote the minimum initial

---

[1] Credit creation by banks saves coin and becomes the origin of the banks' profit (Marx, 1907, 396). The productivity of banks' credit creation, which consists of g, determines the ratio of the banking sector to the total economy.

[2] This model can also explain why unstratified societies consisting of small clans have gathered together after a long process and become stratified chiefdoms, which was the only way to coordinate these originally separate societies. That is, a larger society can enjoy the merits of the division of labor. Service (1962) explains this process by several non-economic factors, but our model can explain it using economic reasoning only.

[3] Sekine (2017) shows mathematically the special characteristics of the great mechanized industries as a technology of increasing return.

cost required for all the costs of operation. Additionally, assume that the size of κ directly reflects the scale of all the elements of production. In this case, we can assume the following type of production function:

$$Y = A(\kappa - \kappa_0)^\alpha, \alpha < 1 \qquad (2.12)$$

Condition $\alpha < 1$ indicates that it has a technology of diminishing return and the efficiency of this production system can be measured by

$$\frac{Y}{\kappa} = \frac{A(\kappa - \kappa_0)^\alpha}{\kappa} \qquad (2.13)$$

With differentiation, we obtain

$$\frac{\partial}{\partial \kappa}\left(\frac{Y}{\kappa}\right) = \frac{A\alpha(\kappa - \kappa_0)^{\alpha-1} \cdot \kappa - A(\kappa - \kappa_0)^\alpha}{\kappa^2} = \frac{A(\kappa - \kappa_0)^{\alpha-1}(\alpha\kappa - \kappa + \kappa_0)}{\kappa^2} \qquad (2.14)$$

To check the sign of this equation, we first calculate the condition that makes it equal to zero. That is,

$$\kappa_0 = (1-\alpha)\kappa \Leftrightarrow \kappa = \frac{\kappa_0}{1-\alpha} \qquad (2.15)$$

Therefore, under the condition $\alpha < 1$, mentioned above,

when $\kappa < \frac{\kappa_0}{1-\alpha}$, $\frac{\partial}{\partial \kappa}\left(\frac{Y}{\kappa}\right) > 0$, therefore, increasing return to scale

when $\kappa = \frac{\kappa_0}{1-\alpha}$, $\frac{\partial}{\partial \kappa}\left(\frac{Y}{\kappa}\right) = 0$, therefore, constant return to scale

when $\kappa > \frac{\kappa_0}{1-\alpha}$, $\frac{\partial}{\partial \kappa}\left(\frac{Y}{\kappa}\right) < 0$, therefore, diminishing return to scale.

This result explains that even if there is a technology with a diminishing return in the form of $\alpha < 1$, there is a certain zone where returns increase, and the higher the initial cost becomes, the larger this area becomes. In this way, the merit of specialization comes from the merit of scale (i.e., the increasing returns to scale) in modern industry. Therefore, the appearance and establishment of the great mechanized industries was indispensable in making the merit of specialization unwavering. Only because of the growth of the great mechanized industries did commodity production become general in human history.

However, the other condition for the productivity of distribution has much closer relations with the machine-based production system. The essential character of this increase in industry started before the Industrial Revolution, when big ships were introduced. For example, Venetian ships trading in the Mediterranean introduced the capital-labor relationship by employing many workers, even before steam engines were invented. After this introduction the productivity of distribution increased dramatically. Hicks (1969) also states that such ships represent the first fixed capital to play an important role in the production process. We need to understand that the role of machines was decisive in the development of a commodity economy. In summary, both the merit of specialization and the rise in the productivity of distribution depend on the introduction and development of what we have called machines (in other words, fixed capital).

2.1.3 Productive and Non-Productive Labor

Incidentally, this understanding of the efficiency and commodity-producing economies results a completely different understanding of productive and non-productive labor. This is because, while there are two definitions of productive labor: "the first definition" and "the second definition," this book uses the first definition that differs from the common understanding.[4] The common opinion in Marxist economics is that productive and non-productive labor differ in terms of whether they are material or non-material labor. In this view, interpersonal services such as cooking, cosmetology, medicine, education, and entertainment services are classified as non-productive labor. However, in our understanding the essential difference between them is whether the sector was necessary from the beginning of human history or not. In other words, services for commerce and finance were not necessary and appeared only after commodity production began. Therefore, the commercial and financial sectors are not productive, and they exist only to support the productive sectors. Although these sectors play a very important role, they cannot exist without the object they support. This is how we can understand the failure of Iceland and the U.S.A. in the global financial crisis in 2008, which specialized in the financial sector. It is also a failure in modern economic theory, which has no categories for the productive and non-productive sector. Until now, no communist-led country has experienced a financial crisis. This might result from the positive effect of Marxist economics.

On the other hand, an important area of service businesses consists of the business-oriented services that support corporate activities, which include designing the manufacturing products and product inspections, which existed before the appearance of commercial and financial sectors, as well as the services for delivery and sales operations that appeared only after the commercial and financial sectors had become an independent

---

[4] The second definition covers all the labor that is useful for capitalists to make a profit under capitalism, and it is called the "historical provision of productive labor" because this definition is effective only in the capitalist era.

part of the social division of labor. Using the distinctions above, the former must be understood as a productive sector and the latter should as a non-productive sector.

However, what is important here is that all these services were originally supplied by individuals (or households) and companies on the basis of their own self-sufficiency. Historically, fisherfolk taught children how to fish, cared for their own and their family's hair, and searched for medicinal herbs. These were all primitive forms of education, beauty, and medical services, but they were not independent industrial sectors. Originally, each manufacturing company engaged in its own designing, inspecting, delivering, and marketing, but these tasks are now specialized, independent industries Therefore, in general, this specialization must be understood not as a development of the service business, but as a specialization of the economy; that is, a development of the social division of labor. Business-oriented services often come about in the form of spin-offs or outsourcing. We regard this phenomenon as an example of the transition from a self-sufficiency economy to a commodity economy and an economy of specialization.[5]

2.2 PRODUCT AND COMMODITY: COMMODITY EXCHANGE IN A CONCRETE AND MATERIALISTIC CONCEPT OF THE HUMAN BEING

2.2.1 Production for Commodity Exchange

Because we discussed the conditions for commodity production, we hereafter assume that the society exists under a commodity economy. Because of this, we have to change Figure 1.1, in which all humans are treated as one group, into a much more concrete type of figure that expresses individual humans. Even if all members engage in productive activity, all necessary products are made by others. This is the typical commodity-producing society, and therefore, we must replace Figure 1.1 with **Figure 2.1**.

The point in this figure is that each agent no longer produces for their own consumption; they produce for others who need their product. In other words, no agent can live without acquiring the products of others, and they do not acquire it as a gift or through theft, then each member should give their own product as compensation. This is commodity exchange.

However, just as in a self-sufficient economy, agents in a commodity-producing society maximize their net utility (utility minus disutility) by balancing the marginal disutility of their labor with the marginal utility of the product that is acquired by production. Because all agents in the society do so, the total net utility in this society must be maximized as a whole. We will now show this mechanism. The readers may skip the proof if desired, or return to it later.

In addition to the above, military forces, police, ideologues for the ruling class to dominate, and managers who simply control workers in the workplace also constitute unproductive labor at a certain stage of class society.

MARXIAN ECONOMICS

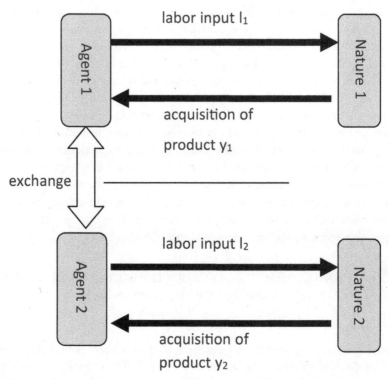

Figure 2.1 Commodity-producing society as a society maintained by product exchange

Here, labeling the two agents' labor inputs l_1 and l_2 and the products made by these productive activities as y_1 and y_2 in Figure 2.1, let us introduce the equilibrium condition as in the former one-agent case in Figure 1.1. Then, for agent 1 in the upper row of this figure,

$$\frac{dD_1}{dl_1} = \frac{dU_1}{dy_2} \cdot p \cdot \frac{dy_1}{dl_1} \qquad (2.16)$$

should hold.

Here, D_1 and U_1 represent the disutility of work and the utility to consume for agent 1 respectively, and p shows that the product produced by agent 1 is exchanged for the product of agent 2 with an exchange ratio of p. This shows that the marginal disutility of labor is equal to the marginal utility of product 2, obtained by exchanging a unit of product 1 produced by agent 1. We show this relation more clearly using the fact that $p = (dy_2/dy_1)$ because substituting (dy_2/dy_1) with p in the above equation (2.16) leads to

CAPITALISM AS A COMMODITY-PRODUCING SOCIETY

$$\frac{dD_1}{dl_1} = \frac{dU_1}{dy_2} \cdot \frac{dy_2}{dy_1} \cdot \frac{dy_1}{dl_1} = \frac{dU_1}{dy_1} \cdot \frac{dy_1}{dl_1} \tag{2.17}$$

This expression is exactly the same as that in the case in Figure 1.1, and also can be simplified as

$$dD_1 = dU_1 \tag{2.18}$$

Furthermore, because this relationship is naturally true also for agent 2,

$$\frac{dD_2}{dl_2} = \frac{dU_2}{dy_1} \cdot \frac{1}{p} \cdot \frac{dy_2}{dl_2} \tag{2.19}$$

should hold. Substituting (dy_2/dy_1) with p in equation (2.19) again leads to

$$\frac{dD_2}{dl_2} = \frac{dU_2}{dy_1} \cdot \frac{dy_1}{dy_2} \cdot \frac{dy_2}{dl_2} = \frac{dU_2}{dy_2} \cdot \frac{dy_2}{dl_2} \tag{2.20}$$

and $dD_2 = dU_2$.

Therefore, we can conclude that the whole of human society, represented as the two agents in Figure 2.1, maximizes its utility by exchanging their outcomes of labor.

In addition, the two agents in this figure produce products that other people need, but if we want to say that this commodity-producing economy is more rational than a self-sufficiency economy without commodity exchange, then there must be a certain condition in the labor productivity of both agents. To show this, first, let us assume that e_{11} and e_{21} are the labor productivities of agent 1 to produce the first and second good, respectively. Likewise e_{12} and e_{22} are the labor productivities of agent 2 to produce the first and second good, respectively. In this case, if agent 1 produces y_2 self-sufficiently using their own labor l_1, they takes $e_{21} \cdot l_1$ amount of y_2, and if they produce y_1 and exchange it with agent 2's y_2, they takes $p \cdot e_{11} \cdot l_1$. Therefore, commodity production is rational for agent 1 only when

$$e_{21} < p \cdot e_{11} \Leftrightarrow \frac{1}{p} < \frac{e_{11}}{e_{21}} \tag{2.21}$$

Likewise, the rational condition for agent 2 becomes

$$e_{12} < \frac{1}{p} \cdot e_{22} \Leftrightarrow \frac{e_{12}}{e_{22}} < \frac{1}{p} \tag{2.22}$$

Therefore, summing up these two conditions leads to the socially rational condition of the commodity economy. That is,

$$\frac{e_{12}}{e_{22}} < \frac{e_{11}}{e_{21}} \Leftrightarrow e_{12}e_{21} < e_{11}e_{22} \tag{2.23}$$

This condition is very interesting because the condition for specialized production (just as in a form of commodity production that is differentiated from a self-sufficiency economy) is not an absolute advantage, as indicated by $e_{21} < e_{11}$ and $e_{12} < e_{22}$, but a greatly weaker condition called comparative advantage. For example, even $e_{21} > e_{11}$, commodity production may be rational if the condition $e_{12} < e_{22}$ holds at a certain level. Likewise, even if $e_{12} > e_{22}$, commodity production may be rational if the condition $e_{21} < e_{11}$ holds at a certain level. This principle is called the "theory of comparative cost," created by Ricardo, and now commonly referred to as comparative advantage theory.[6]

---

[6] Although this model assumes that agent 1 consumes only product 2 and agent 2 consumes only product 1, it is reasonable for both agents to consume both goods. For example, if agent 1 divides all labor into labor for commodity production and for self-consumption at a ratio of $\gamma : 1 - \gamma$, then

$$\frac{dD_1}{dl_1} = \gamma \frac{\partial U_1}{\partial y_1} \cdot \frac{dy_1}{dl_1} + (1-\gamma)\frac{\partial U_1}{\partial y_2} \cdot p \cdot \frac{dy_1}{dl_1} \tag{✻}$$

On the other hand, since the marginal utility obtained from both goods must be equal for agent 1,

$$\frac{\partial U_1}{\partial y_1} \cdot \frac{dy_1}{dl_1} = \frac{\partial U_1}{\partial y_2} \cdot p \cdot \frac{dy_1}{dl_1}$$

Note that substituting this equation onto equation (✻) makes $dD_1 = dU_1$, which we introduced on subsection 1.1.1 of this book as the original interaction between humanity and nature.

Furthermore, because the same equation should hold for agent 2,

$$\frac{\partial U_2}{\partial y_2} \cdot \frac{dy_2}{dl_2} = \frac{\partial U_2}{\partial y_1} \cdot \frac{1}{p} \cdot \frac{dy_2}{dl_2}$$

In this case, the exchange ratio p of both goods must be

$$\frac{\frac{\partial U_1}{\partial y_1}}{\frac{\partial U_1}{\partial y_2}} = \frac{\frac{\partial U_2}{\partial y_1}}{\frac{\partial U_2}{\partial y_2}} = p$$

This means that the exchange ratio of goods between them must be the relative ratio of their marginal utility. If we extend this relation to a multi-agent economy, this relation should become

$$\frac{\frac{\partial U_1}{\partial y_1}}{\frac{\partial U_1}{\partial y_2}} = \frac{\frac{\partial U_2}{\partial y_1}}{\frac{\partial U_2}{\partial y_2}} = \frac{\frac{\partial U_3}{\partial y_1}}{\frac{\partial U_3}{\partial y_2}} = \frac{\frac{\partial U_4}{\partial y_1}}{\frac{\partial U_4}{\partial y_2}} = \cdots = p$$

The difference between price and value is determined in this way, as we saw at the beginning of subsection 1.1.2.

Of course, we can say that human history has been full of robberies and gifts as well as the commodity exchange. It is true that the incessant continuation of wars worldwide means that humans are acquiring interest by force, but we also must acknowledge that that does not mean that all necessary products were taken by force. Even in the Chinese Warring States period starting in 475 BC, most products were exchanged or produced self-sufficiently.

Furthermore, we still have a grant economy inside families as a kind of community. However, even interfamily grants on the basis of love may expect compensation. For example, parents may expect their children to care for them in their old age. Therefore, grants can also be understood as a special and wider type of exchange.

2.2.2 Use Value and Exchange Value

Therefore, the purpose of production has changed from that of consuming the product to that of exchanging it with others. Of course, to be exchangeable, products must be useful. However, what is important now producers is not to be useful but exchangeable. These two different characteristics, usefulness and exchangeability, divide the value of the commodity into use value and exchange value.

Consider the characteristics of the former, such as deliciousness, beauty, comfort, and so on. They are kinds of usefulness called Marx calls use value. However, this use value does not always make a product a commodity. For example, imagine a special photo or family memorial goods. They have a use value for certain individuals but they are not exchangeable because they do not have any use value for others. Another example is air, because air has a decisive use value in the sense that we cannot live without it, but it cannot be a commodity because there is a plenty of air (as long as the air is not polluted). In other words, an important characteristic in a commodity-producing society is its exchangeability, which does not matter in a self-sufficiency society. Focusing on its exchange ratio with other commodities, Marx called exchangeability exchange value. Thus, the era in which use value is required has come to an end and exchange value has become the independent purpose of production; changing the purpose of production completely.

2.2.3 From Exchange Value to Value

Taking this on board, what determines the exchange ratio? We can explain this using Figure 2.1 again. In this figure, each person produces their own products to obtain the products of others, and the problem is how much of their labor should be consumed to obtain the product of others. Thus, if the amount of labor input by agent 1 is larger than that of agent 2, we can say that agent 1 has lost out in the exchange. In this transaction, agent 2 has obtained agent 1's product using less labor, and agent 1 may demand that the exchange ratio is changed because it is unfair. If the exchange ratio does not change, then producers move to the lower row and the amount of product 1 (y_1) decreases. This change in the balance will eventually result in a change in the exchange ratio itself; that is, an increase in the exchange value of product 1 = a decline in the exchange value of product 2.

This conclusion leads us to understand that the amount of labor embodied in products exchanged on a one-to-one basis must eventually become the same. In other words, only the products embodying the same labor can be exchanged one-to-one. This can be expressed by the symbols in Figure 2.1 as follows. The two different products are exchanged at the ratio of $y_2 = p \cdot y_1$, between the two agents, but the first good that is produced exclusively by agent 1 can be expressed as $y_1 = e_{11} l_1$, and the second good that is produced exclusively by agent 2 can be expressed as $y_2 = e_{22} l_2$. Therefore, substituting these two into $y_2 = p \cdot y_1$ yields

$$e_{22} \cdot l_2 = p \cdot e_{11} \cdot l_1 \tag{2.24}$$

Hence, the amount of labor exchanged can be equal only when $e_{22} = p \cdot e_{11}$; that is, p is equal to the ratio of labor productivity between both products, and the ratio is achieved by the labor flow among industries.[7] Exchange value itself changes daily for various reasons, but it has a certain center of gravity, which is determined by the amount of labor expended in the production of each product. Marx calls this central point "value" and says that its substance is labor. This is the basic content of the embodied Labor Theory of Value (LTV).

Marx introduces his explanation by stating that different two products can be equal because both products have a certain common substance. As described above, the decisive condition of exchange is that the products have different use values. Marx asks how, because we do not exchange the same products, do these different products become equal? He answers that these different products have only one common property, that of being the product of labor. That is, this commonality makes exchange possible, and also makes labor the "substance of value" that regulates the exchange ratio.[8]

---

[7] If we substitute the result of the note 6 by p in the equation $e_{22} = p \cdot e_{11}$, we have

$$e_{22} = \frac{\frac{\partial U_1}{\partial y_1}}{\frac{\partial U_1}{\partial y_2}} \cdot e_{11} = \frac{\frac{\partial U_2}{\partial y_1}}{\frac{\partial U_2}{\partial y_2}} \cdot e_{11}$$

which means that it also depends on the subjective tastes of how much the exchangers are satisfied by these goods. However, the generalized exchange ratio p that is realized does not depend merely on the utility function of these two subjects but on the average utility function of the whole society. It is not surprising that subjective utility constitutes the theory of embodied labor. This is because this equation includes subjective utility, which is an important part of the metabolic relation between humanity and nature.

[8] Because this labor is characterized by a total abstraction from its concrete content, such as construction work, hairstyling, sewing, and so on, and the only property left is that of human labor, Marx calls the latter abstract human labor. On the other hand, he calls the concrete content of this labor, such as construction work, hairstyling, and sewing concrete useful labor, which gives commodities their concrete use value.

CAPITALISM AS A COMMODITY-PRODUCING SOCIETY

However, even if this reason is not wrong, we need an additional explanation. Critics of Marx's explanation argue that there are various types of commonality among products. For example, sun power must be an input in a certain process of production, whether it is direct or indirect. We can also say that some types of energy are embodied, and we can find a common property in the fact that all these products are processed by humans, animals, or certain non-living things. Thus, if we want to see labor is a special factor of input, then we need an additional explanation.

In fact, this answer was already provided at the beginning of Section 1 in Chapter 1 of this book where we saw that productive activities are the metabolic relation between humanity and nature, and that humans can spend only their labor. Even if the sun contributes to the productive activities of humans, such effects are free of charge just as a gift from the nature. Humans only consider whether to produce something or not under a given condition. But the sun does not engage in such consideration and simply has an effect on nature. However, "labor" is the human's subjective activity when interacting with nature. The sun's effect is just a given condition for human activity. In this way, we can understand that labor is the only one fundamental feature that differs from all other types of commonalities.

However, my explanation using the principle of marginal utility in modern economics in the same part of this book is also important because I explain that the criterion humans use to decide the amount of product is the necessary amount of labor, and this shows the relationship between labor and utility (or use value). We buy a certain commodity by paying a certain amount of money for it because we think that the necessary disutility caused by giving up that money is less than the utility gained by attaining that commodity. In essence, in this case, we compare the amount of disutility of the additional labor and the amount of utility gained by that commodity. Therefore, this utility is measured in terms of disutility, which is ultimately by the amount of labor expended on it. Traditional Marxists say that value theory should reject the concept of utility, but we do not think so. Only by introducing the concept of utility can we prove that value should be measured in terms of the amount of labor.

2.3 MONEY AS A COMMODITY: PRODUCTION FOR EARNING MONEY AS A CONCRETE AND MATERIALISTIC CONCEPT OF THE HUMAN BEING

2.3.1 Production Not for Commodity Exchange Itself But to Gain Money

Using the methods mentioned above, we explained that the purpose of production is not for the producers' own use but to exchange with others. However, this explanation is not concrete enough because have not explained that money is the more concrete purpose of production. That is, people do not exchange products directly, they do so indirectly, using money. Therefore, they must first receive money as compensation for the commodity that

MARXIAN ECONOMICS

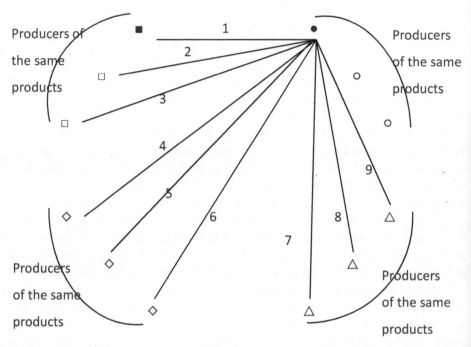

Figure 2.2 Commodity economy with and without money

they sell. In a real capitalist society, money is a universal equivalent that can be exchanged for any commodity. Humans created money as a special commodity in their history and gave it the universal power to be exchanged with anything.

This process of money formation is explained in Marx's *Capital*, Volume 1, Chapter 2, entitled "Exchange." That is, without money, even if producer A finds another producer B who produces a necessary commodity, producer A cannot exchange if producer B does not need the product producer A makes; and vice versa. Therefore, commodities "do not confront each other as commodities, but only as products or use-values."[9] This is the difficulty of showing that use value is not enough for exchange, and this difficulty was solved "before they thought," as in Faust's words "in the beginning was the deed."[10] Money was created before they thought about it by making a certain commodity a universal equivalent.[11]

Therefore, money is a very convenient tool for solving this difficulty and because "convenience" means "rationality" in economic jargon, we can express how and why a commodity

[9] Marx, 1887, 61.
[10] Ibid.
[11] Ibid.

CAPITALISM AS A COMMODITY-PRODUCING SOCIETY

economy with money is better than one without money. For this purpose, we assume the existence of a twelve-person society where three producers produce four kinds of products. Here, we assume that every member knows who produces each kind of product, symbolized as ○, □, ◇, and △.

In this case, if money exists, the producer indicated by the black dot (●) can receive the □ type of product through routes 1, 2, or 3; the ◇ type of product through routes 4, 5, or 6; and the △ type of product through routes 7, 8, or 9. As a result of all these transactions they can enjoy all products, including their own (○). There is no difficulty in getting these three products using money because money is the general equivalent of products and therefore all producers can exchange their products using money. Therefore, each producer needs only three transactions to attain goods from others, and the total number of transactions in this society overall is $3 \times 12 = 36$.

However, without money, the situation is different. If producer ● wants to exchange their products with ■, they can do so only when producer ■ wants an ○ type of product. In other words, if producer ■ has already attained ○ from other producers, then they cannot exchange their products with ■. This is a barter transaction, which differs from one in a money economy. In this case, we assume the probability that producer ● can barter with producer ■ is 1/3. This assumption means that all three producers of ○ equally have 1/3 of a probability to exchange their products with producer ■ successfully. In reality, there is another possibility, namely, that one producer of □ wants to transact with ○, but changes their mind and transacts with ● instead, if the original intended transaction has not been completed. However, it is too complex to calculate this, so we ignore this possibility for simplicity. Under this assumption, the probability of a successful transaction between ● and ■ is 1/3.

However, if producer ● cannot transact with producer ■, then they must go to the next □ producer by route 2 and can exchange successfully with a probability of 1/2. Hence, we have this probability from the first stage, $2/3 \times 1/2 = 1/3$.

Now we need to calculate the third case in which the transaction failed in both routes 1 and 2 and ● goes to route 3. In this case, □'s partner is only ●, and therefore, this transaction must be successful. Then, this possibility must be $2/3 \times 1/2 \times 1 = 1/3$.

Summing up these three cases, we can calculate the expected number of negotiations (transactions) for producer ● as follows:

(1) to obtain a □ type of product:

$$1 \times \frac{1}{3} + 2 \times \left(\frac{2}{3} \times \frac{1}{2}\right) + 3 \times \left(\frac{2}{3} \times \frac{1}{2}\right) = \frac{1}{3} + \frac{2}{3} + \frac{3}{3} = \frac{6}{3} = 2$$

(2) to obtain all other types of products:

$2 \times 3 = 6$

Therefore, for all the individuals in this society to obtain all types of products the number of transactions is:

$$6 \times 12 \times \frac{1}{2} = 36$$

Here, we multiply the left-hand side by 1/2 because both sides of the exchange can obtain their intended objects by one exchange, and we therefore must avoid double counting. In conclusion, the expected number of transactions results in 36 in cases both with and without money.

However, this coincidence is accidental and results from the special condition that the numbers of producers and kinds of products are 12 and 4, respectively. Therefore, we need to generalize this formulation by stating the number of producers and kinds of products as n and m, respectively, and assuming n/m as an integer and that the same number of producers produces each product. In this setting, we can calculate the expected number of negotiations: that is,

in the case with money:

$$n(m-1) \tag{2.25}$$

in the case without money:

$$n(m-1)\left\{\frac{1}{4}\left(\frac{n}{m}+1\right)\right\}^{12} \tag{2.26}$$

---

[12] This is the result of the following calculation. Suppose I am a producer producing a certain good and visit the producers of certain other goods. The possibility that I can transact successfully with the first producer who produces other goods is $1/(n/m) = m/n$. Therefore, the possibility of failure is $1 - m/n$, and in this case, I have to go to the next producer of that good. The possibility that I can transact successfully with this producer becomes $1/(n/m - 1)$ because the number of possible producers left after the first failure is now $n/m - 1$. Therefore, the possibility that I can transact successfully with the second producer is $(1 - m/n) \times 1/(n/m - 1) = m/n$. We can thus calculate the possibilities of a successful transaction as follows:

Successful transaction with the first producer of that good:

$$\frac{1}{\frac{n}{m}} = \frac{m}{n}$$

Successful transaction with the second producer of that good:

$$\left(1-\frac{m}{n}\right) \times \frac{1}{\frac{n}{m}-1} = \frac{n-m}{n} \times \frac{m}{n-m} = \frac{m}{n}$$

Successful transaction with the third producer of that good:

$$\left(1-\frac{m}{n}\right) \times \left(1-\frac{1}{\frac{n}{m}-1}\right) \times \frac{1}{\frac{n}{m}-2} = \frac{n-m}{n} \times \frac{m-2m}{n-m} \times \frac{m}{n-2m} = \frac{m}{n}$$

With this formula, we can understand that the advantage of a money economy over a barter economy depends on whether $\frac{1}{4}\left(\frac{n}{m}+1\right)$ is more than 1 or not. If $n/m = 3$, then both economies' expected number of transactions become equal; if $n/m > 3$, then the money economy becomes more efficient; and if $n/m = 1$ or 2, then the barter economy becomes more efficient.

This result is very interesting because it shows that the larger the society is in size, and thus in the number of individuals (n) compared to the number of products (m), the greater is the necessity of money. However, if this n/m ratio (the number of producers producing each product) is only 2, then situation reverses. In this case, a barter economy that does not use money is better than the commodity economy with money, in the sense that it is more efficient.

We can imagine a type of village economy in a small face-to-face society with only ten families and five types of products. For example, everyone might make their own special products, such as necklaces, bows, axes, stone kitchen knives, and furs, while they all might engage in hunting and gathering food. In this case, necklaces, bows, axes, stone kitchen knives, and furs can be exchanged in a barter economy without using money and this type of exchange is better than exchange using money. Even in simple agricultural villages, we can image there might be only fifty families and twenty types of products. Villages of this size did not need money to exchange products.

On the other hand, the n/m can be small, not only in small villages but also in large societies with a large n and m. Although this may seem strange at first, even in an economy with a large n and large m, we can imagine such a situation. For example, when the ratio is 2, if I fail to exchange with the first producer of a certain good, I can exchange successfully with the second producer of that good. Thus, if the ratio is very small, then the cost of the failure to exchange without money is not serious. Thus, even in a large economy, the ratio n/m is very important. In conclusion, a commodity economy and a money economy are just special forms of different economies, even if we think that they are natural and universal economic systems.[13]

Therefore, we know that the potential for a successful transaction with these producers is same. Then, the number of transactions I can expect to engage in to get that good becomes

$$1\times\frac{m}{n}+2\times\frac{m}{n}+3\times\frac{m}{n}+\cdots+\frac{n}{m}\times\frac{m}{n}=\frac{m}{n}\sum_{k=1}^{\frac{n}{m}}k=\frac{m}{n}\cdot\frac{1}{2}\cdot\frac{n}{m}\left(\frac{n}{m}+1\right)=\frac{1}{2}\left(\frac{n+m}{m}\right)$$

Using these results, we can obtain the expected total number of transactions for the whole society:

$$\frac{1}{2}\left(\frac{n}{m}+1\right)\times(m-1)\times n\times\frac{1}{2}=n(m-1)\left\{\frac{1}{4}\left(\frac{n}{m}+1\right)\right\}$$

[13] The models in subsection 2.3.1 can be converted into a model of corporate financing through financial organizations and a model of direct financing without financial organizations. The former is the conversion from the monetary economy model and the latter is that from the nonmonetary exchange economy model. See Nagata (2020).

2.3.2 Money as a Special Commodity

This discussion has shown in concrete terms the relationship between a commodity economy and a money economy. However, still we have a question: what is money? Knowing how money works and what money is are different questions, and the latter question is related to the question of why people regard money as a universal equivalent? Although the above arguments ignore such questions, we must now answer them on the basis of Marx's special theory of the form of value.

As we explained, value has "substance," but the problem is that it cannot be seen, and therefore, we have to express it by using other commodities. For example:

x units of commodity 1 = y units of commodity 2.

In this case, because the value of x units of commodity 1 is expressed by y units of commodity 2, the former is called the relative form of value and the latter is called the equivalent form of value. This type of expression using equation is called the "elementary or accidental form of value."

Furthermore, x units of commodity 1 can be expressed by other commodities such as

x units of commodity 1 = y units of commodity 2
  = z units of commodity 3
  = v units of commodity 4
  = w units of commodity 5
  ...

This form is called the "total or expanded form of value." However, using this form, we can understand that the value of y units of commodity 2, z units of commodity 3, v units of commodity 4, w units of commodity 5, and so on, also can be expressed by x units of commodity 1 conversely. That is,

$$\left.\begin{array}{l} y \text{ units of commodity 2} \\ z \text{ units of commodity 3} \\ v \text{ units of commodity 4} \\ w \text{ units of commodity 5} \end{array}\right\} = x \text{ units of commodity 1}$$
...

but the meaning of this new form is completely different from the former ones. Here, y units of commodity 2, z units of commodity 3, v units of commodity 4, w units of commodity 5, and so on, change from the equivalent form of value to the relative form of value, and are now expressed as x units of commodity 1 as the equivalent form of value. In this way, x units of commodity 1 has become a form to express all the commodities except for itself. That is, value is fixed here as a form and called the "general form of value," and the commodity on the right-side is called the universal equivalent. Thus, money is a universal equivalent crystalized as a special commodity, called the "money form." The exchange ratio between money and the commodity becomes the price.

This is the logic of Marx's theory of the form of value, and it becomes a twin of the theory of exchange. While the theory of exchange discussed the actual inevitability of the formation of money in the former subsection, the theory of the form of value discusses the theoretical basis of the existence of money. In other words, it explains what money is.[14] With respect to a commodity economy this theory helps us to understand that money was originally a commodity that itself had its own value, and how became a special commodity excluded from the general world of commodities. This is called the theory of commodity money, a commodity that appeared from the world of commodities but is excluded from the world of commodities. In Chapter 1 of this book, I have shown that the means of production is also a part of nature. As with this explanation, money should be understood as part of the set of commodities. Using this analysis, we should not misunderstand the sequence by imagining that gold and silver had appeared as money and then came to express the values of commodities. The true logical order of money and commodity is the reverse: commodities must be expressed in a certain form (the form of value), and so a certain commodity should become money. Commodities are not created by money, but their necessity created money.

2.3.3 The Various Functions of Money

However, it is also true that gold and silver have the characteristics necessary to become money. Because they are expensive (they need much labor), a large amount of value can be expressed by a small amount of these precious metals. They are homogeneous and can be split in any way.

Of course, in modern societies, coins are used only for small amounts money, and we typically use banknotes. Furthermore, some central banks seek potential digital currencies nowadays. However, even if banknotes and digital currencies can be circulated on the basis of a government's creditworthiness, when this goes beyond the border of a country, its reliability declines. Therefore, the only international currency that existed until the 1930s was gold (the gold standard).[15] After President Nixon unilaterally ended the exchange between gold and U.S. dollars in 1971, the dollar standard and the U.S. dollar itself lost international credit. In other words, without the backing of the original currency, namely, gold, the capacity of bills to circulate freely is limited.[16] Based on Marxist economics, China is

---

[14] This interpretation of these two theories was first clarified by Kuruma (1957) in Japan.

[15] Ancient nomadic people used sheep and horses instead of money for exchange with many other ethnic groups. These animals were also a kind of world money.

[16] This situation is developing and the Israeli government uses electronic payments to restrict the exchange of banknotes. This shows that the essential power of circulation does not come from the banknote itself but the value objectively owned by commodities. Money as gold has the ability to express the value of other commodities by its own value.

well aware of this fact, and is now accumulating gold to make the Chinese Yuan the key currency.

It is interesting that these international events reveal the essence of money. Gold is creditworthy as an international money because of its original labor-embodied value. In other words, it can mediate international transactions only on the basis that gold itself has value as a product of labor and can express the value of other commodities quantitatively on that basis. This is because gold has a unified function as the measure of value and the medium of circulation.[17]

In addition, even without gold, but if gold functions ideally, banknotes can be the measure of value and the medium of circulation, but in the case of real gold, its twin functions for hoarding and as a means of payment, respectively, are independent of each other. These twin functions also show that money must be a commodity in nature and a product originally embodying value. This is the key for understanding what money is.

2.4 CAPITAL AS SELF-VALORIZING MONEY: PRODUCTION FOR PRODUCING SURPLUS VALUE AS CONCRETE AND MATERIALISTIC CONCEPT OF ENTERPRISES

2.4.1 Profit-Taking as the Concrete Aim of Commodity Production

As a result of this analysis, we understand that actual commodity exchanges are exchanges mediated by money, and that individual producers first have to obtain money as the universal equivalent before they can exchange. We now turn to the concrete motive for obtaining profit because in the capitalist system, actual producers are basically capitalists and their purpose for engaging in production is to obtain profit. Obtaining money alone is not sufficient; the amount of money they obtain should be greater than the original amount of money. That gap between the original and the obtained money is the profit, which is the purpose of production.

This fact forms a turning point of this chapter because the original purpose of production is to consume (Section 2.1). However, in the capitalist era, people obtain goods for consumption by commodity exchange (Section 2.2). Furthermore, in certain conditions, obtaining money has become a precondition for that commodity exchange (Section 2.3), but all these activities are undertaken for the purpose of obtaining necessary commodities, which differs from the fourth purpose: profit-seeking. For this purpose, use value does not matter at all, and producers pay attention only to value itself. The body of profit is still money, but now its ability to be exchanged is not important and its "value" has become

---

[17] These functions as the measure of value and means of payment correspond to money in the theory of the form of value and the theory of exchange, respectively. See Takeda (1983, 1984).

valuable autonomously and purely on its own account. In this way, money plays a decisive role in changing the purpose of economic activity from obtaining other commodities to taking profit.

However, we cannot obtain profit just by using money. Money is used only to facilitate commodity exchange, because commodity exchange or commodity circulation must be an equal exchange where the value of what you hand over in exchange (e.g., currency) and what you receive (e.g., goods) must be equal. Therefore, an exchange alone cannot create profit. But in reality, there are profits. How is this?

2.4.2 The Appearance that Profit Comes from Circulation

To answer this question, first we must introduce the idea in modern economics that "surplus" can be produced simply through the process of commodity circulation. The concept of surplus consists of the producer's surplus and the consumer's surplus, where the former means "profit + fixed cost" and the latter means the "value of the utility of the goods – the price of the goods." Therefore, the surplus for the producer must be equivalent to profit if we ignore the fixed costs. In this way, modern economics discusses profit as if it is created in the process of commodity distribution.

Given this notion, we now have to investigate their surplus theory, to discover why commodity distribution itself seems to create profit. Then, we must take the consumer's perspective when they want to buy a certain commodity at a certain price without compulsion. In this case, the consumer must think this transaction is beneficial, and therefore, for them,

value of the utility of that commodity > price of that commodity.

The gap between both sides is the consumer's surplus. On the other hand, because producers also want to transact without compulsion, it must be

cost to produce that commodity < price of that commodity.

As with the former inequality, the gap between both sides in this inequality is the producer's surplus, and we must note here that there are two kinds of surplus and profit itself seems to be created only through commodity exchange.[18] Earlier, we said that commodity exchange is an equal exchange and therefore did not produce any profit. However, now we have profit. What happened? Alternatively, which view is correct?

To think about this problem, let us understand the miracle of the two inequalities above. These two inequalities show that, compared with the price of that commodity, on one side,

---

[18] "With reference, therefore, to use value, there is good ground for saying that "exchange is a transaction by which both sides gain" (Marx, 1887, 111).

the value is larger, and the other side, the cost is smaller, but if we combine these two inequalities, we obtain

value of the utility of that commodity > cost to produce that commodity

and can solve the problem by this inequality because this condition shows only that that cost should less than the utility in terms of money. If the price falls between the value of the utility of that commodity and the cost to produce it, any price will work for this transaction. In other words, this condition can be simplified into the formula:

utility of that commodity > cost to produce that commodity.

Otherwise, in the words of the Labor Theory of Value (LTV) explained in Chapter 1,

utility of that commodity > disutility of the labor necessary to produce that commodity.

That is basically same as the condition of the metabolic relation between humanity and nature, in which humans maximize the gap between the utility of the commodity and the disutility of the labor expense described in subsection 1.1.1 and 1.1.2. Thus, we must identify productive activities as a precondition of the existence of surplus (profit). Therefore, even if every profit can be created through the commodity distribution, the basis of profit is the productive activities.

2.4.3 Marx's Explanation: Profit Created from Productive Activities

Marx's way of explaining the origin of profit is somewhat different from the explanation above. Marx's explanation starts from the next formula

$G - G'$ (or $G - G + \Delta G$)

which shows the capitalists' expectation that the final amount of money (G') should be larger than the original (G'). Here, ΔG is the money added; in other words, profit.

However, that final G' cannot be attained directly, it must be attained indirectly through the purchase of the means of production and the labor power, using the original G and the sale of the produced commodity (and the realized G'). Therefore, we should change this formula into

$G - W - G'$ or $G - W - G + \Delta G$

where W indicates the purchased means of production and the labor power. Alternatively, using symbols, where p_m is the means of production and A as the labor power,

$G - W \left\langle \begin{smallmatrix} Pm \\ A \end{smallmatrix} \right. \cdot\cdot P \cdot\cdot W' - G'$

This formula is called the general formula for capital, and it shows that the total amount of value (G) of the original means of production and labor power and the final amount of value (G') of the commodity produced are different. This is because its first $G - W$ as a simple exchange must be an equal exchange, and the final $W' - G'$ is also an equal exchange as a simple exchange. Therefore, the secret of profit creation must be in the productive activities, expressed as "$\cdots P \cdots$" in this formula.

Furthermore, Marx investigates the difference between p_m and A among all factors of production. That is, because only the value of the means of production p_m is preserved, so the change in value must originate from labor power A. Therefore, Marx concludes that there is a difference between the value of labor power and the value created by that labor power. The former consists of the necessary labor to reproduce that labor power; namely, the necessary labor to produce the necessary goods and services to be consumed to reproduce that labor power. This definition is identical as the definition of all other commodities. They include food, clothing, and shelter for the producers and education for the next generation. The latter is the value created by the expenditure of labor power. While the labor power is created in the former case, the labor power creates in the latter case, and generally in the capitalist society, the latter is larger than the former. This gap is what is called surplus value, and at the level of the price it becomes profit. As you can see, even if this explanation is different from the one offered in the previous subsection, the essential content is fundamentally the same.

The existence of this gap has been essential throughout human history since primitive communism came to an end. It can be explained by imaging that the whole of human society needed a certain amount of labor to produce the necessary products (called necessary labor), and if productivity is enough high, then that amount of labor can be less than the potential working time. For example, a slave in a slave system could produce a larger quantity of products than the quantity of products necessary for himself (necessary products). This gap consisted of surplus products, which corresponds to surplus value in capitalist society. Because contemporary capitalist society is also simply another kind of class society, we need to compare necessary labor and working time in all types of class society.

On the concept of the trade of labor power (for instance, in distinction to the trading of slaves), I want to clarify that labor power is just the potential power to work and if the buyer fails to use it efficiently, then they cannot create profit. This is why Marxian economics uses the concept, the "trade of labor power," which is different from the concept of the "trade of labor." Labor itself is not a commodity: it is only the expense of the labor power that is tradable as a commodity. This theoretical correction does not appear in modern economics, which does not distinguish between the "trade of labor power" from the "trade of labor."

Another comment on the term, the "trade of labor power" is that it is different from the "trade of slaves." Workers in capitalist societies are not slaves because they do not sell the property rights of their labor power but only their *jus disponendi* and labor power for a limited time. That is, workers still hold their right to the property of their labor power, and in this sense, capital-labor relations in the capitalist era differ from the relation in the slavery system, where slaves sold themselves to the slave owners. However, also in the capitalist system, the necessary quantity of products for the reproduction of labor power and the quantity of newly produced products by that labor power differ. That is the point.

This chapter has explained the concrete human activities in a commodity-producing society, and that its actual purpose is the profit and its origin is productive activity; that is, labor. Consequently, these productive activities increase the amount of money from its original level through productive activities and commodity exchanges. This is the conceptual progress from money to becoming self-valorizing money. It can also be expressed as self-valorizing value, and this is the first definition of capital. Our discussion of the concept of money now must shift to the discussion of the concept of capital in the next chapter.

3

Capitalism in Industrial Societies: The Qualitative Character of Capitalistic Production and Capital as the Command Over Labor

The previous chapter discussed the characteristics of capitalism as a commodity-producing society and ended with the concept of capital as self-valorizing value. That is, we started from the concept of a commodity and ended with the concept of capital. We found that a capitalist society is not only a commodity-producing society but also one in which capital reigns as the ruler of labor.

This chapter starts by explaining why labor must be ruled and what exploitation is, deduced from the qualitative characteristics of productivity. The previous chapter stated that only a certain quantitative level of productivity could realize commodity-producing societies, but here we explain that the new quality of productivity of great mechanized industries, which was achieved after the industrial revolution, formed the capitalist societies as the industrial societies.

3.1 THE COMMAND OVER LABOR

3.1.1 Another Definition of Capital: The Command Over Labor

The last part of Chapter 2 defined capital as a self-valorizing value, which is understandable because the original meaning of the word "capital" is "a fund," that is, the originally invested money G. It is invested because it increases to G'; that is, it self-valorizes. This valorization is the purpose of capitalists, as Marx discusses in *Capital*, Volume 1, Chapter 4.

However, Marx's *Capital* provides another definition of capital, which is, "the command over labor, i.e., over functioning labor-power or the laborer himself" and "Capital further developed into a coercive relation, which compels the working class to do more work than the narrow round of its own life-wants prescribes" (Marx, 1887, 216). These words are not occasional. This part summarizes the first complete analysis of capital in the last part of *Capital*, Volume 1, Part 3: The production of absolute surplus value.[1] The basis of this

---

[1] In *Capital*, Volume 3, Marx sums up the essence of capital as "self-expanding value" as the result of the "command [of the] the labour of others" (Marx, 1907, 240).

self-valorization lies in the power to command labor, and therefore this characteristic is the definition of capital. The previous chapter discussed the world of commodity exchange and we assumed the perfect equality of all agents. In other words, it mentioned only the miracle that gives profit to capitalists even under the existence of equality and fairness and without fraud or tricks. This chapter discusses the asymmetric power balance between capitalists and workers; that is, the rule over workers by capitalists.

The need for this definition of capital arises because the previous chapter did not explain why the necessary quantity of products for the reproduction of labor power is smaller than the quantity of products newly produced by that same labor power. How does this phenomenon occur? Marx's answer is the command over labor, and his explanation does not contradict the fact that the trading labor power is also an equal exchange. It is because workers, as owners of the commodity of labor power, follow the principle of equivalent exchange when they conduct transactions, and the "coercive relation" takes place outside that exchange. That is, workers' freedom stands only before they enter the factory (they have no freedom inside them) and its coercive relation exists only in the factories. In this way, the gap between G and G' can be explained even under the condition in which capitalists pay the full value of labor power.

3.1.2 Okishio's Proof of Labor Exploitation: Fundamental Marxist Theorem

If we look only at the surface, we cannot understand the difference between the value produced by labor and the value of labor power. This is mainly because the value produced by labor is usually understood only as part of newly added value, of which the other part is understood as being produced by machines. This mistake made by modern economists has already been criticized in Chapter 1 of this book by showing that (1) machines are also produced by labor (and that labor is the only original factor of production)[2] and that (2) the total amount of labor input, including indirect labor to produce machines for its final product is determined by the balance of the utility of that product with the disutility of the expenditure of labor. If we understand these two points, we can avoid the misunderstanding that machines also produce value.

Marx called this type of profit-taking activity "exploitation" and the gap between G and G' "surplus value", in the sense that its fundamental basis is the coercive relation mentioned above. Strictly speaking, it should be called capitalistic exploitation, because exploitation itself occurs not only in the capitalist era, but also in general in all types of class societies.

---

[2] This point was illustrated by Izumi and Li (2005) providing the concept of "total labor productivity." In other words, they proposed the formulation that labor productivity should be the denominator of total labor, both direct and indirect, rather than the denominator of direct labor input only.

CAPITALISM IN INDUSTRIAL SOCIETIES

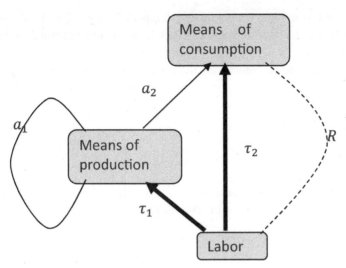

Figure 3.1 Symbols for the fundamental Marxian theorem

Even so, this description is not enough to persuade modern economists. Okishio (1967) therefore formulated a famous theorem called the "fundamental Marxian theorem" (FMT). This modelling was so shocking to modern Japanese economists that he was elected as the president of the biggest Japanese academic association of modern economists.[3] We introduce it here. Note that the following mathematical proof is essentially the same as Marx's explanation discussed at the end of Chapter 2, that exploitation is caused by the gap between the value of labor power and that created by the consumption of labor power. Therefore, readers who are not interested in mathematics can skip this section.

To prove this theorem, we start by setting some symbols in **Figure 3.1**. This figure is clearly based on Figure 1.3, with the difference that the means of production is also used to produce the means of production in this new figure. Concretely speaking, we input $a_1(>0)(<1)$ units of the means of production and $\tau_1(>0)$ units of direct labor to produce one unit of the means of production, and $a_2(>0)$ units of the means of production and $\tau_2(>0)$ units of direct labor to produce one unit of the means of consumption. Furthermore, we symbolize the price of the means of production and the means of consumption as p_1 and p_2, respectively, and the nominal

---

[3] As Professor Masanori Nozawa, my teacher at Kyoto University, was a friend of Professor Okishio, I also attended the joint seminar with both professors every month for two years as a graduate student.

wage per hour (money wage rate) as $w(>0$, equalized among sectors). In this case, the amount of the means of consumption taken by one hour of labor (real wage rate) can be expressed as $R = w/p_2$.

Using these symbols, we can express the conditions for the existence of profit of both sectors as

$$\left.\begin{array}{l} p_1 > a_1 p_1 + \tau_1 w \\ p_2 > a_2 p_1 + \tau_2 w \\ w = R p_2 \end{array}\right\} \quad (3.1)$$

The last equation expresses the relation between the real wage rate and the nominal wage rate, and using this definition, the two inequalities above become

$$\frac{p_1}{p_2} > \frac{\tau_1 R}{1-a_1}, \quad \frac{1-\tau_2 R}{a_2} > \frac{p_1}{p_2} \quad (3.2)$$

Summing them, we obtain

$$\frac{1-\tau_2 R}{a_2} > \frac{\tau_1 R}{1-a_1} \quad (3.3)$$

A further transformation yields

$$1 - R\left(\frac{a_2 \tau_1}{1-a_1} + \tau_2\right) > 0 \quad (3.4)$$

Here, we use the condition $a_2 > 0$.[4] This inequality expresses the fact that the range of the real wage rate is determined by the technological parameters in Figure 3.1, and now we will translate this inequality (3.4) into the key form:

$$1 - R t_2 > 0 \quad (3.5)$$

Here, t_2 is one of the solutions of the following value equations (3.6). That is, when we include the total direct and indirect labor in the production of the means of production and consumption as t_1 and t_2, respectively:

$$\left.\begin{array}{l} t_1 = a_1 t_1 + \tau_1 \\ t_2 = a_2 t_1 + \tau_2 \end{array}\right\} \quad (3.6)$$

---

[4] A strict explanation needs a proof that p_1, p_2, and w have positive solutions, but for simplicity, we omit it here.

Because these two equations have two unknown variables, we can solve t_1 and t_2 thus:

$$\left.\begin{aligned} t_1 &= \frac{\tau_1}{1-a_1} \\ t_2 &= \frac{a_2\tau_1}{1-a_1} + \tau_2 \end{aligned}\right\} \qquad (3.7)$$

Above inequality (3.5) can be obtained from the second equation in (3.7).

Here, we summarize the point of the key inequality (3.5). Because t_2 is the total direct and indirect labor used in to produce the means of consumption, Rt_2 is the amount of labor embodied in the means of consumption, which workers receive as 1 hour of wages. Therefore, inequality (3.5) indicates the amount of labor that workers receive from the capitalists after one hour of work is smaller than one hour. In other words, workers give one hour's labor to the capitalists, but the capitalists' pay them for less than one hour. If this exploitation is sustainable, it must mean that the amount of labor required to reproduce labor power (i.e., its value) is less than the amount of labor expended (the value it has created). This is the FMT (Fundamental Marxian Theorem) which proved labor exploitation.

In addition, dividing by t_2, we can transform inequality (3.5) to the material level as

$$\frac{1}{t_2} - R > 0^5 \qquad (3.8)$$

This transformation has another implication: the quantity of products produced at the material level (e.g., measured in terms of kg/hour) is larger than the quantity of products paid as real wage (e.g., again, kg/hour). In this way, the fact that workers obtain only a part of their production can be expressed at the material level, which is the essential nature of labor exploitation beyond any stage of class society.

Because this illustration addresses the case of average workers, it is possible that marginal workers (whose productivity is same as the wage level indicated as B in **Figure 3.2**) obtain all the products they produce by themselves.[6] If so, then in reality, where there are various levels of workers' productivity, the rational reaction of the entrepreneur should be

---

[5] If we reverse the denominator and numerator, this relation becomes $(p_2/w) > t_2$, which indicates that the amount of labor commanded for the consumption good, expressed by the left-hand side (the amount of labor that this good can buy using its own purchasing power), is larger than the embodied labor in this good. Thus, the exploitation proved by the FMT can be rephrased in various ways.

[6] Here, the quantity of product is measured as the part of product derived from the workers' contribution in the labor process. Additionally, it is possible that workers who are very unproductive are employed in some cases. In this case, there is negative labor exploitation, but this is an exceptional situation if entrepreneurs make a profit.

MARXIAN ECONOMICS

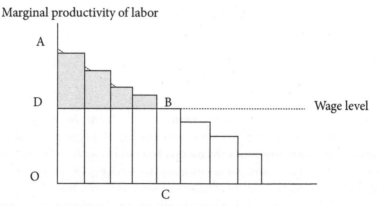

Figure 3.2 A case of diminishing marginal productivity

to employ the most productive workers first, the second-most productive workers second, and so on, and to stop employing them after their productivity falls to the level of the wage. This is the principle of the marginal productivity of labor and is shown in Figure 3.2. In this case, the shaded part called producers' surplus consists of the surplus products created by exploitation. Total employed workers produce the trapezoid ABCO part of the products in this figure but receive only the quadrangle DBCO portion of the products in the form of wages. But even without the principle of marginal productivity, exploitation may occur if workers receive only a portion of what they produced using a different condition, for example, a buyers' monopoly, shown by **Figure 3.3**. In this case, the levels of the workers are the same, and workers are exploited equally or as a whole.

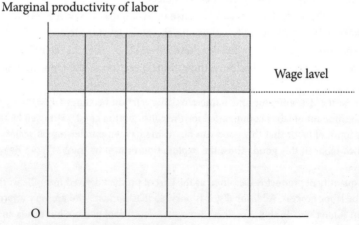

Figure 3.3 A case of constant marginal productivity

The point of this explanation is that the essential condition of capitalism is the existence of profit as a whole.[7] The late Professor Okishio proved that if profit exists as a whole, then it means there is exploitation. Starting from the condition that capitalists take profit, he shows that it is synonymous with the fact that the value of labor power (the value of consumer goods the workers received) is less than the value produced by labor power.[8] Because it is a fact that profit exists in this actual world, this must be a proof of exploitation. Just like the discussion of self-valorizing value from G to G' at the end of the previous chapter, Okishio proves that this self-valorization is the result of the exploitation of labor in the production process. Without exploitation, capital cannot be self-valorized.

Clearly, this exploitation itself has existed in human society ever since the end of the primitive societies. In other words, the ruling class has continued to exploit a portion of the labor product of the ruled class. Therefore, what we showed here is that capitalist society, which is based on equal exchange, is just one of these class societies, and the difference between it and other class societies is just appearance. This has allowed us to capture the nature of capitalism in all of human history.

3.1.3 Analytical Marxism's Class Exploitation Correspondence Principle

Although the existence of labor exploitation was proved in the above section, this explanation is still not sufficient to confirm that the cause of exploitation is a coercive relationship in the workplace. This is because we can explain the cause of exploitation only by introducing a certain coercive relation from outside the model, which shortens the total labor embodied in the consumption goods received by workers as wages per hour (Rt_2). So far, we have shown no more than the proof that the existence of profit is the existence of exploitation.

Therefore, we need to supplement the above explanation using a new model called the "class-exploitation correspondence principle" proposed by Analytical Marxism in the

---

[7] This condition, "as a whole" is also very important. It means that in the case of the original two inequalities in (3.1) one can be an equation but at least one must hold, for example, when $p_1 > a_1 p_1 + \tau_1 w$ and $p_2 = a_2 p_1 + \tau_2 w$. This can be understood by checking the calculations in this theorem. However, this is a condition in which the whole economy is disaggregated into two sectors, and therefore, if we look at individual enterprises, there can of course be some enterprises that have negative profits.

[8] In the sense that Okishio proved no more than this fact, Yoshihara (2008) argues that this theorem should be understood only as a proof of exploitation in general, not of labor exploitation. In my opinion, this criticism is right if we do not accept the labor theory of value. Therefore, to take this theorem as the proof of labor exploitation, the justification of the labor theory of value should complement it. I carry out this task at the beginning of Chapter 1 in this book. Additionally, Analytical Marxism, which is discussed below, does not accept the labor theory of value.

MARXIAN ECONOMICS

Table 3.1 Capital lending and exploitation in Analytical Marxism

	Capitalist			Worker			Total		
	Machine	Labor	Production	Machine	Labor	Production	Machine	Labor	Production
Initial holding	10	1	3	2	1	1	12	2	4
After lending capital	6	1	2.5	6	1	2.5	12	2	5
Rent transfer			+1.49			−1.49			
Final gain			3.99			1.01			

U.S.A. in the 1980s. It explains voluntary capital lending between two classes, one of which is the group of capitalists with much fixed assets (the means of production) and the other is the wage workers with few fixed assets and shows that capital lending causes exploitation.[9] It implies that the exploitation of labor needs a coercive relation between the capitalists and the workers, and details of this relationship are set out below.

For this explanation, we must first show what exploitation is, according to Analytical Marxists. For this purpose, we give a numerical example. In **Table 3.1**, a capitalist and a worker have the same amount of labor power, but different quantities of machines. Each unit of production is assumed to be three and one, respectively, making the total units of production equal to four. But if four of the capitalist's machines are lent to the worker, then each quantity of production must become equal, for example 2.5.[10] In this case, this society totally produces one additional unit of production, and a problem arises as to which group can take that part. Analytical Marxists suppose almost all of it is taken by the capitalists, under the assumption that capitalists are stronger than workers where capital is scarce, and

---

[9] Strictly speaking, Analytical Marxism defines five types of social class: pure capitalists who do not work in a workplace at all, small capitalists who also provide labor to some extent, self-employed people who employ no workers and carry out their work alone, the semi-proletariat who work using their own means of production but also are employed, and the pure proletariat living on a wage income only. In this book, we call the first two "capitalists" and the last two "workers." See Roemer (1980, 1982) and Mayer (1994).

[10] This assumption is not new because it is the same as the diminishing return to capital. In this example, when the number of machines increases from 0 to 2, the marginal productivity is 0.5 (assuming that production using zero machines is 0). When it increases from 2 to 6 it becomes 0.375, and when it increases from 6 to 10 it becomes 0.125.

they called this part of additional acquisition exploitation, because this additional production was created by another person's labor.

One point about this explanation is that workers usually accept this unfair distribution. If you check the numerical examples in this table carefully, you find that they show that it is better for workers to accept this contract than to refuse it. If they refuse it, they can realize only 1 unit of income, but if they accept it, they can realize 1.01 units of income, and it is apparent that the latter is better. Analytical Marxists were thus able to explain successfully such transactions in reality, and theirs is a better explanation than any other to describe what happens in reality as a result of such voluntary contracts.

However, we must note that this unfairness also depends on the balance of power between capitalists and workers. For example, if workers' bargaining power is much stronger, then we can suppose that workers transfer (pay rent) 0.51 units and capitalists should accept this proposal, because to accept this unfair contract is better than to refuse it in this case. If they refuse this contract, then they can realize only 3 units of income, but if they accept it, they can realize 3.01 units of income. This means we need to understand the importance of their relative bargaining power, expressed as the capitalists' command over labor or their coercive relationship with labor, which we are discussing in this chapter.[11]

We must remember that this real and everyday exploitation occurs not when lending capital but in factories. The relation between borrowers and lenders and the relation between capitalists and workers are different. Employed workers are under the direct command of the capitalists and effective command is important to capitalists because this effectiveness directly determines the degree of exploitation. This fact shows that the essence of employed labor is the capitalists' command over it or its coercive relationship with labor. In fact, in *Capital*, part 3 of Volume 1, totally explored the inseparable relation between the command over labor and the nature of the labor process, and first discussed the labor process and the process of producing surplus value separately in Chapter 9, then integrated these two processes in Chapters 8 to 11. Thus, we cannot discuss command over labor separately from the labor process that is equipped with machines. This is employed

---

[11] Analytical Marxists themselves say that the market relationship of downward pressure on wages due to the existence of unemployment determines labor's share, and they avoid using the term "coercive force." However, it is through changes in the balance of power between labor and capital that market pressures, change labor's share, which is one of the coercive forces in any case. Alternatively, these effects on labor's share can be altered without going through the market: for example, by the intervention of the state. However, both kinds of downward pressure constitute the problem that emanates from the balance of power between labor and capital. I argue that, regardless of their perceptions, this is the most important implication of the framework of Analytical Marxists. As for the effects of the balance of power on the growth path, see Addendum V of this book.

labor, and the question here is how these laborers work and what means of production they use. These questions address the quality of productivity. While Chapter 2 of this book discusses only the quantity of productivity, this chapter discusses its quality.

3.1.4 The Nature of Employed Labor in Contestable Exchange Theory

Samuel Bowles and his post-Walrasian academic school focused on the nature of employed labor using game theory. Although Walras was unable to describe the market system fully, it is not enough to understand the nature of capitalism, in which various types of bargaining are used. This school argues that the most important transaction is the buying and selling of labor power, and terms this theory "contestable exchange theory." We just stated that the beauty of Analytical Marxism is that it shows the importance of the power balance between capitalists and workers, but this issue cannot be discussed without understanding the nature of employed labor. The distribution between capitalists and workers shown in Table 3.1 can be discussed by a pure market theory, which determines the equilibrium by indifferent curves. However, contestable exchange theory criticizes such studies for failing to express the nature of employed labor. Therefore, let us move on to the concrete contents of contestable exchange theory, as shown in **Figure 3.4**.[12]

The $v = v^*$ curve in this figure is a worker's special indifference curve that shows that a lower work effort decreases utility by raising the possibility of being fired and a higher effort increases the disutility to work, conversely. These opposite effects make this curve convex left, and that the worker's optimal reaction to wage w^* becomes **a**. This is because if the worker selects any point but **a** to wage w^*, either their disutility to work or the possibility of being fired becomes higher. For these reasons, the worker's optimal response function becomes $e(w, m; z)$. Here, e is the worker's effort to work, m is the capitalist's cost to monitor the worker's labor, and z is the fallback option for the worker can receive when they lose that job (e.g., receiving unemployment compensation).

When the worker reacts in this way, the capitalist must minimize the unit cost of production, which is inclicated as an inclination of the straight line that starts from point m^* on the horizontal axis to the upper right) (m^* is the optimal monitoring cost determined outside this figure). This straight line is the best reaction a capitalist may make to minimize the unit cost shown as the inclination, given the worker's optimal response function $e(w, m; z)$.

Thus, it seems that point **a** is the optimal for both sides, but the contestable exchange theory reveals that this is not true. The shaded part shown in this figure is better than **a** for both of them, and therefore, if workers can stop their members from freeriding, both sides must negotiate (called "contestable exchange" in this theory) rather than undertaking the usual transactions in the market. Also, even if the deal between labor and capital is not binding, it is possible to implement a better agreement in a game-theoretic sense if one

---

[12] Chapter 8 in Bowles (2004) provides a detailed explanation.

CAPITALISM IN INDUSTRIAL SOCIETIES

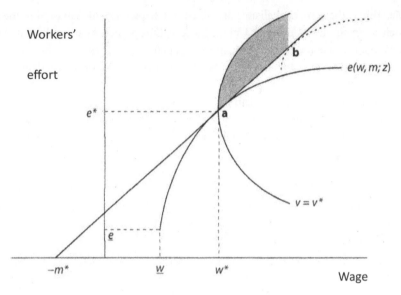

Figure 3.4 Contestable exchange of wage and workers' effort

side can take revenge on the other for their betrayal. In this way Post-Walrasian economics regards this contestable exchange as the nature of employed labor.

In my opinion, this point of view is the same at a deep level as the special Japanese tradition of Marxist economics called Unoist economics. This Japanese school developed a category for the forced commodification of labor power, which implies a fundamental difference between lending capital and employed labor. How much effort employed workers expend is not determined when the labor contract is signed. It is determined in the workplace after the contract. This essence of employed labor is discussed as the imperfectness of the labor contract in modern economics.

3.1.5 The First and Second Definition of Exploitation: What Is a Capitalist?

We can understand the limitations of Analytical Marxism by focusing on the special nature of employed labor compared with contestable exchange theory, but there is a much more important problem that arises from characterizing the capitalist. We have noted that Analytical Marxism defines capitalists as those with much fixed assets; in other words, the concept of social classes is defined by the amounts of assets held. However, in some cases, this definition is not enough to understand the case in which all people are equally exploited by the state. In this case, there is no exploitation among people, but the state directly commands labor and accumulates capital using the surplus value exploited from the people. This is the state capitalism[13] discussed in Chapter 1, which existed in the real world, such as in the former Soviet Union, Eastern Europe before around 1990, Mao's China, Japan, and Germany before 1945, Indonesia before 1967, and India before 1991.[14] In these cases, the state functioned as a capitalist, requiring a more profound definition of the term, capitalist.

In our understanding, capitalists can be understood by their function. As Marx said, a capitalist is only capital personified. In our words, this means a capitalist is just the mouthpiece for realizing the will of capital to exploit labor and accumulate itself; namely, self-valorization. According to Marx:

> The expansion of value ... becomes his subjective aim, and it is only in so far as the appropriation of ever more and more wealth in the abstract becomes the sole motive of his operations, that he functions as a capitalist, that is, as capital personified and endowed with consciousness and a will.[15]

In addition:

> As capitalist, he is only capital personified. His soul is the soul of capital. But capital has one single life impulse, the tendency to create value and surplus-value, to make its constant factor, the means of production, absorb the greatest possible amount of surplus labor.[16]

As we have already explained, "*capital*" has a will to self-valorize, but because capital itself does not have a mouth, someone must advocate its will in human society; that spokesperson is the "*capital*ist." The important point now is not the fact the capitalist earns a good income from surplus value but whether it can lead to capital accumulation or not; in other words, whether it realizes capital's will to valorize or not. Therefore, poor capitalist: capital personified without consuming for its own sake is much better if it functions as the command over labor. Furthermore, bureaucrats, politicians, and ideologues who oppose the workers' movements and promote capital accumulation can also be regarded as a part of the capitalist class. Because bureaucrats and politicians use state organizations for this purpose, this type of ruling system can be called state capitalism; that is, *uklad*. If this *uklad* covers the whole of society, then it can be called state capitalism as a mode of production.

---

[13] A class society in which the state itself becomes the ruler before the capitalist era is defined as a state slavery system and state serfdom in Nakamura (1977).
[14] I made this assertion (Onishi, 1990, 1992) before the collapse of the Soviet Union.
[15] Marx, 1887, 107.
[16] Marx, 1887, 162.

CAPITALISM IN INDUSTRIAL SOCIETIES

Even with this critique of Analytical Marxism, we can agree with its great achievement in its focus on the effects of assets and income disparity. Because Analytical Marxism is relevant to the U.S.A.; a terribly unequal society, it is natural for them to propose such a framework to analyze such disparities. In this sense, we must appreciate their concept of exploitation and thus call it the "second definition of exploitation." Of course, the first definition is different, as we define exploitation as what is sometimes performed by the state and sometimes by private agencies, such as employed managers or poor small capitalists.

3.1.6 Capitalism as a Society in Which the Means of Production Plays a Decisive Role

Another important characteristic of Analytical Marxism is its focus on the means of production. However, while the interest of these theorists is in property in the means of production, our interest is in its technological character. As we note, we should find that the nature of capital is command over labor in the process of production, and through the integrated unity of the labor process and the process of producing surplus value. In other words, the framework of historical materialism is that technology determines the social system.

Therefore, we first need to identify capitalism itself as the system of machine-based production that developed after the industrial revolution, which finalized the era of tool-based production. From the basic viewpoint of this book, which focuses on the importance of the means of production, we now proceed to the next step to discuss the nature of capital as the command over labor based on machine-based production, because now we focus on the labor process or process of production. However, our explanation of machine-based production starts from a comparison between it and tool-based production in the previous period.

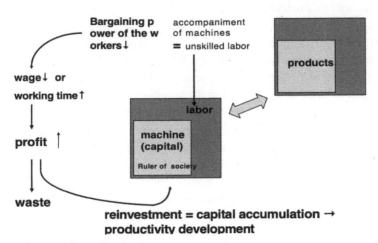

Figure 3.5 Mechanism of capitalist development

For this purpose, let us recall the characteristics of tool-based manufacture before the industrial revolution. In that period, there was no way to increase productivity without leveling up the skills of the craftsperson, because the decisive factor to determine productivity was not tools, but skills. Imagine a situation in which a great craftsperson produces beautiful goods but bad craftsperson produces only bad goods using the same tools. This difference does not come from the quality of the tools, because both workers use the same tools, but from the difference in skill. Thus, the most important task in these societies was to brush up skills, and for this purpose, those societies formed a special social system.

The most important institution among them in the feudal human relationship was the apprenticeship between craftsperson and masters, under which workers repeated the same work thousands of times every day under the absolute control of their masters. Because it was only by such obedient repetition could their skills be improved, this special human relation was established and maintained, and its own ideologies were developed for this purpose. In East Asia, this special ideology was crystalized as the Confucian spirit (or the neo-Confucian principle that emphasized imitation and habituation[17]), and elders were respected as highly skilled craftspersons, ending up with formulating a kind of seniority system because the elders were the most productive. This is completely different from modern societies in which we have a retirement system and blame elders for holding onto their positions for too long. Therefore, the non-existence of a retirement age had a special technological base in tools before the industrial revolution.

This transformation of the system occurred not only in terms of human relations in the workplace but also in the size of the craft workshops and marketplaces. For example, the apprenticeship system could not train too many craftsperson as this would hinder the deep and intimate human relationship that was necessary between the masters and the craftsperson, and this nature limited the size of the craft. Readers might imagine small craft workshops in the feudal era but cannot imagine any huge factories of craftspeople. Such small craft workshops are technically conceptualized as "small-scale modes of production" in Marxist economics. Furthermore, this necessity to keep a craft workshop small in size also needed restrictions on market competition because market competition drives the divergence of craft sizes. Stronger crafts become larger and weaker crafts go bankrupt through market competition. It is very difficult to keep small craft workshops that are equal in size; therefore, all feudal societies formulated their own restrictions on competition using the guild system, which was a voluntary agreement among craftspeople to keep the peace among themselves and keep their size of production small.

However, this beautiful and peaceful relation ended with the appearance of machines. Machines broke the skill base of production by making it useless, and from that time the quality

---

[17] Chapters 5 and 6 in Kakiuchi (2015) explain this logical transformation from the original Confucianism.

and quantity of products have been determined not by feudalistic skills but by the quality and quantity of machines. Skilled workers have been replaced by unskilled workers who have little bargaining power against the capitalists, as it has become very easy for capitalists to fire them. Unlike the former skilled craftsperson, whom even absolute masters could not fire, unskilled workers can be replaced by many substitutable workers waiting in the labor market. Their weak bargaining power, which is the result of them becoming only the accompaniment of machines (in *Capital* and *The Communist Manifesto*) worsens their working conditions, such as their wages and working times (via the law of pauperization)[18] and increases profit, most of which is invested in machines. This is capital accumulation (self-valorization), and it results in an increase in productivity (economic development). In this way, machines act as the rulers of society, and their self-valorization—their expansion as their own movement—becomes the objective task of society. Because machines themselves are called capital usually, societies whose task is the self-valorization of machines can be redefined as societies whose task is the self-valorization of capital. Thus, we must call our society after the industrial revolution a "*capital*ist society;" in which the primary task is capital accumulation.[19]

However, this self-valorization of machines realizes economic growth;[20] so that now, total production is not determined by the workers' degree of skill (or human development) but by the quality and quantity of machines in capitalist society. This causal relationship is due to the nature of capitalism, in which only capital accumulation can realize economic development. From its birth, capitalism has developed by building this self-valorizing mechanism.

Therefore, we assert that the technological nature of the new means of production (machines) after the industrial revolution necessitated the self-valorization of capital, and, for this purpose, it also necessitated bad working conditions, a decline in labor's share and so on. Strong command over labor is the necessary condition for all these purposes.

This book is discussing the technological determination of the society as a whole, and this historical materialism is now concretely explained as the qualitative character of machines as a new technology which makes capitalism inevitable. The command over labor is the

---

[18] However the bad working conditions must be enough, to reproduce labor power as a class because, for capital to continue production, it is necessary for the total population of the working class (labor power population) must not decrease. For this purpose, the wages for workers as a class must cover the cost of raising the next generation. This bottom line is called the subsistence wage.

[19] Marx opposes the notion that the nature of capitalist production is the accumulation for the consumption and says that the aim and compelling motive of capitalist production originally was not consumption, but "the snatching of surplus-value and its capitalisation, i.e., accumulation" (Marx, 1894, 306).

[20] "[T]he scale of the process of production (productive power development remaining the same) depends on the mass and volume of the means of production which a given quantity of labour-power can cope with" (Marx, 1894, 62).

essential element of the capitalistic relation of production and also the inevitable result of the quality of capitalistic productivity.

In addition, we want to clarify the difference between our understanding and the ordinary understanding of capitalism, in terms of the market and private property. Japanese Marxist economics has already developed to recognize that capitalism is differentiated from the market system or the system of private property, but among ordinary Japanese people and those in many other countries, the market and private property are seen as the fundamental characteristic of capitalism. In our opinion, there is no better term than "capitalism" to express a society in which capital accumulation is the primary task; in short, a society that exists for capital. If someone wants to characterize a market-oriented society, they should call it "marketism," while the appropriate name for a society based on private property would be a "private property society." They must not use the term "capitalism." Nevertheless, capitalism has a close relation with the market system and private property, and therefore, we need to clarify its relation with the market system first, although we have discussed it in terms of a commodity exchange society in the previous chapter.

In the previous chapter, we assumed that the capitalist society is fully covered by the "commodity production"; in other words, it is a society where there is equal exchange in the markets. Furthermore, it is true that markets promote capital accumulation as the essential task of capitalism. For example, severe market competition among capitalists decreases wages, accelerating workers' competition, while it also promotes capital accumulation, improving productive efficiency.

However, the market is not the definition itself of capitalism, but it simply promotes capitalism, and can be regard as capitalistic only when it promotes capital accumulation. Though capitalist society is a commodity-producing society, the characteristic captured by the term capitalism is not the market but that of capital accumulation. Thus, capitalism can exist in societies with limited markets or without markets at all, such as in the case of state capitalism, if it promotes capital accumulation.

Another difference in the way of understanding capitalism is the focus on private property, and we agree that this is also a very important condition for capital accumulation, especially in the early stage of capitalism. For example, let us imagine the very primitive stage of capitalism in which property rights are ambiguous. In this case, capitalists could not establish their right to command labor, and therefore, enlightenment thinkers declared that property rights which ensures capitalists to command labor. In addition, and in modern history, property rights in the so-called socialist system in the Soviet Union, Eastern Europe, and Mao's China were also ambiguous, and therefore, in these countries the state functioned as the capitalist instead of private capitalists themselves.

However, we must again say that the existence of private property itself is not the definition of capitalism: it simply promotes capitalism in certain conditions, and it can be regarded as capitalistic only when it promotes capital accumulation. For example, although consumers' cooperatives or private universities and corporatized national universities do not have their own special owner, all of them are managed under fundamentally the same principle as other capitalistic private companies, such as by cutting wages and transferring major revenue for building new stores and new departments. In general, we must identify the definition itself from the characteristics and extract the essence of the situation accurately.

3.1.7 Dialectic Understanding of Capital and Capitalism: Definition—Relation—Concrete Entity

The last discussion in this section takes on another disagreement with our definition of capitalism. Such critics argue that our definition does not differentiate between capital and machines. They say that capital is not a substance but a relation, and the term capital includes not only the means of production (machines) but also labor power.

However, in response to the first criticism it might be enough to say that our explanation also focuses on the relations of production; in other words, command over labor based on the technological character of machines. This is the fundamental relation of capitalism, and we see machines not just as a substance but as a coercive power, which is the essential basis of exploitation, as we discussed above. Historical materialism is a theory that states that social relations are determined not by ideal creatures but by material substances.

On the other hand, the second criticism of our definition is very important, and it raises the problem of how to understand the dialectical methodology. Even if both machines and labor power are included in capital, the dialectical method aims to determine which is fundamental. Thus, this criticism is deeply related to the problem of dialectics. We therefore discuss the essential methodology of dialectics below.

In general, the dialectical methodology can be expressed as follows. Suppose that there is something A that society generally recognizes. However, this A does not exist only as A, because it can demonstrate its existence as A in relationship with others. Otherwise, this A cannot remain an A for a long, but continuously is being transformed. Therefore, this A can be understood wholly only when it is understood as a relation that includes a *non-A* in the former case, or as an identical universality = "essence" of itself including the *non-A* that A has just become in the last one minute, in the latter case. Our recognition of this substance proceeds as a process in which

$$A - non\text{-}A - A'$$

Therefore, even if the recognition of A is not perfect, it is the original determinant without which this recognition process cannot begin. In other words, this recognition process can be understood as

definition—relation—concrete entity.

That is, the definition is the original words used to provide an explanation in one sentence. We can find many such examples in dictionaries, but they are not enough to understand wholly or concretely because these few words are selected by a certain process of abstraction and omit other aspects of that object. Therefore, in this case, to understand the term wholly and concretely, we need to know the relation between the first words (definition) and the other aspects that have been omitted. If this relation can be understood as an entity, then this final goal must be the whole and concrete recognition as a concrete entity.

Our recognition of capitalism is the result of similar processes of recognition. Our starting point in this recognition is the occurrence of capitalism by the industrial revolution, and therefore, capitalism is defined as a machine age. This understanding corresponds to the general characterization of the industrial age by modern economics, and basically reflects its primary influences.

However, we need to do more than recognize capitalism as the machine age because it includes the relation between machine and labor (the capital–labor relation), which has changed after apprenticeships were dismantles and direct producers ceased to exist. Therefore, the complete and concrete recognition of the term must cover these two primary and secondary factors and integrate them as a concrete entity.

Actually, this analysis of capitalism as the integration (sublation) of these primary and secondary factors can be explained in a much concrete way because this capital–labor relation has the following structure: "thesis—antithesis—synthesis"; in other words, "A—non-A—A'". This is because that capital–labor relation increases profit and valorizes capital, which is definition of capital itself. Thus, A is redefined as A', and we can extend this method of redefinition into a complete and concrete understanding of the capitalist economy, including the non-capitalistic sector, or a more complete and concrete understanding of capitalist society including its superstructure. These relations are shown in **Figure 3.6**.

In this figure, three syntheses are indicated:

(1) Capital cannot stand alone and necessitates wage labor as its own obverse. Then, only by being integrated with the latter it can formulate capital-labor relation as the concrete entity in the direct production process.
(2) However, this capital–labor relation does not cover the whole capitalist economy. In other words, even if we can say the Japanese capitalist economy was already established in the pre-war period, most Japanese were engaged in non-capitalistic sectors such as agriculture during that period. That is, even if its advanced sector, which exhibited capital–labor relations, was the minority in terms of its population, it is possible that it determined the fundamental structure of society, and sometimes exploited such backward sectors. Thus, the concrete entity of the capitalist economy can be understood only by integrating the capitalistic sector as the principal factor with the non-capitalistic sector as the secondary factor.

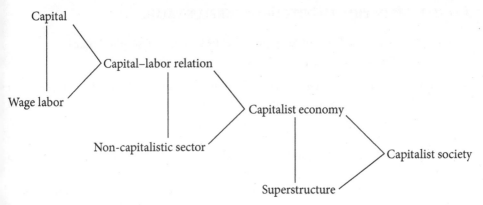

Figure 3.6 Logical structure of the analysis of capital and capitalism

(3) However, the above analysis is still not enough to understand the whole of capitalism concretely because the capitalist system consists of not only the economic system but also a social system. Therefore, for a whole and concrete understanding, we must also understand its superstructure, such as its ideology and political system. In summary, what should be understood here is that a dialectical recognition has a stratified structure, and the recognition of its structure proceeds as:

$$A - non\text{-}A - A' - non\text{-}A' - A'' - non\text{-}A'' - A'''.....$$

This is Hegelian dialectics, in which the same categories repeatedly appear in the different strata, such as "being" in the chapter on being; "determinate being as such" in the chapter on determinate being, "being-for-self as such" in the chapter on "being-for-self," "pure quantity" in the chapter on quantity, "absolute ground" in the chapter on ground, and so on. A can be defined again and again as A', A'', A'''... reiterated in the different dimensions.

This explanation is just on the special category of capitalism, but it is worth noting that everything can be defined and redefined repeatedly in different dimensions in dialectical logic, unlike in formal logic. While is true that *Capital* states that capital includes both the means of production and labor power, this is a definition of capital along just one dimension. However, it has a different definition in different dimensions. That is, the original definition before analysis and the new definition after the analysis.

Additionally, this descriptive methodology, based on the 3–3–3 structure, is also applied in this book. For example, the structure of this book, starting with Chapter 1, then Chapters 2, 3, 4, and 5, then Chapter 6, has the form of "universal, special, general". Chapters 2, 3, and 4 in this book correspond to the internal dialectical structure in the first volume of *Capital*. I invite you to examine the contents of this book from this viewpoint.

3.2 CHANGES IN THE MAGNITUDE OF SURPLUS VALUE

3.2.1 The Production of Absolute Surplus Value: Production of Surplus Value by Prolongation of Working Time

In Section 3.1 we outlined the essential understanding of capital and capitalism. This section discusses changes in the magnitude of surplus value as the result of capitalists' command over labor and labor exploitation. For this explanation, first we provide two special Marxist terms: constant capital and variable capital.

Here, the term constant capital expresses the means of production. Marx used this term because the original value of the means of production and the value transferred to its products are the same (constant). For example, when we use 1 kg of flour to make a cake, the exact value of 1 kg of flour is transferred to the cake. However, the value of the product produced is greater than the total value of the means of production and labor power bought by the capitalist. This is because the value added by labor power to the means of production is larger than the value of its actual labor power, as we saw in the previous section. Marx thus used the term "variable capital" for labor power, as it is a kind of capital. Of course, the gap between its original value and added value is the surplus value.

Then, to explain the relations among constant capital, variable capital, and surplus value, Marx introduced three symbols c, v, and m, respectively. Although surplus value is expressed as "s" in English textbooks, here, we use "m" (Mehrwert), according to Marx's original use in German. In this case, the value of a commodity is $c + v + m$, the profit rate is $m/(c + v)$, and the rate of surplus value (exploitation rate) is m/v. However, these basic rates can also be expressed using Okishio's terminology, as explained in the previous section. Okishio's two sector value equations are

$$\left. \begin{array}{l} t_1 = a_1 t_1 + \tau_1 \\ t_2 = a_2 t_1 + \tau_2 \end{array} \right\} \tag{3.9}$$

Here, $a_1 t_1$ and $a_2 t_1$ represent constant capital (c) and τ_1 and τ_2 are $v + m$, but the latter parts can be divided into v and m using Rt_2, which indicates the value of the hourly wage rate (see subsection 3.1.2 again). Therefore,

$$\left. \begin{array}{l} t_1 = a_1 t_1 + \tau_1 R t_2 + \tau_1 (1 - R t_2) \\ t_2 = a_2 t_1 + \tau_2 R t_2 + \tau_2 (1 - R t_2) \end{array} \right\} \tag{3.10}$$

This is the complete form of $c + v + m$ in Okishio's value equations, and in this case, the profit rates of these two sectors become

$$\left.\begin{array}{c}\dfrac{\tau_1(1-Rt_2)}{a_1t_1+\tau_1Rt_2}\\[2mm]\dfrac{\tau_2(1-Rt_2)}{a_2t_1+\tau_2Rt_2}\end{array}\right\} \quad (3.11)$$

respectively, and their rate of surplus value becomes

$$\dfrac{1-Rt_2}{Rt_2} \quad (3.12)$$

which is identical for both sectors.[21] Therefore, Okishio's model is a sophisticated extension of Marx's model, and it shows that there are two ways to increase surplus value ($1-Rt_2$); that is, by cutting R and decreasing t_2.

On this issue, Marx first focused on the way to cut the real wage rate (R) under the given t_2: by prolonging working time in general. While every capitalist wants to cut wages directly, this method is limited because the working class as a whole need to reproduce by consuming a certain amount of goods. In other words, while the reproduction cost of the working class is basically fixed, working hours are relatively flexible. For example, even if a capitalist prolongates working hours by 20 percent, this does not increase the reproduction cost of labor power directly. This 20 percent extension means a 17 percent cut in the hourly real wage R, but its direct effect is just to reduce the workers' free time. Marx regarded this method of valorizing surplus value as the essential and most fundamental method, which he called the production of absolute surplus value.

In fact, our Japanese society is a rare example of very long working hours. While the average working time in Japan is falling as a result of the increase in the use of part-time workers, the real working time of full-time workers is totally different from that in European countries, and Japanese employees work from early morning to late at night, way beyond their physical capacity. Having free time or not is a decisive difference between being a wage worker and a slave, because by nature wage workers sell their labor only for a limited number of hours, while slaves must provide it throughout their entire lives. However, this means that if wage workers have no free time they are equivalent to slaves. Hence, Marx emphasized that the real wealth is having freely disposable time outside that needed for direct production.

Therefore, workers press for shorter working hours to avoid being slaves, while capitalists contrive to prolong their working hours. This is the class struggle over working hours,

---

[21] Marx introduces a theoretical assumption that the rate of surplus value is equalized among industries and calls this rate the "general rate of surplus value." Behind this description, we can suppose an inter-industrial movement of labor power as a result of workers' competition, which equalizes the rate of surplus value (Marx, 1907, 131).

which are flexible and indeterminate, and Marx explained the real history of such class struggles from the fourteenth century in England. Surprisingly, the first law on working time in the world was the Statute of Laborers in 1349 in England. This law tried to lengthen working hours by compulsion, while modern labor laws compulsorily shorten them. In the fourteenth century law the state represented capitalists' demand directly, and England persisted in this path, enacting similar factory acts in 1496 and 1562 to lengthen working hours.

Centuries later, the English factory acts were promulgated to reduce working hours, and the first effective law to do so was passed in 1833.[22] A workers' rebellion and the development of the labor movement promoted this change. In other words, laws are not simply a way of regulating the relationship between capital and labor from outside society; they are formed and changed to reflect the real balance of power between capital and labor. The present situation in Japan, in which laws on labor standards have little effect, also reflects the real balance of power. Without changing this real balance there will be no reduction in working time.

On this point, Marx added an important fact: regulations on child labor in certain industries also lead to similar regulations in other industries as a result of the pressure of capital to seek equal exploitation between industrial conditions.[23] In other words, when workers want to improve their working conditions, it is not necessary to enlarge all industries at once, but enough to start with individual industries or individual companies. This first breakthrough for the improvement of labor conditions can occur through the activities of the labor union at the level of the enterprise or industry. As a result of the improvement in some companies and industries, capital that accepted the new conditions will now apply pressure on other capital as a labor ally. Workers' praising some companies or industries for their advanced working conditions aims for this effect.

3.2.2 Production of Relative Surplus Value: Producing Surplus Value by Improving Productivity

The previous section discussed one way to increase surplus value, and we now move on to the next method: decreasing t_2 (the necessary labor time to produce one unit of consumption goods) in Okishio's model. This decrease reduces Rt_2 and then increases $1 - Rt_2$.

However, if this is the case, then what does reduce t_2? For an answer, we must go back to the second equation in (3.7):

$$t_2 = \frac{a_2 \tau_1}{1-a_1} + \tau_2 \tag{3.13}$$

---

[22] The five factory acts issued from 1802 to 1833 officially aimed to limit working time, but in reality, they remained a dead letter due to their restricted application to parish apprentices only, as well as the lack of money laid aside to supervise their enforcement. See Marx, 1887, 83.

[23] Marx, 1887, 271.

This equation indicates that there are many ways of changing t_2, not only via the parameters of productivity in the means of consumption sector (a_2 and τ_2) but also in the means of production sector (a_1 and τ_1). In summary, the improvement in productivity in any sectors decreases t_2 and thus increases surplus value.[24] Marx called this method to increase surplus value the production of relative surplus value.[25]

In an actual economy, how does productivity rise? Some types of technological progress lead to the growth of productivity, and Marx lists and analyzes three ways in which this occurs: by cooperation among workers, the division of labor, and machine-based production.

The most important effect of cooperation comes from a kind of saving, for example, the collaborative use of a workplace, storehouse, container, tool, or installation. There also are some special types of work that can be conducted only by using combined labor, such as lifting heavy weights, turning a winch, or removing an obstacle. Furthermore, emulating and stimulating animal spirits, which also are the result of cooperation, can heighten the efficiency of each individual worker, for example, two people carrying bricks side by side.

However, instead of this simple kind of cooperation achieved by performing the same work side by side, work efficiency can be further enhanced if the division of labor is introduced into manufacturing. For example, hand skills can be improved by specializing individual work, reducing the interruption time of doing different work at the same workplace, improving efficiency by allocating workers with different degrees of expertise and personal characteristics to appropriate work, and specializing and improving laborers' tools.

However, these methods are available only at the level of organizational modification, and the introduction of machines resulting from the industrial revolution was absolutely necessary for the full development of productivity. This is because machinery productivity, measured in terms of the degree to which the machine replaces the human labor power, became dramatically higher than it was previously. In *Capital*, Marx gives the example of self-acting mule spindle machines, which shortened the time to spin 366 pounds of cotton from the 27,000 hours it took prior to the machine's introduction to 150 hours. In this case, even if this machine can spin only 366 pounds of cotton during its lifetime (in the extreme assumption that it will break after 150 hours of use), if this machine can be produced for less than $27,000 - 150 = 26,850$ hours = $2,685$ working days ≈ 9 working years labor time, its introduction is productive. The improvement in productivity is obvious.

---

[24] Although any improvement in both sectors affects t_2 in this model, it is different in the luxury goods sector because no improvement in this sector reduces the embodied labor of workers' consumption goods (wage goods). See the three-sector model, which includes the luxury goods sector in note 32 in subsection 4.4.3.

[25] Note that this is a devaluation of the means of consumption for workers. In other words, the devaluation of luxury goods has no effect on the value of labor power.

3.2.3 Production of Absolute and Relative Surplus Value: Real Subjection of Labor to Capital

Incidentally, it is not possible to distinguish clearly between absolute surplus value and relative surplus value. According to the previous explanation, if surplus value is added by prolongating working hours or strengthening labor intensity, it produces absolute surplus value, and if we add it by decreasing the reproductive cost of labor power due to the rise in productivity, it produces relative surplus value. However, the latter is also subject to forced labor over the working hours required to reproduce that labor power, and the former necessitates enough labor productivity for the labor paid as wages to be less than the working hours expended by the workers.

We can explain this point by the fact that three ways to improve productivity also strengthen the command of capital over labor by technological necessity. For example, cooperation itself necessitates a certain command over many workers, and strengthens labor intensity. The division of labor forces the workers in the latter process to follow, without delay, workers in the previous process. This also strengthens labor intensity. Finally, the introduction of expensive machinery required companies to maintain high levels of use and long working hours because they must use the machine sufficiently before it becomes outdated.

However, the most important point is that the required cooperation and division of labor was actually realized only by the use of machines. Let us imagine that the machinery system is also a kind of cooperation and division of labor, because it gathers many workers in one factory and makes them work together and take responsibility for different processes. Thus, the technical necessity (Marx, *Capital*) of cooperation and the division of labor could be established only by using machines.[26]

---

[26] This relation is clearly defined by Sekine (2017), who provides a unique production function to show the special characteristics of the great mechanized industries. To discuss the relationships among workers, such as their cooperation and the division of labor, we must specify the internal structure of labor power L besides the means of production K. According to Sekine (2017), we can define L_k as labor power allocated to the k_{th} process (here, the assumed number of processes is n), and the production function is

$$Y = AK^\alpha (\prod_{k=1}^{n} L_k^{\alpha_k})^\beta$$

Here, we assume that $0 < a_i < 1$ for $a_i (i = 1, ..., n)$, $0 < \alpha, \beta < 1$ for α and β, and a constant return to scale; that is, $\alpha + \beta = 1$, where a_i is the elasticity of production to labor input for each process. In this production function, even if the same amount of total labor $L = \sum_{k=1}^{n} L_k$ is the input, the manner of labor allocation among the processes determined by the form of the organization determines the amount of production Y. In particular:

In this way, cooperation and division of labor could cover the whole society only after the appearance of machines in the industrial revolution,[27] and thus, Marx called the rule over labor after becoming a technical necessity the "real subjection of labor to capital," and before becoming a technical necessity the "formal subjection of labor to capital."[28] This is because the destruction of skill and the generalization of unskilled labor by machines provided more employment opportunities for women and children and reduced wages per capita, as well as reducing the reproductive cost of labor power. Although simple cooperation did not have this characteristic, the division of labor in manufacturing transformed workers into detail laborers on a large scale. Further progress toward industrialization using large machines makes most laborers unskilled. The characteristic of labor becoming unskilled is decisive for our understanding of capitalism, in contrast to the skilled labor in the feudal mode of production.[29]

---

(1) No matter which L_k is zero, $Y = 0$. This indicates that labor must be used in every process; in other words, it shows the technical necessity of the division of labor. Consequently, for example, a workers' strike at any stage can easily stop the whole production process.

(2) The marginal productivity of individual labor

$$\frac{\partial Y}{\partial L_i} = \alpha_i \beta A K^\alpha \left(\prod_{\substack{k=1 \\ k \ne i}}^{n} L_k^{\alpha_k \beta} \right) L_i^{\alpha_i \beta - 1} = a_i \beta \frac{Y}{L_i}$$

diminish gradually because $a_i \beta - 1 < 0$ under the condition of a_i and β defined above. This also shows that labor does not improve qualitatively as a result of training. Unlike feudal skill, it represents unskilled labor in large mechanized industries.

(3) If labor is allocated optimally, then the marginal productivity of additional labor in each process must be equal; and this leads to the general condition $a_i/L_i = a_j/L_j$. Therefore, $L_i/L_j = a_i/a_j$, which indicates that the allocation of labor is determined by technology as a kind of technical necessity.

[27] In Marx's evaluation, "the capitalistic mode of production ... now appears as one unique type of mode of production" by large mechanized industries (in *Marx and Engels*, 1994, 106). In other words, though capitalism previously existed as a *uklad*, it could not become "one unique type of mode of production" before the industrial revolution. Thus, capitalism as a regime must be understood as the society that followed the industrial revolution.

[28] Marx, 1887, chapter 16.

[29] "All fully developed machinery consists of three essentially different parts, the motor mechanism, the transmitting mechanism, and finally the tool or working machine" (Marx, *Capital*, 1887, 259). However, this book emphasizes working machines because we think the lack of skill in labor power is the most important condition for capitalism. Thus, "whether the motive power is derived from man, or from some other machine, makes no difference in this respect" (260). The dramatic development of steam engines and their application to the production process were also a result of the simplification of work due to the development of these working machines. Nevertheless, we cannot ignore the

However, this question may arise: is skilled labor not predominant now? In fact, we are also hesitant to describe modern labor, in which clerical labor is the major activity, as unskilled. Nowadays everyone uses personal computers and the internet freely in office work. We cannot imagine a clerical worker who cannot use them. However, in reality, those office workers are nothing more than the modern type of unskilled worker.

This is because the use of a personal computer in office work is another means of the mechanization of clerical labor, because it is a machine that can be applied to anything. Thus, in reality, every type of business activity has been simplified to the point at which it can be done using Microsoft's Excel, Word, and access to the internet. Thus, work has become easy. A considerable part of contemporary non-regular workers[30] is engaging clerical labor, because this kind of work has become unskilled office work by the mechanization of clerical labor.

We can also conceptualize it as a problem of fundamental differences between the feudal type of skilled labor and modern skilled labor. As we saw in the previous section, feudal skill is the type of skill that can be learned only by repeating the same work for many years during an apprenticeship. The word "skilled" includes that meaning. For example, skill is a thing that can only be acquired by craftsperson who have made tens of thousands of the same kitchen knives as blacksmiths, and this is the skill of hands. It is not related to the brain.

However, modern skilled labor is completely different. We do not need long years to learn to use Word and Excel, and do not need an apprenticeship. However, workers must graduate at least from high school because they need to be literate and able to calculate numbers to some extent. This is the decisive difference. In the feudal era, workers needed to enter an apprenticeship to obtain solid and specific feudal skills, but they did not need the education provided in the modern school system. The modern school system aims to form general ability and would not be useful for skilled work in the feudal mode of production, when it was more important to become apprenticed to a master from childhood than to go to school. Thus, the term "skilled" is suitable only for characterizing feudal skill, and modern skill should be expressed using different words such as "high working capacity." Thus, the essence of "modern skill" is not that it is skilled but that it is unskilled.

---

huge influence of the development of machines after they were coupled with the steam engine. For example, steam engines enabled factories to be located in cities and towns and formed the technical foundation of the conflict between urban and rural areas.

[30] This is a uniquely Japanese concept of workers who are discriminated against. They constitute a class of unskilled, low-waged workers who are treated like an appendage of regular workers.

Therefore, becoming unskilled, which became general after the industrial revolution, weakened the workers' bargaining power by eliminating skilled workers and devaluing the reproduction costs of labor power and wages. However, it also created the need to acquire modern knowledge, so that the necessary attendance of children at school historically increased these costs. As capitalism is a system that humans needed in a certain historical stage to increase productivity, the result has been an increase in per-capita income, and along with this, wages have also increased. Thus, the devaluation of the wage as work has become deskilled has been limited. Nevertheless, the general wage reduction due to the non-regularization of office work in Japan currently is a typical example of wage reduction due to deskilling.

However, what is important about deskilling labor is not just its effect on productivity and wage trends: it also liberated workers from the solid, persistent feudal division of labor. The school system gave workers the ability to match the constantly changing labor processes by educating them to read, write, and calculate. According to Marx and Engels (1910), the essence of capitalism is ceaseless change. This is because machines have become central to the production process because machines evolve constantly. Capitalism was also indispensable for nurturing general abilities that can handle anything, not for unflexible skilled labor. This was a big change, and to show it, read the following sentences by Marx:

> [M]odern industry, by its very nature, therefore necessitates variation of labor, fluency of function, *universal mobility of the laborer*... at all points, modern industry, on the other hand, through its catastrophes imposes the necessity of recognizing, as a fundamental law of production, variation of work, consequently fitness of the laborer for varied work, consequently the greatest possible development of his varied aptitudes. It becomes a question of life and death for society to adapt the mode of production to the normal functioning of this law. Modern Industry, indeed, compels society, under penalty of death, to replace the detail-worker of to-day, grappled by life-long repetition of one and the same trivial operation, and thus reduced to the mere fragment of a man, by the *fully developed individual*, fit for a variety of labors, ready to face any change of production, and to whom the different social functions he performs, are but so many modes of giving free scope to his own natural and acquired powers.[31]

The modern school system was established to answer the above question of life and death, which Marx also emphasized in *Capital*. The factory acts mentioned in the previous

---

[31] Marx, 1887, 319. Italicized parts are by the author.

section included an educational clause from the beginning, and this shows that the modern school system was formed to address the requirements of the factories. Marx says that this brings to workers a broader perspective and even the ability to bring about social change. Indeed, schools can even teach what is happening now in the world, the rights of workers, and even knowledge of scientific economics, even if schoolteachers are teaching anti-worker ideologies under the control of state power.

Following on from this, on the unexpected human development caused by the use of unskilled labor in great mechanized industries, Marx commented:

> However terrible and disgusting the dissolution, under the capitalist system, of the old family ties may appear, nevertheless, modern industry, by assigning as it does an important part in the process of production, outside the domestic sphere, to women, to young persons, and to children of both sexes, creates a new economic foundation for a higher form of the family and of the relations between the sexes.[32]

Here, the view of Marx (and Engels) extends to understand that the participation of women and children in the social networks of labor liberates them from the old family system. The economic conditions for women's liberation from their status of subordination to their husbands to become autonomous, and for youth liberation from the feudal system of seniority (elders' rule) have been provided by the industrialization. It is also noteworthy that Marx writes, "However terrible and disgusting the dissolution, ... of the old family ties may appear..." this "new economic foundation for a higher form of the family and of the relations between the sexes" forms only by dismantling of the old family system, and that he suggests that the pain and suffering of this development must be accepted. In many cases, as Marx says, this progressive change is also "terrible and disgusting," and that, facing this transformation, people tend to mourn or rebel. However, we should not be captured only the destructive aspect of the participation of women and children in production, because this forms the conditions for the next stage. In my opinion, progress in the market economy and globalization have the same nature as that of widening participation in the labor process to include women and children. It is important to characterize these historical roles from a longer perspective.

In this way, the real subjection of labor to capital also creates "fully developed individuals" as one of the elements that may overcome capitalism as itself.[33]

---

[32] Marx, 1887, 320.

[33] The Institute for Fundamental Economic Science, which was established in 1968 and has a history over half a century, is a research group that emphasizes this point.

3.3 INDUSTRIAL REVOLUTIONS AND CAPITALISTIZATION IN NON-MANUFACTURING SECTORS

3.3.1 The "Industrial Revolution" in the Construction Industry and Its Capitalistization

The industrial revolution, which inevitably requires capitalism as machines appear, did not only affect the manufacturing sector. Although I have so far described the manufacturing industry, this revolution occurs in turn at different times and in other sectors. Thus, we now turn to the industrial revolution in other sectors.

The first of these is the construction industry whose form before the industrial revolution focused on carpenters. The means of labor for carpenters are typically, chisels, saws, hammers, and planes, and in this case, there is a difference in quality between the products made by skilled and unskilled craftspeople, even if both of whom use same tools. In other word, skill is the most important element here because carpenters use only tools. Therefore, an apprentice system was established, and the craft size remained small in scale. In other words, there was no huge workshop of carpenters. Additionally, due to the lack of the technical necessity to cooperate and for the division of labor to start, a single carpenter could say "I made this house."

However, over time the nature of the labor process in construction work changed completely due to the spread of electric tools and the progress of in-plant production such as the two-by-four (2 × 4) method and by the change from wooden to steel frame or reinforced concrete buildings. Briefly, this constitutes the dismantling of skill and the deepening of the division of labor. For example, see **Table 3.2**. This merely one example from one construction workers' union with which I have been connected, showing the organization rate by occupation. It also shows that the ratio of carpenters in the construction industry decreased dramatically over this period and the division of labor deepened. After "construction work" has turned to the collaborative work of many workers in this way, no one can say "I made this house."

Even more interesting is that construction workers as a whole changed from being craftsperson to being workers in this process, which seems to have increased their participation ratio in the labor unions. As shown in **Figure 3.7**, the ratio of skilled workers in the construction industry and the union participation ratio in this region are inversely correlated, with one exceptional region shown in the northeast part of this figure. This means that construction workers are strengthening their class consciousness in line with the process of being deskilled, typically shown as the ratio of carpenters to all the construction workers. I created this figure in 1992, and as a result, I advised the union to characterize itself as a workers' union rather than a carpenters' union. This recommendation played a major role in increasing its membership. In any case, the changes from skilled to unskilled workers in the labor process, and thus the progress to the capitalist mode of production also took place in the construction industry.

MARXIAN ECONOMICS

Table 3.2 Trends in the proportion of union members by job category of the Fushimi branch of the All Kyoto Construction Workers' Union (%)

Category	2010	2005	2000	1995	1990	1987	1980
Carpenter	14.5	16.7	19.0	20.4	25.5	32.1	42.5
Electrician	5.0	4.9	5.7	6.0	6.8	8.3	7.3
Plumber	5.0	5.2	5.1	4.5	4.6	3.7	4.4
Formworker	4.2	4.5	4.5	4.7	4.1	3.1	
Interior construction worker	4.0	4.4	4.6	3.9	3.4	2.2	
Plasterer	3.5	4.0	4.4	4.7	5.6	6.0	5.4
Painter	5.2	4.1	4.6	3.8	5.3	4.1	1.7
Civil engineering worker	4.3	3.9	4.7	4.7	5.5	6.3	8.4
Dismantling worker	3.5	2.4	1.1	0.8	0.9		0.2
Steeplejack	2.5	2.2	1.8	0.7	0.5		
Operator	2.0	2.1	1.8	2.0	1.3	3.5	
Office worker	2.6	1.7	1.6	2.6			
Ironworker	2.0	2.3	3.1	3.1	4.1	4.6	4.4
Carrying worker	2.3	2.1	1.4	1.4	0.1		
Building designer	2.3	2.1	1.0	1.1	0.4		
Chipping worker	2.2	2.0	0.8	0.6	0.2		
Earth worker	1.1	1.3	1.6	2.1	3.3		3.3
Tinsmith	2.3	1.7	1.4	1.3	1.0		
Gardener	1.1	2.0	1.9	1.8	1.0		
Household appliance placer	1.8	1.4	1.8	1.3	0.8		
Waterproof placer	1.9	1.4	0.8	0.6	0.2		0.2
Reinforcing-bar placer	1.5	1.3	1.4	2.2	1.5		0.8
Building material constructor	1.2	1.3	1.0	1.1	1.1		0.6
Supervisor	1.2	1.5	1.0	1.0	0.9		
Washer	1.9	1.4	1.3	1.3	1.5	1.3	
Air conditioner plumber	1.9	1.6	1.7	2.0	1.4	1.9	3.8
Helper	1.0	1.3	1.0	0.7			
Telephone placer	1.6	1.2	2.1	1.9	1.3	1.7	0.6
Machine placer	1.1	1.1	1.0	1.0	0.9		0.6
Building contractor	0.9	1.1	1.4	1.5	0.9		
Real estate agent	1.6	1.0	0.8	0.5	0.5		
Roof tile placer	0.9	0.9	1.0	1.3	1.6	1.6	2.3
Light gauge worker	1.3	0.8	0.7	0.6	0.3		
Window frame worker	0.9	0.8	0.8	0.7	0.8		

(Continued)

Table 3.2 (Continued)

Category	2010	2005	2000	1995	1990	1987	1980
Tiler worker	0.5	0.7	0.9	0.9	1.0		
Exterior construction worker	0.6	0.7	0.7	0.4	0.1		
Hardware placer	0.9	0.7	0.4	0.7	0.1		
Woodworker	0.8	0.7	0.3	0.7	0.8		1.9
Signboard worker	0.5	0.6	0.7	0.8	1.0		
Gas plumber	0.8	0.6	0.6	0.6	0.3		
Pavement constructor	0.5	0.5	0.4	0.2	0.5		
Exterior designer	0.3	0.4	0.4	0.3	0.6		0.6
Picture framer	0.5	0.4	0.3	0.5	0.5		
Bricklayer	0.3	0.4	0.3	0.5	0.4		
ALC worker	0.4	0.3	0.3	0.4	0.4		
Lath placer	0.2	0.3	0.2	0.2			
Roofing constructor	0.3	0.5	0.7	0.4			
Stone mason	0.2	0.4	0.4	0.1	0.2		0.6
Welder	0.2	0.3	0.3	0.2	0.6		
Joiner	0.1	0.3	0.4	0.1	0.1		0.2
Surveyor	0.5	0.2	0.2	0.2	0.2		
Cutter	0.1	0.2	0.3	0.3	0.3		
Furniture maker	0.2	0.2	0.3	0.4	0.2		
Tatami maker	0.2	0.1	0.1	0.2	0.1		0.4
Glass worker	0.1	0.1	0.2	0.2	0.3		0.8
Concrete worker	0.0	0.1	0.0	0.1	0.1		0.2
Heat insulator placer	0.1	0.0		0.0	0.1		
Shutter placer	0.0	0.1	0.1	0.1	0.1		
Road constructor	0.1	0.1	0.1	0.1	0.1		0.2
Kiln constructor	0.1	0.1	0.1	0.1	0.3		
Boring worker	0.0	0.0		0.1	0.1		
Incubator constructor	0.0	0.0	0.1	0.2	0.2		
Tent worker	0.0	0.1	0.1	0.1			
Well driller	0.1	0.0	0.0	0.0	0.1		
Refrigerator placer	0.0	0.0	0.1	0.1	0.1		
Sanitation and moth-proofing constructor	0.1	0.0	0.0	0.1			
Stainless placer	0.0	0.0	0.0	0.0	0.1		
Others	0.6	3.0	2.7	2.7	4.1	19.7	8.4
Total	100	100	100	100	100	100	100

Source: Internal documents of the Fushimi Branch of All Kyoto Construction Workers' Union.

MARXIAN ECONOMICS

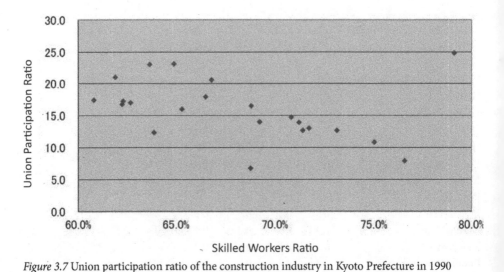

Figure 3.7 Union participation ratio of the construction industry in Kyoto Prefecture in 1990

Source: The ratio of skilled workers is calculated by the number of skilled and production process workers, and laborers, divided by the total number of construction workers in each district of Kyoto prefecture in 1990. These numbers are taken from the 1990 Japanese Census. The union participation ratio is calculated by the total number of all Kyoto Construction Workers' Union members divided by the number of construction workers in each district in Kyoto prefecture in 1990. The former figures come from the internal documents of the union and the latter from the 1990 Japanese Census.

3.3.2 The "Industrial Revolution" in Hospitals and Its Capitalistization

A similar process is taking place among hospital doctors and medical engineers. Even if they have an opposite social status to construction workers, their similarity becomes clear if we examine the relationship of skilled to unskilled work and remember that formerly town doctors conducted their work using only a stethoscope and syringe. Of doctors who use the same stethoscope, some may diagnose the patients correctly while others make a misdiagnosis.[34] Similarly, of two nurses using the same syringe, one may use it without causing pain but the other may cause pain. As these differences arise not from the differences between stethoscopes or syringes but from difference in skill, only a thousand repetitions of examinations of a thousand patients can polish their skill. Thus, all faculties of medicine in universities have very rigorous hierarchical system, similar to the feudal apprenticeship system.

However, the situation is completely different in a modern hospital (or hospital networks linking local hospitals), because now it is not a question of what kind of doctor is in that

---

[34] Japan has a special word for unskilled doctors: *yabu*.

hospital, but whether computed tomography scanners and other state-of-the-art medical equipment are available. Hence, examinations and treatments currently are decisively dependent on machines, which reduces the importance of skill. Of course, in special fields like cardiac surgery, skill still plays an important role, but this will be true only until new, state-of-the-art machines of cardiac surgery appear.

In this new situation, various new technologists are being developed. For example, when I entered Kyoto University as a student in 1975, the faculty of medicine had a nursing school, which later changed its name to the Medical Technology Junior College, the Department of Health Science of the Faculty of Medicine, and finally the Department of Human Health Science of the Faculty of Medicine, which trains mainly laboratory technicians, physiotherapists, and occupational therapists as well as nurses. This means that the current medical labor is a collaborative work as a segmented division of labor system, so that no one can say "I healed this patient." This process shows that each individual work has become unskilled detailed labor, mainly that of operating machines (in the meaning described in this text). Of course, since the modern hospital is a huge organization, someone is made the employer and the others the employees, which is the capital-wage labor relationship. This contrasts with past "town doctors" consisting only of doctors and nurses. In other words, this change also transformed capitalist mode of production in this industry.

3.3.3 The "Industrial Revolution" in the Commercial Sector and Its Capitalistization

In terms of capitalistization, these hospital systems are very similar to the supermarket chains that developed from individually managed small retail stores. The small retail store cannot be described as a tool-based system, but supermarket chain systems, with visually appealing stores and strong networks with commodity producers, can be described in this way. As a result, small retail stores that lack these features need to find and use other advantages such as the merchants' own special talents and personal connections, through which they can keep customers in their village. My father was an owner of a small store in a small village and was skilled in finding customers and setting the best prices for each customer, using all his human relations and detailed knowledge of his customers and their needs. This kind of talent is decisively based on a deep personal interdependency that has been broken by capitalism, especially in urban areas. This talent can be classified as a kind of skill obtained over a long period of "training."

We may ask whether the primarily important origin of this kind of productivity lies within or outside the individual. As we saw, in my father's case, it was internal, but in the case of supermarket chains, it lies outside the workers. The first section of this chapter discussed the nature of capitalism in which the objects to be accumulated were the machines external to human beings because machines have become the most important origin of productivity, while the skill internal to a human being was the most important origin of productivity in the feudal era. This shows that we can categorize the transition from small stores to supermarket chains as the transition from the feudal type of system to the capitalist system. Needless to say, in this process, the size of stores and the scale of the division of labor has grown, and the capital–labor relationship has developed.

3.3.4 The "Industrial Revolution" in the School System and Its Capitalization

The last case is the conversion of the school system from the *terakoya* schools in the Edo era to the present modern school system in Japan. In Edo era Japan, local temples were a kind of community center and the monks taught letters, calculation, and morality to the children. These old types of school system differ fundamentally from the present one in the sense of the tools or facilities used. For example, in the *terakoya* system, monks used only brushes and ink to write letters and the Japanese abacus to calculate numbers. These things were just tools, and the students' educational achievements did not depend on the quality of these tools but solely on the monks' skill. Thus, the origin of the productivity of these schools lay inside the monks who taught all subjects (so they were able to say "He is my pupil").

However, after the industrial revolution, the modern school system incorporated sophisticated textbooks and various types of facilities. Some contemporary private universities in particular appeal to young high-school students with their beautiful cafés or restaurants on campus, and if these cafés and restaurants are very attractive to students, we may say that they function as a kind of machine as the origin of their productivity (measured here in terms of its attractiveness). This leads some universities, in which working conditions of the staff members are very bad, to build new campuses and new departments rather than paying higher wages to the staffs. This is capitalism. In other words, money is used primarily to accumulate capital, not to empower the teachers' ability to teach.

Therefore, not only the manufacturing sector but also other industries had experienced or are now experiencing a transition from a feudal type of system to a capitalist one. Even though these transitions took place at different times in each sector, the nature and the cause of the transition is identical. Technological changes in the decisive element of productivity from that of skill to that of machines have caused these transitions in all industries.

4

The Growth and Death of Capitalism: Accumulation Theory: A New Quality Created by Quantity

4.1 THE BIRTH, GROWTH, AND DEATH OF CAPITALISM: THE MARXIAN OPTIMAL GROWTH MODEL

4.1.1 Formulation of the Issues

In the previous chapter, I explained capitalist exploitation in terms of the qualitative character of capitalist productive forces, and Marx followed up those explanations in the first volume of *Capital* with a discussion of the long-term tendencies of capitalism. The law of capitalism, which makes capital accumulation the primary task is thus an explanation of the various tendencies resulting from capital accumulation, which can also be called the accumulation theory. And since Marx states at the end of this explanation that "capitalist production begets, with the inexorability of a law of Nature, its own negation"[1] "and that "the knell of capitalist private property sounds,"[2] this issue here is to discuss the growth and death of capitalism. Discussing the growth of capitalism (via accumulation) leads to discussing the death of capitalism, in historical materialist terms.

Now, it must be recalled that Marx stated in the preface to the first edition of *Capital* that (a society) "can neither clear by bold leaps, nor remove by legal enactments, the obstacles offered by the successive phases of its normal development."[3] This sentence is also indicated by the statement: "The country that is more developed industrially only shows, to the less developed, the image of its own future,"[4] indicating that industrial development or capitalistic development is necessary before the stage of future society. In other words, only the development of capitalism can prepare for the transformation to the post-capitalist mode of production. The question here, therefore, is how capitalism paves the way for the transformation of the mode of production. This chapter explains this problem by discussing the growth and death of capitalism. Showing its death is a

---

[1] Marx, 1887, 542.
[2] Ibid.
[3] Marx, 1887, 7.
[4] Marx, 1887, 6–7.

consequence of its growth. In other words, since capitalism is "a system in which capital accumulation is the primary task," this chapter explains that the necessity of capital accumulation itself will disappear as a result of its progress.

To explain this, this chapter also relies on mathematics to a large extent. In the previous chapter, the proof of exploitation was presented as Okishio's FMT, but here it is necessary to discuss the core proposition of historical materialism, the death of capitalism, in mathematical form to confront contemporary mainstream economics. In *Socialism: Utopian and Scientific* (1880), Engels states that Marx's theory is composed of the theory of surplus value and historical materialism. Therefore, the theory of historical materialism also needs to be proved, in the same way that Okishio proved theory of surplus value. My research group is working on this issue and has fulfilled this task through the framework of the Marxian optimal growth theory, which was formed in 2002 and has developed since then.

Since this chapter discusses the tendencies of capitalism, I also discuss the law of the falling rate of profit and the increase in the relative surplus population discussed in Marx's *Capital*. However, while the former is also derived from the Marxian theory of optimal growth, the latter leads to a different conclusion. Since these theories are all quantitative in nature, a certain use of mathematics is needed. However, we have made considerable efforts to ensure that even beginners can read and understand them. If you still find it difficult, you can skip the mathematical part and read only the text. Even if you skip the mathematical formulas, the conclusions are clearly presented in each section.

Then, the first task is to describe the nature of capitalistic production. In the previous chapter, we presented it as a society of machines that developed after the industrial revolution, and broke away from a society of tools, and that difference between these two societies can be conceptualized as follows. That is, while the accumulation of tools before the industrial revolution did not result in a higher capacity for production, the accumulation of capital after the industrial revolution did. This is because, while giving a second or third hammer to a feudal craftsperson who used tools did not result in an increase in their production capacity, an increase in the number or sizes of machinery used by a single worker in modern industry alone leads to an increase in their production capacity. This relationship also can be expressed in terms of the elasticity of production with respect to the means of production, with this elasticity having a value of zero in the former case and a positive value in the latter. When expressing this elasticity in a production function, labor input serves as a factor of production in addition to the means of production, we can express this in the form of a Cobb-Douglas function, commonly used in modern economics, as follows:

$$Y = AK^\alpha L^\beta \quad (\alpha = 0 \text{ before the industrial revolution,} \qquad \qquad (4.1)$$
$$\alpha > 0 \text{ after the industrial revolution}).$$

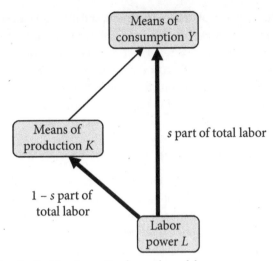

Figure 4.1 Redefinition for the Marxian optimal growth model

For the moment, Y represents the production of the means of consumption, A represents the total factor productivity (a technological coefficient), K is the means of production (capital stock), and L is labor input. In this case prior to the industrial revolution, the amount of K required to maximize production Y was zero because $\alpha = 0$, while after the industrial revolution, it became a quantity of some size because $\alpha > 0$. It goes without saying that in the latter case, an increase in K leads directly to an increase in Y.

Another important consideration is to redefine the symbols of roundabout production in **Figure 4.1** differently from Figure 3.1 in which a part of means of production also is used to produce the means of production. However, here for simplicity, we assume that machines are produced only by using labor.[5] Figure 4.1 shows the resulting relation between Y, K, and L. The key point here is that use of labor power L is split into two sectors, dividing the activities of the human being into the production of the means of production and the production of the means of consumption. In this figure, s has a value of $0 \leq s \leq 1$, which defines the portion of labor power s that is diverted to the production of the means of consumption, and the portion $1 - s$, to the production of the means of production. In this scenario, the production function of the means of consumption becomes

$$Y = AK^{\alpha}(sL)^{\beta} \qquad (4.2)$$

I simplify the other production function of the means of production as

$$\dot{K} + \delta K = B(1-s)L \qquad (4.3)$$

---

[5] We can return to the original assumption shawn in Figure 3.1 in subsection 3.1.2. However, in essence the result is identical.

As seen above, this is the simplest linear function, and does not take into account the use of machinery in the production of machinery. Here, K represents the stock of the means of production, $\dot{K}$ is the amount of increase in K over a single period (e.g., one year),[6] B is labor productivity, and the value of δ is $0 < \delta < 1$. For example, if the means of production depreciate and are written off through their use over 20 periods, then $\delta = 0.05$, and since this component must be also produced in the sector producing the means of production, δK is added on the left-hand side.

4.1.2 Optimal Capital Stock Targeted after the Industrial Revolution

The other issues that arise here are the ratio (s) of the total labor power allocated to the sector producing the means of consumption, the ratio ($1 - s$) allocated to the sector producing the means of production, and the issue of the equilibrium capital labor ratio (that is, the volume of the means of production used per capita). The solution to this problem can be sought by considering the following situation in the case of optimal allocation: with an equilibrium at the starting point, in the event of a slight increase (ΔL) in total labor power under these conditions, the increase in the sectors producing the means of consumption and the means of production will both have the same effects on the final objective of producing the means of consumption. To derive these conditions, the effect on the right-hand side of the figure (the direct addition of ΔL to the sector producing the means of consumption) can be calculated as follows:

$$\frac{\partial Y}{\partial L} = \beta A K^\alpha L^{\beta-1} \qquad (4.4)$$

In this calculation, both s and $1 - s$ are ignored because here we consider the case of adding ΔL to both, and we are concerned only with the function form. On the other hand, the result on the left-hand side of the this figure (i.e., the results of the indirect contribution of ΔL to the production of the means of consumption through an increase in the production of means of production) becomes:

$$\frac{dK}{dL} \cdot \frac{\partial Y}{\partial K} = B\alpha A K^{\alpha-1} L^\beta \qquad (4.5)$$

However, we also need to take into account the fact that, because the effects on productivity here function as an increase in machinery and equipment available for use over the long term, we must also consider the cumulative results of these effects as well. These effects do not appear in the current period but begin in the next period and then continue steadily. For this reason, using a subjective discount rate for future utility (or the time-preference rate) expressed as ρ using a value such as 0.1, we must employ a calculation such as:

---

[6] This can be expressed mathematically as $\dot{K} = dK/dt$ (here, t represents time).

$$\frac{B\alpha AK^{\alpha-1}L^{\beta}}{1+\rho} + \frac{B\alpha AK^{\alpha-1}L^{\beta}}{(1+\rho)^2} + \frac{B\alpha AK^{\alpha-1}L^{\beta}}{(1+\rho)^2}\ldots \quad (4.6)$$

However, we also must consider that the effect will decrease in each period, because the additional K that ΔL creates will also depreciate each year. Therefore,

$$\frac{B\alpha AK^{\alpha-1}L^{\beta}}{1+\rho+\delta} + \frac{B\alpha AK^{\alpha-1}L^{\beta}}{(1+\rho+\delta)^2} + \frac{B\alpha AK^{\alpha-1}L^{\beta}}{(1+\rho+\delta)^3}\ldots \quad (4.7)$$

In this case, according to the formula for the sum of an infinite geometric series:

$$\frac{B\alpha AK^{\alpha-1}L^{\beta}}{1+\rho+\delta} \cdot \frac{1}{1-\dfrac{1}{1+\rho+\delta}} = \frac{B\alpha AK^{\alpha-1}L^{\beta}}{\rho+\delta} \quad (4.8)$$

Hence, the result on the right side of the figure is equal to equation (4.4):

$$\beta AK^{\alpha}L^{\beta-1} = \frac{B\alpha AK^{\alpha-1}L^{\beta}}{\rho+\delta} \quad (4.9)$$

This equation can be rewritten simply as

$$\beta(\rho+\delta)K = B\alpha L \quad (4.10)$$

However, another matter must be given further consideration: the above calculation does not account sufficiently for the fact that the total capital K decreases autonomously due to depreciation in each period. If we account for this, the labor in the amount of $\delta K^*/B$, which is needed in each period to continue maintaining the optimal capital K^{*7}, must be deducted from total labor. Thus, we must rewrite equation (4.10) as

$$\beta(\rho+\delta)K^* = B\alpha\left(L - \frac{\delta K^*}{B}\right) \quad (4.11)$$

For this reason, it is clear that the ultimate optimal capital–labor ratio $(K/L)^*$ will be

$$\left(\frac{K}{L}\right)^* = \frac{B\alpha}{(\alpha+\beta)\delta + \beta\rho} \quad .[8] \quad (4.12)$$

---

[7] This is derived from the production function of the sector producing the means of production.
[8] Kanae (2008) and others show that by substituting the realistic assumption that the means of production is used in the sector producing the means of production as well,
$Y = AK_c^{\alpha_2}L_c^{\beta_2}$, $\dot{K}+\delta K = BK_k^{\alpha_1}L_k^{\beta_1}$

This calculation result contains much information. First, it demonstrates the fact that when $\alpha = 0$; that is, under the technology of the feudal system before the industrial revolution, this ratio was zero. Strictly speaking, when $\alpha = 0$ and $K = 0$, the production function $Y = AK^{\alpha}(sL)^{\beta}$ cannot be defined mathematically,[9] so K must be assumed very close to zero. This must be in agreement with the real-world. This is because in our theoretical framework, the most important task during the feudal period is to improve craftspeople's skills, and accumulating tools was unnecessary. Here, we have verified this thesis mathematically.

The per capita consumption $(Y/L)^*$ resulting this new technology is usually greater than the per capita consumption $(Y/L)^-$ of the old technology, but this is not always true. To show this problem, let us introduce both types of per capita consumption as follows:

$$\left(\frac{Y}{L}\right)^* = A\left(\frac{B\alpha}{\delta + \beta\rho}\right)^{\alpha} \tag{4.13}$$

$$\left(\frac{Y}{L}\right)^- = AL^{\beta_0 - 1} \tag{4.14}$$

where, for simplicity, we assume the constant return to scale: $\alpha + \beta = 1$ of the new technology, and $\alpha = 0$ and $\beta = \beta_0$ ($0 < \beta_0 < 1$) for the old technology. Therefore, the adoption of new technology is rational only in the following case:

$$L > \left(\frac{\delta + \beta\rho}{B\alpha}\right)^{\frac{\alpha}{1-\beta_0}} \tag{4.15}$$[10]

This is because of the assumption that the old technology has diminishing returns, where the smaller that L is in the old technology, the larger the per capita production

---

and the optimal capital–labor ratio is

$$\left(\frac{K}{L}\right)^* = \left(\frac{\alpha_1}{\rho+\delta}\right)^{\frac{\alpha_1}{1-\alpha_1}} \frac{B^{\frac{1}{1-\alpha_1}}}{\delta}\left\{\frac{\alpha_2\beta_1\delta}{\alpha_2\beta_1\delta + \beta_2(\rho+\delta-\alpha_1\delta)}\right\}^{\frac{\beta_1}{1-\alpha_1}} L^{\frac{\alpha_1+\beta_1-1}{1-\alpha_1}}$$

we can derive this by substituting $\alpha_1 = 0$ and $\beta_1 = 1$ into the optimal capital–labor ratio used in this text. It also is clear that this equation has largely the same characteristics as in this text. However, the fact that $(K/L)^*$ here is a multiple of $L^{\frac{\alpha_1+\beta_1-1}{1-\alpha_1}}$ means that the optimal capital–labor ratio is an increasing function, constant, or decreasing function of labor input (or population) if the production function of the means of production is under increasing, constant or diminishing returns to scale, respectively.

[9] 0^0 cannot be defined mathematically.

[10] This optimal solution can be held easily for individuals with a lower time preference rate ρ, but individuals with a larger ρ want to maintain the old technology. The Luddite movement was a resistance movement carried out by such workers in the early stage of the industrial revolution.

(of the means of consumption) becomes. Pomeranz (2000) argues that access to cheap coal and other foreign resources was essential for the British industrial revolution, because this condition more easily satisfied (4.15) by increasing productivity-related parameters such as B and α in the means of the production sector in England. However, after the late nineteenth century, this became common in many countries, including Japan.

Second, a further increase in α or decrease in β will cause the optimal capital–labor ratio to rise. Since coefficients α and β indicate whether a greater contribution to production is made by capital or labor inputs in the production function of the means of consumption, our ultimate interest is the ratio between α and β. In fact, it is said that in the macro economy in general, constant returns to scale apply, in which doubling both capital and labor would result in double the production as well. Since this means that $\alpha + \beta = 1$,[11] an increase in α means a decrease in β. Moreover, since the effects of the input of labor in the means of production sector become relatively large in this case, a greater production of the means of production occurs, and thus, the optimal capital–labor ratio (capital per worker) will rise as well.

Third, the effects of A and B are of some interest. We see that an increase in B leads to greater production of the means of production and a high capital–labor ratio in the same way as α because it makes the input of labor producing the means of production more efficient. However, A shows no such relationship. This is because an increase in A has exactly the same effect on the direct input of additional labor to the sector producing the means of consumption (illustrated as a straight upward arrow from L in Figure 4.1) and the indirect input of additional labor through the additional production of machines (this is illustrated as the arrows from L to K and from K to Y in Figure 4.1).

However, since A has a direct effect on the production of Y as shown in equation (4.2), per capita consumption Y/L increases every year at the ratio $(1+\eta_A)\cdot(1+\eta_B)^\alpha$, when η_B and η_A denote the rate of increase in B and in A, respectively. This is because the increase in the productivity of the means of production sector affects the production of the means of consumption indirectly through the roundabout production system. In this way, if such a roundabout production system is deepened by generating more production sectors that specialize in producing the means of production, a further synergistic effect of productivity increase is generated. For example, $(1+\eta_A)\cdot(1+\eta_B)^\alpha \cdot(1+\eta_C)^{\alpha\beta}\ldots$. Note that deepening the roundabout production system makes technological innovation more important, even if this has a diminishing effect.

---

[11] For example, assuming a doubling of both K and L in the production function of the sector producing the means of consumption, $Y' = A(2K)^\alpha (s2L)^\beta = 2^{\alpha+\beta} AK^\alpha (sL)^\beta = 2^{\alpha+\beta} Y$. It is clear that the new production Y' will double of the original Y when $\alpha + \beta = 1$. In addition, when the economy displays uncertainty, this optimal capital–labor ratio must rise to prepare for this uncertainty. See Part II in Kanae (2013).

MARXIAN ECONOMICS

The next important topic is the effect of the depreciation rate δ. Because a higher depreciation rate causes a loss of the accumulated means of production at that rate, the effects of the accumulation of K decrease. Clearly, the result is the same effect as a decrease in α or a drop in B.

The last subject we consider here is the rate of time preference. Unlike other variables, this indicates the effect of the subjective factor of people's preferences. For example, preferences for investment differ among ethnic Chinese and ethnic natives in Southeast Asia, and between the Han and ethnic minorities in China. This also leads to differences in these groups' economic, and thus social and political, status.[12] Here, this fact is expressed as differences in the optimal capital–labor ratio. A higher rate of time preference—that is, the discount rate—"discounts" the effects of production through capital accumulation from the next year, and leads to a lower optimal capital–labor ratio.

We also can calculate the ratios of the allocation of labor to sectors producing the means of consumption and producing the means of production at the point where the above capital–labor ratio is optimal. Here, because under the above production function for the sector producing the means of production, $\dot{K} = 0$ must be held, and therefore

$$B(1-s)L = \delta K^* \tag{4.16}$$

Thus, changing this equation by substituting the K^* derived above will lead to

$$1 - s^* = \frac{\delta \alpha}{(\alpha+\beta)\delta + \beta\rho} \tag{4.17}$$

Because all the parameters on the right side are positive, s^* is positive and smaller than one.

4.2. CONDITIONS FOR REPRODUCTION WITHOUT ACCUMULATION

4.2.1 Marx's Simple Reproduction Scheme

The model described above divides total social production into the sectors that produce the means of consumption and the means of production, expressing the operation of society as a whole as the relationship between them—an idea that, incidentally, began with Marx's reproduction scheme. In modern economics it was the late Professor Hirofumi Uzawa of the University of Tokyo who developed the two-sector growth model in the 1960s, admitting that his idea came from Marx's reproduction scheme. That is, the Marxian optimal growth model described above is also based on that idea that began with Marx and was exported into modern economics.

---

[12] In my opinion, this is a main cause of ethnic conflicts in capitalism. For example, see Onishi (2008, 2012).

Nevertheless, there are differences between the two. While the Marxian optimal growth model used direct measurement in material terms, Marx's model used value terms. The Marxian optimal growth model is set in material terms because whatever the case, material terms are necessary to express the fact that the accumulation of machinery is effective for production. While this model using material terms is rewritten in value terms of labor inputs in subsections 4.2.2 and 4.4.1, we explain the reproduction scheme first. This makes the conditions required for reproduction clear.

There are two types of schemes for reproduction: a simple scheme without capital accumulation and an extended scheme that incorporates capital accumulation. We describe the former first because it is the basic model, and its starting point is the division of value into $c + v + m$. Since Marx called the sector producing the means of production sector 1 and the sector producing the means of consumption sector 2, we can use these notations to start by expressing $c + v + m$ for the two sectors, as shown below:

$$W_1 = c_1 + v_1 + m_1$$
$$W_2 = c_2 + v_2 + m_2 \quad (4.18)$$

Here, W_1 and W_2 represent the value of total output in a single year in each sector, and as we know, we can derive the following simple reproduction condition from the equilibrium condition between the total supply of the means of consumption and the total demand of them shown in (4.18):

$$v_1 + m_1 = c_2 \quad (4.19)$$

From this condition alone, the same scale of production will continue in the current, next, and subsequent periods without accumulation, because capitalists consume all the surplus value generated in those periods.

4.2.2 Explanation Using the Marxian Optimal Growth Model

The next discussion concerns the relationship between this simple reproduction scheme and the model in section 4.1: the Marxian optimal growth model. In particular, since the calculation of the optimal capital–labor ratio in subsection 4.1.2 was a calculation of the optimal equilibrium ratio, this indicates the conditions for simple reproduction. For this reason, if we plug K^* and s^* into the production functions for the two sectors above and substitute $\dot{K} = 0$, as the capital–labor ratio is constant here, we can compare them with the reproduction scheme below:

$$W_1 = c_1 + (v_1 + m_1) \quad \delta K^* = 0 + B(1 - s^*)L$$
$$W_2 = c_2 + (v_2 + m_2) \quad Y = A(K^*)^\alpha (s^* L)^\beta \quad (4.20)$$

Furthermore, rewriting them to express K^* and s^* as L yields

$$\left.\begin{array}{ll} W_1 = c_1 + (v_1 + m_1) & \delta K^* = 0 + B\left(\dfrac{\delta\alpha}{(\alpha+\beta)\delta+\beta\rho}\right)L \\ \\ W_2 = c_2 + (v_2 + m_2) & Y = A\left(\dfrac{B\alpha L}{(\alpha+\beta)\delta+\beta\rho}\right)^\alpha \left\{\left(1-\dfrac{\delta\alpha}{(\alpha+\beta)\delta+\beta\rho}\right)L\right\}^\beta \end{array}\right\} \quad (4.21)$$

These forms directly indicate the amount of labor input to each sector. That is, if we recall that the means of production in the consumption goods sector on the second line are produced only by labor (that is $\alpha = 0$), then we can see that all inputs are labor. These two equations express this fact.

This is particularly important because the formulae on the right can be rewritten as the labor input, and when doing so, it largely is rewritten in the form $c + v + m$. This rearrangement is shown in **Table 4.1**. As seen above in sector 1, the sector producing the means of production, for simplification, we assume that no means of production are used. Thus, the value of c_1 is 0 here. In addition, the value corresponding to c_2 is K^* multiplied by δ, based on the concept that in each period, only a portion of K^*—that is, δK^*—is depreciated. Moreover, we cannot decompose $v + m$ into v and m here because we need to define m in the first definition of exploitation later in subsection 4.3.3.

We should also note that in this table, total $v + m$ in both sectors is L. Thus, the total value added in this period through both sectors is equal to the total labor input. The LTV (labor theory of value) argues that total labor input is itself total value added. This is expressed in Table 4.1.

In conclusion, the most important point in this subsection is that the condition $v_1 + m_1 = c_2$, derived from the above simple reproduction scheme, is satisfied. This is because both the left-hand and right-hand sides of this equation are $(\delta\alpha/\{(\alpha + \beta)\delta + \beta\rho\})L$ in Table 4.1. Hence, the Marxian optimal growth model leads to the same conclusions as the simple reproduction scheme.

Table 4.1 Decomposition of labor input in the optimal state of the Marxian optimal growth model

	c	$v+m$
Sector 1	0	$\left(\dfrac{\delta\alpha}{(\alpha+\beta)\delta+\beta\rho}\right)L$
Sector 2	$\left(\dfrac{\delta\alpha}{(\alpha+\beta)\delta+\beta\rho}\right)L$	$\left(1-\dfrac{\delta\alpha}{(\alpha+\beta)\delta+\beta\rho}\right)L$

4.3 CONVERSION OF SURPLUS VALUE INTO CAPITAL

4.3.1 Marx's Extended Reproduction Scheme

While we derived the conditions for the continuation of production, the simple reproduction above is no more than a hypothetical state introduced temporarily for explanatory reasons. The nature of capitalism is change, and capitalism cannot be envisioned with constant reproduction in practice. For this reason, Marx conceived of the following extended reproduction scheme. That is, partially following the preceding expression:

$$\left. \begin{array}{l} W_1 = c_1 + v_1 + m_1(c) + m_1(v) + m_1(k) \\ W_2 = c_2 + v_2 + m_2(c) + m_2(v) + m_2(k) \end{array} \right\} \quad (4.22)$$

Here the surplus values of both sectors m_1 and m_2 are newly invested in c and v in addition to capitalist consumption $m_1(k)$ and $m_2(k)$. Then, this time, unlike in the case of simple reproduction, we need a new $m_1(c)$ and $m_2(c)$ from sector 1 as materials, and $m_1(v), m_2(v), m_1(k)$, and $m_2(k)$ from sector 2. For this reason, the supply-demand coincidence conditions for the means of production and the means of consumption become, respectively:

$$\left. \begin{array}{l} c_1 + v_1 + m_1(c) + m_1(v) + m_1(k) = c_1 + m_1(c) + c_2 + m_2(c) \\ c_2 + v_2 + m_2(c) + m_2(v) + m_2(k) = v_1 + m_1(v) + m_1(k) + v_2 + m_2(v) + m_2(k) \end{array} \right\} \quad (4.23)$$

Cleaning up both sides of these equations yields:

$$v_1 + m_1(v) + m_1(k) = c_2 + m_2(c) \quad (4.24)$$

Thus, our task is to compare this condition with the simple reproduction condition $m_1 + v_1 = c_2$, and for this purpose, we add $m_1(c)$ on both sides, as follows:

$$v_1 + m_1(v) + m_1(k) + m_1(c) = c_2 + m_2(c) + m_1(c) \quad (4.25)$$

This can be rewritten as

$$v_1 + m_1 = c_2 + m_2(c) + m_1(c) > c_2 \quad (4.26)$$

The last sign of inequality comes from the assumption $m_2(c) + m_1(c) > 0$, which is a very basic condition of extended reproduction. Therefore, we can check whether or not this condition holds for the Marxian optimal growth model under extended reproduction.

4.3.2 Explanation by the Marxian Optimal Growth Model

Next, we address the condition of extended reproduction in this model where the surplus value is used to purchase labor power and the means of production, instead of being consumed in its entirety by capitalists. For this reason, we assume that capitalist consumption is zero; that is, $m_1(k) = m_2(k) = 0$. This represents the simplest case of extended reproduction.

While this assumption may appear extreme, it is not. This is because capitalists are only capital personified, and even if they live at the minimum level, they can work well as capitalists. Alternatively, they sometimes work harder than general workers do. However, the problem is the fact that they do not have to be rich in order to command labor despotically, exploit workers, and work as hard as they can to increase capital. Thus, they are also a kind of employee that represents the will of capital; namely, the true rulers of the companies. Therefore, we can assume no consumption by these representatives of capital; that is $m_1(k) = m_2(k) = 0$. This assumption does not matter if capitalists work well as representatives. Marx himself also made this assumption, saying: "in order not to complicate the formula, it is better to assume that the entire surplus-value is accumulated"[13] and, "In so far as the capitalist merely personifies industrial capital, his own demand is confined to the means of production and labor-power."[14]

Nevertheless, this expression gives rise to the question of what the purposes of surplus value and exploitation are. If surplus value is not intended to enrich capitalists, then truly it must be to enrich capital itself, or to accumulate. If we concur with the understanding of capitalism described above, then surplus value alone makes economic development after the industrial revolution possible. That is, this can be said to be performing a role for the entire society; and in fact, several ideologues argue that it is necessary to secure funds for capital accumulation for this reason, and politicians and government officials continue to support this argument. There are thus legitimate grounds for exploitation in the sense of productivity.

Two more factors to introduce in the Marxian optimal growth model are the assumption that the population size is fixed and that the entire population provides labor power. This is because, while the population size is a very important variable in economics, it is not easy to set as an endogenous variable. While it is easy to model the mechanism by which labor power is absorbed into and ejected from the labor market through the process of the business cycle, our interest is now in the basic long-term tendencies that run through these changes. For this reason, in the Marxian optimal growth model we introduce the assumption that even in extended reproduction, there is no change in the total labor power (which occurs only in a transfer between two sectors), and only the means of production are added. In other words, we assume $v_1 + v_2 + m_1(v) + m_2(v) = L$; that is, $m_1(v) + m_2(v) = 0$. In this case the extended reproduction scheme above becomes

---

[13] Marx, 1894, 45.
[14] Marx, 1894, 68.

$$W_1 = c_1 + v_1 + m_1(c) + m_1(v)$$
$$W_2 = c_2 + v_2 + m_2(c) + m_2(v) \qquad (4.27)$$
$$m_1(v) + m_2(v) = 0$$

That is, the entire exploited surplus value here is used for net investment $m_1(c) + m_2(c) + m_1(v) + m_2(v)$ under the condition that total labor is constant and there is no increase in workers' wages. In fact, capitalistic development has been a process of converting surplus value into capital. Instead of the constant conditions of simple reproduction seen above, this is a growing economy, in the process of accumulation or growth of capital.

Bringing this closer to the Marxian optimal growth model, we can express it as follows. The calculation of the equilibrium (steady state) in the preceding section was derived from the assumption that $\dot{K} = 0$, and there is no guarantee that the economy can reach this state immediately. It can be obtained only after long-term capital accumulation.

For example, we know that, in general, economic growth rates are lower in advanced countries than in developing countries. This is because capital accumulation rates are decreasing toward the state $\dot{K} = 0$. Japan's net investment rate is shown in **Figure 4.2**. However, this is not the case in developing countries. According to Onishi (2016), China's capital accumulation as of 2009 had reached only 11% of the steady-state level. Rough calculations by Onishi (1998a) show that as of 1994, South Korea, Taiwan, the Philippines,

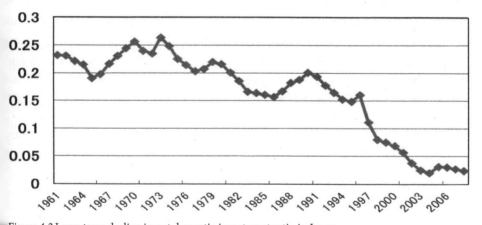

Figure 4.2 Long-term decline in net domestic investment ratio in Japan

Source: Statistics Bureau, Ministry of Internal Affairs and Communications, *Annual Reports of National Economic Accounts*. Note: The net investment ratio is calculated as (gross domestic fixed capital formation + net increase in inventories − gross domestic depreciation)/net domestic production. It is calculated using net domestic production instead of gross domestic production to match the 1 − s of the Marxian optimal growth model.

MARXIAN ECONOMICS

and Indonesia reached the levels of 36%, 21%, 39%, and 51%, respectively. These facts mean that apart from the advanced countries that have reached the level of zero growth, the process of capital accumulation continues, and for this reason, developing countries are not in the condition of simple reproduction; they are still in the condition prior to that of a steady state. However, to simplify the issue, we described simple reproduction prior to extended reproduction, in reverse historical order.

In this case, the issue here is how to derive this process of accumulation or growth, which we formulate as the issue of maximizing the production of the means of final consumption, using the two production functions introduced above. However, considering diminishing marginal utility of consumption goods, we identify the level of instantaneous utility to the human being at any moment (instantaneous utility) as $\log Y$. This is because marginal utility diminishes in this form. In addition, converting the sequence of utility continuing into the future to its present value using the discount rate ρ, which expresses the preference between the future and the present, we can rewrite utility as

$$U = \int_0^\infty e^{-\rho t} \log Y(t) dt \qquad (4.28)$$

This is called intertemporal utility. Here, e is the base of the natural logarithm (Napier's constant) and (t) appended to Y indicates that this calculation accounts for the fact that Y varies over time. Using this form, the content of the integral sign on the right-hand side can be understood by thinking about it in terms of the following discrete system.

When, for example, $\rho = 0.1$, the discounted present value of instantaneous utility at time t, $\log Y(t)$ can be expressed as $\left(\dfrac{1}{1+0.1}\right)^t \log Y(t)$, but this calculation holds only when that discount rate is assumed to be annual interest. However, strictly speaking, it should be calculated at more frequent rates, for example, a half-year's rate is $\rho/2$, a 4-month rate is $\rho/3$ a 3-month rate is $\rho/4$, and so on. Therefore, if we take an unlimited frequency of the content of the integral sign on the right-hand side of the above utility function, it becomes

$$\lim_{n\to\infty}\left(\dfrac{1}{1+\dfrac{0.1}{n}}\right)^{nt} \log Y(t) = \lim_{n\to\infty}\left(1+\dfrac{0.1}{n}\right)^{-nt} \log Y(t)$$

$$= \lim_{n\to\infty}\left\{\left(1+\dfrac{1}{\dfrac{n}{0.1}}\right)^{\dfrac{n}{0.1}}\right\}^{-0.1t} \log Y(t) = \lim_{m\to\infty}\left\{\left(1+\dfrac{1}{m}\right)^m\right\}^{-0.1t} \log Y(t) = e^{-0.1t} \log Y(t) \quad (4.29)$$

Here, we considered $m = n/0.1$ and we used the definition of e (Napier's constant) at the last equal sign. Then, we need to sum up (integrate) each year's discounted present value $\log Y(0)$ at time 0, $\log Y(1)$ at time 1, $\log Y(2)$ at time 2, $\log Y(3)$ at time 3, $\log Y(4)$ at time 4, and so on, in the form of a continuous variable. Using this calculation, we can derive the above intertemporal utility function U.

Therefore, the issue we face is to maximize the intertemporal utility U under the conditions of the two production functions identified above. That is,

$$\begin{aligned} &\max U = \int_0^\infty e^{-\rho t} \log Y(t) dt \\ &\text{s.t. } Y(t) = AK(t)^\alpha (s(t)L)^\beta \\ &\dot{K}(t) + \delta K(t) = B(1 - s(t))L \end{aligned} \quad (4.30)$$

Here, "s.t." means "subject to," which indicates that the following two equations of $Y(t)$ and $\dot{K}(t)$ are the constraint conditions, and the terms $Y(t)$, $K(t)$ and $s(t)$ indicate that Y, K, and s vary over time. Thus, the ratio $[s(t): 1 - s(t)]$ at which the total labor power is split into two production sectors is a control variable of humanity overall. Thus, this model is called the Marxian *optimal* growth model: the issue is formulated as an optimization problem in the growth process.

We now attempt to work out this model in practical terms. Since the issue is a conditional maximization problem of intertemporal utility while satisfying certain conditions, we employ the following current Hamiltonian value:

$$\begin{aligned} H_c &\equiv \log Y(t) + \mu(t)[B\{1 - s(t)\}L - \delta K(t)] \\ &= \log A + \beta \log s(t) + \beta \log L + \alpha \log K(t) + \mu(t)B\{1 - s(t)\}L - \mu(t)\delta K(t) \end{aligned} \quad (4.31)$$

with the first-order conditions of the optimization being

$$\frac{\partial H_c}{\partial s} = 0 \Leftrightarrow \frac{\beta}{s} - \mu BL = 0 \quad (4.32)$$

$$\frac{\partial H_c}{\partial K} = \rho\mu - \dot{\mu} \Leftrightarrow \frac{\alpha}{K} - \mu\delta = \rho\mu - \dot{\mu} \quad (4.33)$$

$$\frac{\partial H_c}{\partial \mu} = \dot{K} \Leftrightarrow \dot{K} - \delta K = B(1-s)L \quad (4.34)$$

and the transversality condition. Here, we omit the term (t) for Y, K, s, and μ, for simplicity. The first equation leads to

$$\frac{\dot{\mu}}{\mu} = -\frac{\dot{s}}{s}, \quad \mu = \frac{\beta}{sBL} \quad (4.35)$$

Substituting this into the second first-order condition gives

$$\frac{\alpha}{K} \cdot \frac{sBL}{\beta} = \frac{\dot{s}}{s} + (\rho + \delta) \qquad (4.36)$$

which can be transformed further to derive

$$\dot{s} = \frac{BL}{K} \cdot \frac{\alpha}{\beta} s^2 - (\rho + \delta)s = s\left\{\frac{BL}{K} \cdot \frac{\alpha}{\beta} s - (\rho + \delta)\right\} \qquad (4.37)$$

Because it is a $0 < s < 1$ process analysis, in this equation, $s \neq 0$. We first substitute $\dot{s} = 0$ into equation (4.37) to obtain the expression

$$s = \frac{(\rho + \delta)\beta}{\alpha BL} K \qquad (4.38)$$

which satisfies the condition $\dot{s} = 0$. In addition, the subsequent substitution of $\dot{K} = 0$ into the production function of the sector producing the means of production gives

$$B(1-s)L = \delta K \qquad (4.39)$$

The intersection of (4.38) and (4.39) is a steady point satisfying both $\dot{s} = 0$ and $\dot{K} = 0$. Solving this gives

$$\left(\frac{K}{L}\right)^* = \frac{B\alpha}{(\alpha + \beta)\delta + \beta\rho}, \quad 1 - s^* = \frac{\delta\alpha}{(\alpha + \beta)\delta + \beta\rho} \qquad (4.40)$$

This exactly matches the steady state eqilibrium derived by different method in subsections 4.1.2 as (4.12).[15]

However, we have not only confirmed that the steady state is the same as in the preceding section, but we can also identify an important property in the process of accumulation or growth. This is because when we derive K^* and s^* as the intersection of the equations $\dot{s} = 0$ and $\dot{K} = 0$ in **Figure 4.3**, we also can investigate the dynamics in the four segments separated by the lines $\dot{s} = 0$ and $\dot{K} = 0$, as shown in Figure 4.3. That is, above the line $\dot{s} = 0$, s is increasing, while it is decreasing below the line, and in the area to the right of the line $\dot{K} = 0$, K is decreasing, while it is increasing in the area to the left of the line. These are depicted in this figure by bidirectional arrows. What is important here is that we know that when we start from the rational assumption that K is initially less than K^*, the process of accumulation or growth toward K^* must be a saddle path that rises as it moves to the right. This indicates that the percentage of labor allocated to the sector producing the means of

---

[15] The calculations above are based mostly on Yamashita and Onishi (2002).

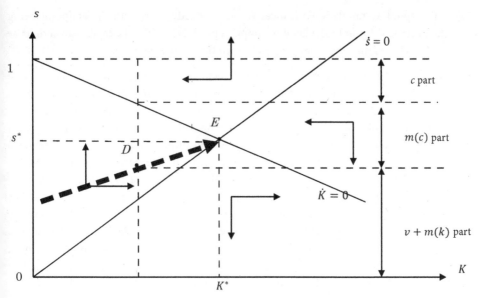

Figure 4.3 Dynamics of capital accumulation toward a steady state[16]

consumption[17] increases during the process of capital accumulation or growth; in other words, the percentage of labor allocated to the sector producing the means of production decreases.[18] This conclusion is the opposite of the law of the preferential growth of sector 1, argued by Lenin (1977).[19]

Thus, the process of capital accumulation or extended reproduction can be summarized as the following two results:

(1) The ratio s of total labor used in the production of the means of consumption rises in the process of capital accumulation. Put another way, the ratio $1 - s$ used to produce the means of production decreases.

---

[16] Ibid.

[17] While Marxian economics distinguishes labor power from labor itself, we examine not the allocation of labor power, but that of labor. This is because we are interested in the amount of labor actually consumed. However, if we assume that all capitalists use their labor power efficiently, then these allocations of labor and labor power are identical.

[18] The above diagram is based on Onishi and Tazoe (2011).

[19] Note that this conclusion holds only in the case of no technological change. For example, if B (labor productivity in sector 1) increases, then the possibility of the preferential growth of sector 1 arises.

(2) This capital accumulation advances toward a steady state, with the endpoint being the same as the point calculated in simple reproduction. That is, capitalism should be understood as the long-term process toward this steady state.

These two conclusions cannot be reached using the reproduction scheme in value terms. This is because while the reproduction scheme can introduce the condition for replenishing materials between $v_1 + m_1$ and c_2, or between the two sectors in value terms, the human object of utility maximization was not formulated. Until now this optimization behavior has been treated as if a society is a collection of completely homogeneous individuals and is deliberately managed by an individual (a "representative individual"). This notion needs to be replaced with the optimization behavior of various individuals, each having independent utility, and with the profit-optimization behavior of various companies, each having their own individual production functions. In the terminology of modern economics, this needs to be developed not as a social planning model but as a competitive market model. To this point we have described only the former model because, under conditions in which there are no externalities, no irrationality and no incomplete information with perfect competition, the solutions to both models are identical.

4.3.3 Böhm-Bawerk's Agio Theory of Interest

Incidentally, following the first-order condition shown in (4.33), we have a very important eqution for understanding exploitation:

$$\frac{\alpha}{K} - \mu\delta = \rho\mu - \dot{\mu} \tag{4.41}$$

This is because it can be transformed into the following equations, which clearly express profit:

$$r_k - \delta = \rho - \frac{\dot{\mu}}{\mu} \tag{4.42}$$

or

$$r_c - \delta = \rho - \frac{\dot{\mu}}{\mu} \tag{4.43}$$

Here, r_k and r_c are the real rental prices of capital in terms of the means of production and the means of consumption, respectively. This transformation can be obtained by the following method:

(1) Because α/K in (4.41) is $(\partial \log Y/\partial K)$, it can be expressed as $(\partial \log Y/\partial Y) \times (\partial Y/\partial K)$. Remember that $(\partial \log Y/\partial Y)$ is the price of the means of consumption (p_c) in terms of utility (differentiated instantaneous utility) and $\partial Y/\partial K$ is the real rental price of capital goods in terms of the means of consumption (r_c).

(2) Here, the current value Hamiltonian in the previous subsection shows μ as the shadow price of capital goods, which is the price of the means of production in terms of utility. Therefore, if $r_c \cdot p_c$ is divided by μ, it becomes r_k. In this way, we can achieve equation (4.42).

(3) On the other hand, if we disregard the difference between the price of the means of consumption and the means of production, (4.41) can be transformed into equation (4.43). This assumption is the same as the assumption in the one-sector model, which disregards the difference between the two sectors.[20]

Next, let us look at equations (4.42) and (4.43), where the real rental prices of capital r_k and r_c consist of depreciation, time preference, and the inflation rate of the price of the means of production. If so, the part of depreciation that corresponds to Marx's constant capital c, the right side of equations (4.42) and (4.43), should be surplus value. That is, the real rental price of capital r_c is the sum of constant capital c and surplus value m in Marxian theory.

We need to show clearly that the Marxian exploitation theorem also holds under the product exhaustion theorem of neoclassical economics. For the production function in the consumer goods sector, if the sum of the two elasticities of the factors of production equals one; that is $\alpha + \beta = 1$, then the return to scale becomes constant, and the sum of the capitalists' income (the real rental price of capital r_c × amount of capital) and workers' income (the real wage in terms of the means of consumption × amount of labor power) becomes the amount of net product. That is:

$$r_c K + RL = \frac{\partial Y}{\partial K}K + \frac{\partial Y}{\partial L}L = \alpha \frac{Y}{K}K + \beta \frac{Y}{L}L = (\alpha + \beta)Y = Y \qquad (4.44)$$

In this relation, it seems as if there is no surplus value for capitalists, as all the products are exhausted. However, we can see that r_c or r_k includes Marx's constant capital c and surplus value m.[21] Here, the profit rate, which in Marx's definition is $m/(c + v)$, becomes

$$\frac{\left(\rho - \frac{\dot{\mu}}{\mu}\right)K}{\delta K + RL} \qquad (4.45)$$

---

[20] Onishi and Kanae (2015) explain the difference between (4.42) and (4.43) without assuming (3).
[21] Because equation (4.44) shows that α and β are the capital share and labor share, respectively, if they change, these shares also change. Therefore, a rising trend of growth in the capital share in the post-war period, shown in Chapter 6 in Piketty (2013), may be caused by a further rise in α by new industrial revolutions in non-manufacturing sectors. See section 3.3 in this book, which discusses industrial revolutions in non-manufacturing sectors.

Furthermore, the more important fact is that the surplus value in the right side of equations (4.42) and (4.43) include time preference ρ. These equations therefore become

$$r_k - \delta = \rho \qquad (4.46)$$

or

$$r_c - \delta = \rho \qquad (4.47)$$

in their stable states (although equation (4.47) needs the above procedure (3)). In this situation, all surplus value comes from the time preference, a problem that was first discussed by Böhm-Bawerk (1890) and then later by Negishi (1985), Chapter 9. Negishi knew Okishio's FMT and agreed with this proof, but also criticized it, stating that this proof was just an interpretation using the LTV. In this context, Negishi supports Böhm-Bawerk's agio theory of interest (namely, that profit and interest are created by the time preference), and our equations (4.46) and (4.47) also seem to support their premise. At the very least, we should recognize Böhm-Bawerk's and Negishi's theory that an important part of profit is created by time preference, as a result of the rational behavior of economic agents. In this way, individual preference, which evaluates the present over the future, creates profit as price minus cost.

However, this is not the essence of profit. Let us remember Okishio's FMT, which does not take into account the period[22] and time preference in explaining exploitation (surplus value). For him, exploitation is just the gap between what workers produce and what workers receive, without concern for any type of time preference. Therefore, what Böhm-Bawerk and Negishi proposed should be understood merely as the mechanism by which surplus value is allocated. In fact, later in subsection 5.3.1 we discuss the intersectoral allocation of surplus value guided by the organic composition of capital (c/v) when surplus value is transformed into actual profit. We also discuss the allocation of profit between the producers of present goods and producers of future goods. The point here is that we must identify the difference in the dimension of value in terms of input labor from the dimension of price, where profit is actually allocated to different agents.[23]

Equations (4.42) and (4.43) contain one more important element; that is, the second term of the right-hand side as the second part of profit. Recall the following equations (4.35) in the last subsection 4.3.2, renamed as (4.48):

---

[22] The temporal single-system interpretation (TSSI), which we discuss later in subsection 5.3.2, is a period-concern approach and therefore cannot accept the FMT. This means that the TSSI cannot regard ρ as a part of exploitation. It is true that Marx included a time structure in his circulation theory by setting the initial cost as "capital advanced." However, in my opinion, the time structure should be incorporated differently, for example, using differential equations.

[23] This problem was clarified first by Morimoto (2011).

$$\frac{\dot{\mu}}{\mu} = -\frac{\dot{s}}{s}, \quad \mu = \frac{\beta}{sBL} \tag{4.48}$$

which indicate that $(-\dot{\mu}/\mu)$ is positive because s is constantly increasing, and at a final point will become zero. In other words, when capital accumulation is necessary before reaching the steady state, a part of the profit should be taken by someone and another part by the rentiers. Here, the "someone" must be the entrepreneur who engages to achieve the social necessity of capital accumulation. This part of the gain will disappear when capital accumulation becomes unnecessary, which means that profit is created by the necessity of capital accumulation. In subsection 3.1.5, I call this exploitation the "first definition of exploitation," which is generated to match the necessity of capital accumulation and will disappear when capital accumulation becomes unnecessary. Therefore, in contrast, the profit gained by the time preference can be regarded as rest of the surplus value which I named "second definition of exploitation" in subsection 3.1.5.

These equations can explain that Piketty's gap, in Piketty (2013), between the growth rate of property income and that of labor income (Piketty's $r > g$ using our terminology in Addendum II, can be expressed as $\tilde{r} > g$; here, g is the growth rate of total income) is caused by the second definition of exploitation. Although Piketty calls the GDP growth rate g and assumes it is the same for the growth rate of labor income, it is better to assume it is the same as the growth rate of consumption. If so, then the following equation in Addendum II

$$\frac{\dot{Y}}{Y} = \tilde{r} - \rho \tag{4.49}$$

can be rewritten using Piketty's notations as

$$g = r - \rho \tag{4.50}$$

That is,

$$r = g + \rho \tag{4.51}$$

Thus, the interest rate is larger than the economic growth rate by ρ, and therefore the gap between r and g is basically constant, as Piketty shows. In this way, we can solve the secret of Piketty's inequality.

However, a much more important point is that this gap ρ is a very significant part of surplus value. Equations (4.46) and (4.47) show that ρ is the surplus value in the stable state. Piketty's critique of capitalism, showing that r is larger than g, should be understood as critique of the acquisition of interest payments as a kind of compensation for waiting time and a critique of the second definition of exploitation. Piketty regards himself as a non-Marxist, stating that his theory has no relation to Marx's exploitation theory.

However, in reality, his most important finding ($r > g$) comes from Marx's theory of exploitation. U.S. Marxist David Harvey criticized Piketty by stating that the latter should discuss exploitation itself (Harvey, 2014). My criticism of Piketty is identical.

In addition, note that the gap between r and g becomes much more conspicuous when both r and g decrease together, as we typically see in advanced economies. If the growth rate of labor income (g) is sufficiently high, then the complaints of the working class are not considered serious, even if the growth rate of property income (r) is also higher. However, if only the rentiers' income grows, without a rise in labor income, then workers' complaints become significant. In this case, the possibility that workers can raise their social status through "success stories" becomes very limited. This might be why Piketty's critique of capitalism has a much stronger influence than other liberal critiques in advanced countries. In one of his lectures, Piketty laments that his critique was not welcomed in developing countries. I think the difference between advanced and developing countries is explained by the lower r and g in advanced economies. In other words, the difficulties in advanced countries are much deeper than in young economies, where capitalism still works well.

4.4 GENERAL LAW OF CAPITALIST ACCUMULATION: THE END OF CAPITALIST ACCUMULATION

4.4.1 Value-Term Expression of the Marxian Optimal Growth Model

We consider it is more important to replace this Marxian optimal growth model with a value term model than with a competitive market model. Marx called this social planning model the "Robinson Crusoe story," and said that it incorporated all the essential determinants of value.[24] To achieve our aim we first derive the total amount of direct and indirect labor inputs to the means of production (t_1) and means of consumption (t_2). Then, we calculate a table similar to Table 4.1 for a model of the accumulation or growth process.

To calculate t_1 and t_2, the values (labor input) per unit of output in both sectors, we follow the expression in Okishio's theorem in the last chapter, which gives us

$$\left. \begin{array}{l} t_1(\dot{K}+\delta K)=(1-s)L \\ t_2 Y = t_1 \delta K + sL \end{array} \right\} \quad (4.52)$$

The right sides express labor inputs, and the left sides express how much labor embodied in both goods in the macro level.

---

[24] Marx, 1887, 50.

Solving this pair of simultaneous equations gives

$$\left.\begin{array}{l}t_1 = \dfrac{(1-s)L}{\dot{K}+\delta K} = \dfrac{(1-s)L}{B(1-s)L} = \dfrac{1}{B} \\[2ex] t_2 = \left(\dfrac{\delta K}{B} + sL\right) / (AK^{\alpha}(sL)^{1-\alpha}) = \left(\dfrac{\delta}{AB}\right)k_2^{1-\alpha} + \left(\dfrac{1}{A}\right)k_2^{-\alpha}\end{array}\right\} \quad (4.53)$$

In the second equation, for convenience of calculation we introduce a new definition, $k_2 \equiv K/sL$, and assume constant returns to scale ($\alpha + \beta = 1$). While it is clear from the first equation that t_1 is a constant expressed in technical parameters only, we require analysis only on t_2. It varies as a function of k_2. We next need to investigate the dynamics of k_2. Since this is a complex calculation, I place it in the footnotes. The results show that k_2 increases over time and thus t_2 decreases over time.[25] A decrease in t_2 means a continual decrease

---

[25] This calculation is as follows. First, differentiating the second equation in (4.53) for k_2 gives

$$\dfrac{dt_2}{dk_2} = \dfrac{k_2^{-\alpha}}{A}\left\{\dfrac{(1-\alpha)\delta}{B} - \alpha k_2^{-1}\right\}$$

Since this expression becomes 0 when $k_2 = \hat{k}_2 \equiv \alpha B/(1-\alpha)\delta$, then when $k_2 < \hat{k}_2$, the increase in k_2 decreases t_2, and the opposite result is attained in the opposite case. However, on which side is k_2 in reality? For this analysis, we look again in detail at the equation (4.37) in subsection 4.3.2:

$$\dot{s} = s\left\{\dfrac{BL}{K}\cdot\dfrac{\alpha}{\beta}s - (\rho+\delta)\right\}$$

We know from the phase diagram in Figure 4.3 that s increases; in other words, $\dot{s} > 0$. Therefore, inserting this condition into this equation will give us an analysis of the dynamics of k_2. For this purpose, first deriving k_2^* when $\dot{s} = 0$ gives

$$k_2^* = \dfrac{\alpha B}{(\rho+\delta)(1-\alpha)}$$

This is k_2 at the target of capital accumulation, or in a steady state. However, because, as noted above, $\dot{s} > 0$ in the economy, then

$$\dfrac{BsL}{K}\cdot\dfrac{\alpha}{\beta} - (\rho+\delta) > 0$$

We can rewrite this condition as

$$k_2 < \dfrac{\alpha B}{(\rho+\delta)(1-\alpha)} = k_2^*$$

ultimately,

$$k_2 < k_2^*$$

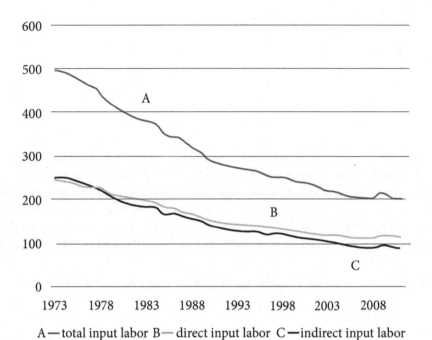

A—total input labor B—direct input labor C—indirect input labor

Figure 4.4 Improvement in input labor per unit in the Japanese economy (in terms of hour/standardized million yen in 2000)

in the labor needed to produce the means of consumption (i.e., a decrease in value) and the increased efficiency of production. This supports our position that accumulation is a rational choice for human societies. However, as mentioned in section 4.3, there is an upper limit to the improvement of productivity, and therefore, the speed of improvement in productivity gradually decreases toward a steady state. We illustrate this situation for the Japanese economy in **Figure 4.4** (see Tazoe, 2015).

Furthermore, **Table 4.2** attempts to calculate the composition of value $c + v + m$ in the same form as Table 4.1 based on labor input, or value, in each sector in the growth process. Since we assume $m_1(k) = m_2(k) = 0$, this decomposes into $c + v + m(c) + m(v)$. However, note that we can neglect $m(v)$ as a whole society because $m_1(v) + m_2(v) = 0$.

This shows that k_2 is in the process of increasing over time.

In addition, because $\hat{k}_2 \equiv \alpha B/(1-\alpha)\delta$ is clearly larger than k_2^*, we have

$$k_2 < k_2^* < \hat{k}_2$$

This shows that in the entire range in which k_2 can move, t_2 decreases. That is, labor productivity increases.

Table 4.2 Decomposition of labor input in the growth process of the Marxian optimal growth model

	c	v	m	Total
Sector 1	0	$(1-s)L$	0	$(1-s)L$
Sector 2	$\dfrac{\delta K}{B}$	$\beta t_2 Y = \beta\left(\dfrac{\delta K}{B} + sL\right)$	$sL - \beta\left(\dfrac{\delta K}{B} + sL\right)$ $= (1-\beta)sL - \beta\dfrac{\delta K}{B}$	$\dfrac{\delta K}{B} + sL$
Society as a whole	$\dfrac{\delta K}{B}$	$\beta\left(\dfrac{\delta K}{B}\right) + (1-s+\beta s)L$	$(1-\beta)sL - \beta\dfrac{\delta K}{B}$	$L + \dfrac{\delta K}{B}$

The steps of this calculation are shown in the notes.[26] From this we can see that the labor input in the growth process, but not only a stationary state, can be converted to $c + v + m$ in this simple form. While Marxian economics developed a useful system of equations known as the reproduction scheme, until now it has been expressed in value terms only. However, we can rewrite labor input structure as the form of $c + v + m$ even if we are employing a modern economics model, as long as this problem has been formulated as a problem of labor allocation. The human behavioral objective of maximizing utility can be expressed only in modern economics models and the resulting dynamics of the model can be analyzed in the form $c + v + m$.

The fact that this has the same format as Table 4.1 means, naturally, that the condition $v_1 + m_1 > c_2$ of extended reproduction can be confirmed in Table 4.2 as well. In fact, the condition $v_1 + m_1 > c_2$ holds because

$$v_1 + m_1 - c_2 = (1-s)L - \frac{\delta}{B}K = \frac{\dot{K}}{B} > 0 \qquad (4.54)$$

Furthermore, identifying the conditions satisfied by the growth process in this way requires knowing what stationary state comes after this growth. To derive the state at the

---

[26] (1) The total of both sectors or total value, can easily be plugged into this table. (2) Component c is the next easiest because it is not present in sector 1 (since there is no input of capital K) and because in sector 2, it can be represented as the amount of labor $\delta K/B$ needed to cover each year's depreciation δK in K. (3) Component v is calculated using the assumption of the marginal productivity of labor equalizing wage. (4) The last calculation is for the component m by subtracting the components c and v from the total value produced in each of the sectors. (5) Both sides of the table match after these calculations.

$c + v + m$ level, K^* and s^* derived in the preceding section must be substituted in Table 4.2. The results in **Table 4.3** are identical to those in Table 4.1. That is, the economy grows to address simple reproduction and stops there.

Incidentally, if we express the entire process, including the endpoint of growth, as the form of $c + v + m$, we can identify two tendencies: the rise in the organic composition of capital and the falling rate of profit. They are expressed by the dynamics in each of c, v, and m in Figure 4.3, in which the thick broken line points to the upper right, although $m(k)$ as a part of m is added to v in Figure 4.3, unlike in Table 4.3 in which all parts of m are added together. Therefore, as a kind of summary, let us clarify the characteristics of these dynamics, as follows.

(1) The total c and total v in Table 4.3 show a rising trend of the c/v ratio, called the organic composition of capital by Marx. Because the rising tendency s and K/L rise

$$\frac{c}{v} = \frac{\frac{\delta K}{B}}{\beta\left(\frac{\delta K}{B}\right) + (1-s+\beta s)L} = \frac{1}{\beta + \frac{B(1-s+\beta s)}{\delta} \cdot \frac{L}{K}} \tag{4.55}$$

under the assumption $0 < \beta < 1$.[27]

---

[27] This point is closely related to the evaluation of the Cambridge capital controversy. Many mathematical Marxists prefer a linear production function in line with Marx, and thus support the arguments of Joan Robinson and others, who criticized the neoclassical production function in this controversy. But this is a mistake because it fails to understand that this debate is aimed at Keynesian distribution theory and has the opposite implication for Marx's theory of the falling rate of profit. Marx's theory discusses the effect of a rise in the capital–labor ratio, as described here, and not only does it argue for the substitution of capital and labor, but also that a rise in the capital–labor ratio reduces the profit rate, which is exactly the opposite of the post-Keynesian school. This controversy was developed by the post-Keynesians as a criticism of Samuelson, who argued that (1) even if linear production functions is used, neoclassical production functions that allow for substitution between factors of production can be derived by arranging multiple production functions, and (2) in that case, a rise in the capital–labor ratio would lead to a decline in the rate of profit. In his Chapter 4 Pasinetti (1977) summarizes this as: (1) the post-Keynesians did not oppose substitution between factors of production, and (2) a rise in the capital–labor ratio does not necessarily lead to a falling rate of profit. In other words, Samuelson's argument alone does not allow us to admit that a rise in the capital–labor ratio leads to a falling rate of profit, and this in itself makes sense. However, the problem is that the profit rate in developed countries is actually lower than that in developing countries, and thus the neoclassical production function is a better representation of this reality, as is Marx's falling rate of profit theory. Furthermore, in Chapter 4 of Lenin (1933), the high-profit rate in less developed countries, which leads to the export of capital from highly developed countries, is due to the scarcity of capital. Therefore, the criticism of Samuelson by Robinson and others is rejected by the reality,

Table 4.3 Substitution of K and s for K and s in Table 4.2

	c	v	m	Total
Sector 1	0	$\dfrac{\delta\alpha}{(\alpha+\beta)\delta+\beta\rho}L$	0	$\dfrac{\delta\alpha}{(\alpha+\beta)\delta+\beta\rho}L$
Sector 2	$\dfrac{\delta\alpha L}{(\alpha+\beta)\delta+\beta\rho}$	$\beta((1-s)+s)L = \beta L$	$\left\{\dfrac{\beta(\delta+\rho)}{(\alpha+\beta)\delta+\beta\rho}-\beta\right\}L$	$\dfrac{\delta\alpha L}{(\alpha+\beta)\delta+\beta\rho}+\dfrac{(\beta\delta+\beta\rho)L}{(\alpha+\beta)\delta+\beta\rho}=L$
Society as a whole	$\dfrac{\delta\alpha L}{(\alpha+\beta)\delta+\beta\rho}$	$\left\{\dfrac{\delta\alpha}{(\alpha+\beta)\delta+\beta\rho}+\beta\right\}L$	$\left\{\dfrac{\beta(\delta+\rho)}{(\alpha+\beta)\delta+\beta\rho}-\beta\right\}L$	$L+\dfrac{\delta\alpha L}{(\alpha+\beta)\delta+\beta\rho}$

(2) On the other hand, $m(c)/(c + v + m(k))$ in Figure 4.3 shows a trend to decrease and finally reaches zero. This is due to the gradual disappearance of the need for capital accumulation, but to determine the trend of the other part of surplus value (the $m(k)$ part), it is necessary to examine the overall movement of surplus value in Table 4.2. The trend of the m_2 part depends on whether $(1 - \beta)sL$, which is increased by the trend increase in s, or β ($\delta K/B$), which is increased by the trend increase in K, changes more significantly. However, because Figure 4.3 shows increasing speed of K is larger than that of increasing speed of s, m_2 tends to shrink.

(3) The value term falling rate of profit can be proven using Table 4.2. $m_2/(c_2 + v_2 + m_2)$ and $m/(c + v + m)$ fall, because $c_2 + v_2 + m_2$ and $c + v + m$ rise while m (or m_2) tends to shrink, as implied in Table 4.2. This confirms that the profit rate ($m/(c + v)$) falls. Marx explains this law based on the rising trend of the organic composition of capital to increase. On the other hand, using the Marxian optimal growth model we reason that the marginal productivity of capital decreases in the process of the rising capital–labor ratio. As this is caused by a rising capital–labor ratio, Marx's reasoning and ours are basically same. Note that this is the law of the falling rate of profit in value terms (in terms of embodied labor), and its price term law is explained in note 4 in Addendum III. The falling rate of profit is the most important law of tendency in Marxian economics. In this way, the Marxian optimal growth model can prove this law.

(4) However, the $m(c)$ part of surplus value decreases toward an ultimate value of zero. Among these four characteristics, (1) and (3) are the same as in Marx, but Marx did not arrive at conclusion (4). In other words, conclusion (4) is a special characteristic of the Marxian optimal growth model. However, it must be said that this conclusion is a new discovery resulting from new developments in the following three aspects of Marxian economics.

First, while Marxian historical materialism argues that capitalism remains legitimate up to a certain historical stage, it also must be able to explain how this legitimacy will disappear in the future. This must include arguing simultaneously for the legitimacy of exploitation or gaining surplus value for a certain period and for the future disappearance of this legitimacy. The above four conclusions show this clearly. To the best of my knowledge, no prior research, including that by Marx himself, has developed

---

even if it is not wrong in logic. This is also why this book uses the neoclassical production function in general.

Related to this problem, it should be noted that although the real wage rate increases as the capital–labor ratio increases, the former increase is not the cause of the latter one. This follows from the original relationship that an increase in the capital–labor ratio reduces the total labor required to produce the final product, and it is also consistent with the individual interests of capitalists, although the calculation is not shown here.

this decisively important argument persuasively. This is because, while there have been many descriptions of exploitation as a form of injustice or to prove it exists, there are no explanations why it is legitimate in one period and that legitimacy disappears in the same analytical framework.

Second, we need once again to be clear that the framework of the Marxian optimal growth theory can provide this new breakthrough. This is because to make the above argument for legitimacy requires arguing for what a society needs as a whole, and for this purpose, it is useful to set an objective function to maximize utility over time by a representative individual. Otherwise, it is not possible to identify, for example, the purpose of this capital accumulation via exploitation, and it can only be condemned as an injustice. In this way, the simple $c + v + m$ model is insufficient to lead to some of the conclusions of historical materialism, and we need a physical level of a different explanation to identify the purpose of capital accumulation clearly. Thus, the Marxian optimal growth model is formulated on the physical level and as a model of labor (= value) allocation model for the two sectors. Each column in Tables 4.2 and 4.3 can be calculated only with this type of labor allocation model.[28]

Third, we need to recognize that there is an upper limit of capital accumulation that must not be exceeded. In fact, the conclusion that the $m(c)$ part will become zero depends on acknowledging this. Furthermore, we need to recognize that the production of machinery, which is of decisive importance, ultimately depends on labor. As we see repeatedly in Figures 4.3, 4.8, and 4.13, the decision whether to produce final products (means of consumption) with direct labor on the right-hand side of the figures or with indirect labor on the left-hand side hinges on efficiency. The key point here is that to arrive at the optimal value $(K/L)^*$ means that any additional capital accumulation will give excessive weight to production through the indirect labor on the left-hand side of this figure, which would be inefficient; that is, over-accumulation. Thus, economic rationality requires accumulation to stop at some point in time, and for this reason, growth and accumulation will stop as well. Thus, the conclusions of Marx and the others who did not consider this issue of a halt in growth must be revised.

Finally, in addition to the above three points, note that even if $m(c)$ becomes zero, it does not mean that $m(k)$, as the remainder of m, also becomes zero. This was discussed in

---

[28] Contrary to Steedman's claim that a physical-level model is sufficient to explain exploitation and that the labor theory of value is redundant (Steedman, 1977), our physical model represents the labor theory of value itself because it shows the optimal decisions of the agents to determine labor input as an interaction between humans and nature. Steedman's model looks unrelated to the labor theory of value because it is not designed to express the human optimization behavior of labor expense. Nagaura (1985) has similar criticisms using an inequality approach model that has an explicit objective function.

subsection 4.3.3 as a result of the decentralized market model explained in Addendum II. According to this result, as we saw that the $m(k)$ part remains positive at the steady state, and therefore, m cannot be zero. This corresponds to the existence of surplus value in the case of simple reproduction. Therefore, if we want to abolish m, the logical consequence is that we need a certain force that comes in from outside the free market. As a whole Marx's *Capital* insists that exploitation exists inside the equal exchange that takes place in commodity production, but if this is so, it cannot achieve the abolition of exploitation without a certain force external to the market, which may violate the principle of equivalent exchange. Because the object of abolition here is $m(k)$ as a source of capitalist private consumption, this does not shrink the economy.

This makes us ask: what might be this force outside the market? I suspect it is the power of trade unions in the private realm, as they control capital but without political power, but if this relationship does not mature, then we need a certain kind of governmental intervention such as minimum wage regulations, which does not allow any discretion in administration.

For example, in Figure 3.2, we saw that a forced wage increase may make companies cut employment to maximize their profit. Therefore, social productivity should become sufficiently mature to allow such a wage increase. In other words, sufficient capital accumulation should have been achieved (and the society should have arrived at a zero-growth economy). In this case, a reduction in the total labor demand will not cause a problem because the wage increase will not have to disturb workers' enough earning with a reduction in their working time, if this can be shared properly among. In this case, the capitalists rather than the working class will pay the cost of a wage increase. But because the capitalist class can be expected to resist this, the working class needs to have enough political and social power to force it through. Hence, Karl Marx insisted on the need for a political revolution.[29]

4.4.2 The Organic Composition of Capital and the Law of Increase in the Relative Surplus Population

This final point among four points in the previous subsection is, in a sense, a critique of the assumption of many Marxist economists that there can be a limitless increase in the

---

[29] Mason (2015) proposes the interesting and simple concept of "information goods" as a type of means of production. First, he states that this type of good also needs labor to produce it, according to the labor theory of value. Second, there is no depreciation in the process of production where information goods are used as the means of production, and this characteristic can be introduced into Table 4.3 by setting $\delta = 0$. In this case, production in sector 1 and the constant capital part in sector 2 disappear. Thus, all the labor is used to produce consumption goods ($s^* = 1$). This view is compatible with our image of a communist society, rather than the case in $\delta > 0$.

organic composition of capital. And because of this assumption, Marxian economics proposes the law of the increase in the relative surplus population, which means that unemployment rates will increase due to the increase in the organic composition of capital. However, this assumes a limitless increase in the organic composition of capital. This argument starts by replacing $v + m$ with L because it represents total labor input, and here c/L can be understood as a kind of organic composition of capital. In this case, L can be rewritten as

$$L = \frac{L}{c} \cdot c \tag{4.56}$$

This shows that L is determined by L/c, or the inverse of the organic composition of capital and total capital c. Because our expectation is that the former will decrease while the latter will increase, in the end, the issue is the size of the relationship between this rate of decrease and increase. Thus, investigating this issue by focusing on the constraints on the annual increase in c gives

$$\Delta c \leq m \leq v + m = L \tag{4.57}$$

meaning that

$$\frac{\Delta c}{c} \leq \frac{L}{c} \tag{4.58}$$

This equation shows that the rate of increase in c itself is deeply related to L/c. That is, it cannot exceed L/c and a sufficient decrease in L/c (i.e., an increase in the organic composition) will lead at some point to a decrease, bringing about a downward trend in L. This is discussed more rigorously below. Here, assuming for convenience in manipulating the equations that

$$\frac{L}{c} \equiv v \tag{4.59}$$

then

$$\frac{dL}{dt} = \frac{d}{dt}\left(\frac{L}{c} \cdot c\right) = c\frac{dv}{dt} + v\frac{dc}{dt} \tag{4.60}$$

From the constraint on $\Delta c/c$ (4.58) derived above,

$$\frac{dL}{dt} = c\frac{dv}{dt} + v\frac{dc}{dt} \leq c\frac{dv}{dt} + v(cv) \tag{4.61}$$

This can be rewritten as

$$\frac{dL}{dt} \leq c\left(\frac{dv}{dt} + v^2\right) \tag{4.62}$$

The accepted argument in Marxian economics is that since the first term inside the parentheses in (4.62) is negative and the second, v^2, approaches zero, the left-hand side must eventually become negative.[30]

However, in our reasoning thus far, we have not concluded that v will approach zero. This is because there is an upper limit to capital accumulation. Thus, we cannot agree with the above accepted argument; namely, that the law of increase in the relative surplus population argued for by Marx depends strongly on the extent to which the capital–labor ratio will advance, and for this reason, its ultimate accuracy can be judged only by a model that clearly identifies and accounts for the behavior of economic agents. Put another way, the fact that the propriety of this argument has not been determined until now is due to the lack of a model like the Marxian optimal growth model.

Thus, while the extremely important issue of trends in the unemployment rate must be studied more rigorously, it is desirable to discuss this issue with a focus on a cause other than an increase in the organic composition of capital. For example, European countries, which are considered to have higher unemployment rates than Japan, have developed systems of unemployment insurance. As their societies have advanced to the point of developing such systems this suggests that enriching by the unemployment insurance system the unemployment rate could be increased. However, if such a causal relationship can be shown to work, it would mean that the high unemployment rates in those countries result from workers' choice and are therefore not a serious issue.

While the unemployment rate varies with economic fluctuations, the factors embodied in these trends center on cyclical fluctuations include the degree of development of job placement and job training systems, in addition to the state of the unemployment insurance systems. I believe that the basic trend in unemployment or surplus population is a function of such systems.

4.4.3 Falling Rate of Profit and the Shibata-Okishio Theorem

We argued above for the importance of understanding the various tendencies put forward by Marx to account for the choices made by economic agents. Okishio (1961) offers a related discussion on the law of the falling rate of profit, known as the Shibata-Okishio theorem. We show its argument, because to assess it we need to understand it clearly.

First, the following expressions can be derived by introducing the equilibrium average rate of profit r to the variables p_1 and p_2, defined as prices in both sectors; a_1 and a_2, defined as input coefficients into both sectors; τ_1 and τ_2, defined as the direct labor needed in both sectors; and R, defined as the real wage rate per labor unit:

---

[30] The above analysis is based on Chapter 4, Section 3 of Okishio (1977).

$$(1+r)(a_1 p_1 + R\tau_1 p_2) = p_1$$
$$(1+r)(a_2 p_1 + R\tau_2 p_2) = p_2 \qquad (4.63)$$

We assume here that p_2 is fixed, since fixing either of the prices will cause no essential difficulty because the issue is the relative price of the two goods. Next, we assume that new technology (a_1', τ_1') is adopted, leading to a new equilibrium rate of profit. This results in the following equations:

$$(1+r')(a_1' p_1' + R\tau_1' p_2) = p_1'$$
$$(1+r')(a_2 p_1' + R\tau_2 p_2) = p_2 \qquad (4.64)$$

However, since here p_2 is fixed,

$$p_2 = (1+r)(a_2 p_1 + R\tau_2 p_2) = (1+r')(a_2 p_1' + R\tau_2 p_2) \qquad (4.65)$$

This second equality shows that there are only two possibilities: (1) $r' < r$ and $p_1' > p_1$, or (2) $r' > r$ and $p_1' < p_1$. Thus, our problem is to identify which is correct when capitalists introduce new technology to realize a decrease in production costs. For this purpose, the new technology (a_1', τ_1') should be incorporated into the equation for sector 1, namely:

$$(1+r')(a_1' p_1' + R\tau_1' p_2) = p_1' \qquad (4.66)$$

which results in change in relative price and profit rate, and can be transformed into

$$(1+r')(a_1' p_1 + R\tau_1' p_2) + (1+r')a_1'(p_1' - p_1) = p_1' \qquad (4.67)$$

Since we assume that capitalists introduce the new technology to reduce their production costs, the following applies:

$$a_1 p_1 + R\tau_1 p_2 > a_1' p_1 + R\tau_1' p_2 \qquad (4.68)$$

Substituting this inequality into equation (4.67) yields

$$(1+r')(a_1 p_1 + R\tau_1 p_2) + (1+r')a_1'(p_1' - p_1) > p_1' \qquad (4.69)$$

Subtracting the state under the old technology in sector 1 from both sides gives

$$(r' - r)(a_1 p_1 + R\tau_1 p_2) + (1+r')a_1'(p_1' - p_1) > p_1' - p_1 \qquad (4.70)$$

This can be rewritten as

$$(r' - r)(a_1 p_1 + R\tau_1 p_2) > [1 - (1+r')a_1'](p_1' - p_1) \qquad (4.71)$$

The fact that $[1 - (1+r')a_1']$ in this equation is positive can be derived from the equation for the new technology in sector 1. This is because dividing both sides of the equation for the new technology in sector 1 (first equation in (4.64)) by p_1' and rearranging the equation leads to

$$(1+r')a_1' + (1+r')R\tau_1'\frac{p_2}{p_1'} = 1 \tag{4.72}$$

and this can be transformed to

$$1-(1+r')a_1' = (1+r')R\tau_1'\frac{p_2}{p_1'} > 0 \tag{4.73}$$

Thus, the equation

$$(r'-r)(a_1p_1 + R\tau_1p_2) > [1-(1+r')a_1'](p_1' - p_1) \tag{4.74}$$

shows that of the above two possibilities [(1) $r' < r$ and $p_1' > p_1$, or (2) $r' > r$ and $p_1' < p_1$)], the first is not acceptable, leaving only the second possibility. This means that the rate of profit will increase, contrary to Marx's argument. This is the content of the Shibata-Okishio theorem.[31]

However, as seen above, we reached a different conclusion; namely, that the rate of profit decreases in the Marxian optimal growth model. Tracing the cause of this difference, we can see that in the Shibata-Okishio theorem the real wage rate R is constant. This assumption differs from the conclusions of the Marxian optimal growth model. In other words, in this case as well, the key point is whether the movements of various variables are derived from a model that explicitly incorporates the behavioral principles of economic agents. While the Shibata-Okishio theorem surpasses the previous theory of a falling rate of profit in that it reflects the behavioral principles of capitalists when new technology is introduced, inevitable changes in long-term wage rates were beyond its scope. In fact, Okishio (1965) in Chapter 3, Section 4 and Okishio (1997, 2001) clarifies that the most important condition for the falling rate of profit is the increase in the real wage rate.[32] This is why we introduced the Marxian optimal growth model, which also leads to a rising trend in the real wage rate.

---

[31] The above analysis is based on Chapter 4, Section 6 of Okishio (1977).

[32] To show this relation, Okishio (1965) introduces a three-sector model that includes the luxury goods sector, as follows:

$$(1+r)(a_1p_1 + \tau_1Rp_2) = p_1$$

$$(1+r)(a_2p_1 + \tau_2Rp_2) = p_2$$

$$(1+r)(a_3p_1 + \tau_3Rp_2) = p_3$$

Here, the subscript "3" indicates the luxury goods sector, and the first two sectors are, as with this subsection, the means of production sector and the means of consumption sector, which determine p_1/p_2 and r independently of the third sector. p_3 is determined by p_1/p_2 and r, which are determined by the first two sectors. Hence, the socially equalized profit rate is also determined by technological conditions and real wage rate in the first two sectors. Conversely, under given technological conditions, only the real wage rate determines the profit rate.

4.4.4 Post-Capitalist Society as a Zero-Growth Society

We have examined the general tendencies in capitalist accumulation in various ways. Finally, it seems natural to discuss the situation at the end of such tendencies. As mentioned above, this situation is a steady-state society with zero net investment and this is our image of the post-capitalist society. It is same as that of Schumpeter (1942), who also described socialism as a steady state.[33] We define capitalism as a society that exists for the purpose of capital accumulation or as one in which the most important task is capital accumulation, so that by definition, a society in which there is no capital accumulation is a post-capitalist society—that is, a society that is based on socialism or communism. Such a society can be said to be "human centered" in the sense that all net product other than depreciation is diverted for consumption by humans directly.[34] **Figure 4.5** depicts feudalism, capitalism,[35] and post-capitalism. That is, feudal society was ended by the industrial revolution, and then capitalist society began. However, once capital accumulation largely ceases, the system can no longer be referred to as capitalism. In ordinary terminology, the only choices are socialism or communism. That is why we argue that the Marxian optimal growth theory proves the death of capitalism. A key point here is the recognition that there is an upper limit to capital accumulation.

However, this explanation requires several additional points. The first concerns the breadth of zero growth and the possibility of a partial recovery of growth. What this means is that the term "zero growth" used here does not necessarily refer to a growth rate that is completely 0%. It allows certain low growth rates to exist. As seen in the steady-state capital–labor ratio calculation above.

---

[33] In the sense that it can be established only by the success of capitalism, this conceptualization is in agreement with Hegelian dialectics. Marx states that Hegelian dialectics "includes in its comprehension and affirmative recognition of the existing state of things, at the same time also, the recognition of the negation of that state, of its inevitable breaking up." See Marx, 1887, 15.

[34] Tazoe (2011) defines this society more correctly as one where all the net product can be consumed directly by the people except for the additional capital investment generated by the technological change.

[35] In this figure, the speed of capital accumulation after the industrial revolution changes from high to middle to low, depending on the conclusion of Figure 4.3, because the saddle path of $D \rightarrow E$ shows a decline in the accumulation rate indicated by $1 - s$ (in terms of labor power allocation), which rises to the right. However, in a more complete two-sector model, such as the one presented in Table 5.5 in Chapter 5, the accumulation path is not as simple as this, and numerical calculations confirm that the speed of accumulation can follow an S-curve path that changes from low to high to low. This may be due to the effect of the acceleration of investment accompanying the increase in the capital stock that is incorporated in the means of production function.

MARXIAN ECONOMICS

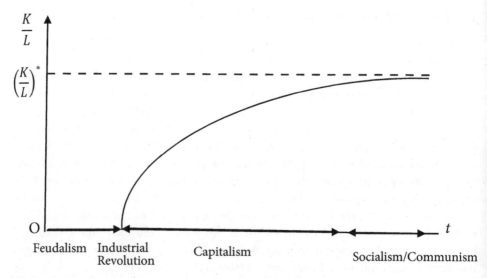

Figure 4.5 Capital accumulation over time since the industrial revolution

$$\left(\frac{K}{L}\right)^* = \frac{B\alpha}{(\alpha+\beta)\delta+\beta\rho} \tag{4.75}$$

An increase in total factor productivity B or technological parameter α in sector 1 leads to growth corresponding to the resulting increase in the target level of capital accumulation. Schumpeter (1942) sometimes says that technological innovation is the only method of extending the life of capitalism. It is thus that this argument can be understood. Even so, however, after the target has been achieved, this type of capital accumulation occurs only by changing the target level itself, so it differs from the original sense of capital accumulation by aiming at a target. This is because it is no more than accumulation due to external causes, which would not have occurred without technological changes.[36]

Thus, the zero-growth societies envisioned here (i.e., advanced countries) do not, strictly speaking, have exact growth rates of 0%. For example, the real growth rate in the

---

[36] According to Tazoe (2011) again, the correct expression of this issue should be that if technological progress is labor-intensive in the means of consumption sector, there must be an increase in the steady-state capital stock of a whole society commensurate with the rate of technological progress. Thus, we can say that while the Japanese economy appears to have shifted to zero growth beginning in 1990, the bubble at the end of the 1980s is understood to have been an artificial prolonging of growth. Labor in the steady state capital labor ratio must be measured in terms of effective labor.

U.S.A. over the years 2000–2020 was 1.7%, and its population grew by 0.9% over the same period. These figures can be converted to a roughly 0.8% per-capita growth rate. We could call this growth rate zero growth. Since the U.S.A. differs from other advanced countries in the sense that it contains a type of developing country internally, which results from immigration and other means, it is natural that its growth rate is higher than the average growth rate in other advanced countries. Put another way, despite such conditions, its real per-capita growth rate was no more than roughly 0.8%, and this type of society can be referred to as a zero-growth society.

Nevertheless, the U.S. economy has continued to be wrongly characterized as strong for a long time. A typical example was that the Clinton years were praised for bringing about a new economy despite the fact that it was a period during which a strong dollar policy led to a loss of industrial competitiveness.[37] The U.S.A. continued to carry out various measures to maintain an artificial growth rate that differed from its actual growth potential. For example, the strong dollar policy during the new economy period was intended to return dollars to the U.S.A., making it possible at first to avoid a shortage of money. Subsequently, the development of economic bubbles such as the IT bubble and the subprime mortgage bubble, in addition to the wars in Afghanistan and Iraq, can be understood as other forms of artificial economic stimuli. These unsustainable policies began to break down in 2008.[38]

In looking at the U.S. economy in this way, it can be seen that it resembles the Japanese economy in many aspects. While the Japanese economy appears to have shifted to zero growth starting in 1990, the bubble at the end of the 1980s is understood to have been the result of an artificial prolongation of growth. Since a long-term decrease in the profit rate also entails the disappearance of profitable targets of investment, it was the investors' orientation toward such targets, even if they were unreasonable, that induced the bubble. In addition, while Japan did not start a war, wasteful public investment, popularity-seeking fiscal expenditures (tax cuts for the rich and making the use of expressways free of charge have same effects on the budget deficit), and the artificial exchange rate policies of Abenomics are understood to have been unreasonable economic stimuli as well.

In addition, the reason why such an effective demand policy is necessary is that the excess supply of investment goods leads to a decline in the price of investment goods, which in turn leads to an expansion of the aggregate supply of consumption goods (Y) under a constant population size, but the aggregate demand for consumption goods shrinks due to the decline in real wages caused by the general decline in the price of investment goods. In other words, the situation in which various types of effective demand policies must be overexploited is one brought about by over-accumulation.

---

[37] On this issue, see Part 1 of Onishi (2003a).
[38] While there are some differences, this is the basic understanding of the Institute for Fundamental Economic Science (2011).

Thus, the second point we must add concerning this zero-growth society is that although these countries need to accept zero growth and form a completely different society suited to such conditions, they have not been able to do so, and instead created several massive problems. This shows how difficult systemic change is, and the considerable effort such a transition requires from the citizenry. For example, the U.S.A. had a strong movement opposing several wars, seeing it as the greatest waste, and this led to the 99 percent movement. Japan also had similar movements against several wasteful public projects. Starting with opposition to the construction of a movable dam on the Yoshino River in Tokushima Prefecture in 1999–2001, there have been various highly visible movements opposing public projects, such as the construction of a new bullet train station in Shiga Prefecture and new airports in Kobe and Shizuoka. Several movements have succeeded in blocking these projects. These movements are beneficial to society as a whole in that they stop accumulation. This can be seen in the case of expressways in **Figure 4.6**. Here, the plots indicate the year when construction was completed for all the expressways in Japan at the time the figure was drawn, along with the number of vehicles using the expressways per day and per kilometer. It shows that while the first expressways were necessary and are still used efficiently, those constructed recently are not used very much. While this would not be a problem if expressways were gifts from God, in reality, they are built by human labor. For this reason, their utility must be measured against the cost of their construction. This balance has broken down to a considerable extent. Clearly, this state of affairs perfectly embodies our theory that the cessation of capital accumulation is necessary at some point.

What we must note on this point is the effect of the nuclear accidents in Japan resulting from the Great East Japan Earthquake in 2011. This event simultaneously unmasked the various problems of nuclear power and forced reductions in the demand for electricity, which forced us to change our lifestyles. It can be said that recognition of the costs of nuclear power has grown. Changes in people's consciousness always arise from such shocks.

I am aware of the fact that this change is being advanced with an anti-science bias, including that of antinuclear powers, and an ecological bias, as this is how revolutionary social changes occur. As is clear from our discussion, ending over-accumulation is not an act that opposes productivity; instead, it favors productivity. That is, it seeks to optimize the cost-benefit ratio. People tend to represent this in their consciousness merely as opposing extravagance. This representation is mistaken. While our own misunderstanding as social scientists cannot be allowed, we know that a broad variety of people tend to have such perceptions. This allows us to see that it is perfectly conceivable that the red-green political coalition in Germany could come into being. From our point of view, this is a battle seeking to convert away from a capital-centered society or to abolish capitalism.

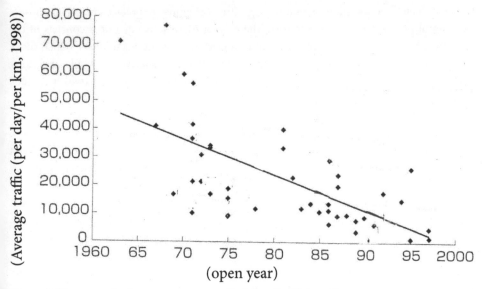

Figure 4.6 The newer the Japanese expressway, the lower its efficiency[39]

However, this brings us to our third point. The forces of resistance are strong, as we seen in the fact that such demands cannot be realized unless they are supported by powerful movements. This is because, no matter how wasteful public projects may have been, to continue with such public construction is in the interest of the construction industry, and the successive Liberal Democratic Party of Japan administrations represent this interest. It is a fact that political face-offs in rural areas took the form of conflicts between construction capitals and residents.

In addition, in some respects, an economic bubble can be said to be the result of actions by active investors (in other words, the "capitalistic personality") who could not tolerate conditions in which there were no targets for investment. They continued to search daily, through their securities companies, for profitable investments, and this pressure induced a bubble by attaching high value to even slight profitability. Alternatively, they induced government to make bubble-promoting policies by welcoming artificial government

---

[39] *Source:* Prepared based on the 1999 Annual Report of the Japan Highway Public Corporation. Japan Ministry of Finance Policy Research Institute (2001).
Notes: (1) Where the expressway opened in parts, "open year" indicates the year the first segment opened. (2) Average traffic (per day/per km) represents the average number of vehicles on the average day on the entire route, calculated as the sum total of distance travelled by vehicles using these expressways (vehicles · km) divided by the total length of these expressways (km).

policies, such as a low interest rate policy. Thus, economic bubbles can be said to be a product of pressure by active investors. The core of these active investors consists of the wealthy, who have high levels of orientation toward investment. Such forces, which make up only one portion of society, have damaged the interests of society as a whole, including that of social stability. This also appears in the form of conflicts among forces within society, and broadly, as class struggle.

Of course, this conflict unfolds inside companies as well because the shift from investment to consumption at the macro level also includes wage increases at the micro level. Thus, the capital–labor relationship must also change. To realize full employment even under conditions of zero growth we need to shorten working hours using methods such as work sharing. Obviously, this constitutes class struggle. On this subject, Marx argues that in the society of the future, growth in productivity would entail not an increase in surplus labor but an increase in free time. This is the position we take when we refer to such a society as a zero-growth society.

In addition, it should be pointed out that the condition of zero growth is important to overcome these various resistances. For that purpose, we can replace the player of the model in Chapter 1, Section 3 with two players: the capitalist class and the working class, as shown in **Table 4.4**. Here, it is assumed that the workers' cooperation with capitalists although the latters' hostility to them increases the gains by a rate of g from the original levels of S_w and S_c, respectively. This represents a situation where the outcome of economic growth trickles down to both classes at the same rate.

However, there is another way for the working class to gain, not from this trickle-down system, but by fighting against the capitalist class. In **Table 4.4**, as in Chapter 1, Section 3, we assume that a workers' fight generates F amounts of income transfer from the capitalist class to the working class, and that capitalists' concessions have the same effect.

Which attitude does each class choose in such a payoff structure? Looking first at the capitalist class, it would not be beneficial to concede to the working class, whatever attitude the latter choses (this strategy is called the "dominant" one). Because the interests of both classes are inherently unreconcilable, the capitalist class instead chooses to be

Table 4.4 Payoff structure of the two classes by their chosen attitudes

		Choice of the capitalist class	
		Concession	Hostility
Choice of the working class	Fight	$S_w + 2F, S_c - 2F$	$S_w + F, S_c - F$
	Cooperation	$S_w + F, S_c - F$	$(1+g)S_w, (1+g)S_c$

hostile as a strategy. Therefore, the working class must choose between one of the two combinations on the right side of this table; that is, $S_w + F$ or $(1 + g) S_w$ if we look only at the workers' payoffs. It is a choice of g gained by cooperation or a choice of F taken by fighting. In short, the greater the value of F, the more advantages gained by fighting, and the smaller the g, the more workers will fight against capitalists. What is important here is that mature capital accumulation makes g nearly zero. Thus, we can understand why a high growth economy leads to workers' cooperation and why a zero-growth economy is politically unstable.

However, this brings us to our fourth point, which is that, as society reaches zero growth and is able to divert considerable wealth to consumption, the quality of the output generated by the human empowered in this manner will change. As muscle power can to some extent be replaced by machines, it is possible that computers will end "nerve labor,"[40] and if such conditions are realized, then the only important labor left for humans will become that of designing, broadly defined, and decision-making. Market pressure may make it possible that only the companies that maintain and promote such human capabilities will survive in competition. That is, the targeting of investments in capitalism that once adhered to the slogan "invest in machines, not people" will change to a post-capitalism approach to the slogan "invest in people, not machines."

In fact, thinking about this thoroughly, one might recall that one implication of our definition of capitalism is that there can be no true change in the mode of production without a change in the main root sources of productivity. If only skill is important, then various social resources will be concentrated in its formation. However, if machinery is the most important, then various social resources will be concentrated in its accumulation. Of course, there is an upper limit, and after reaching this target value, something other than machinery must be of greater importance. If so, then this "something other than machinery" clearly must be the human ability to fulfill the roles of designing, broadly defined, and decision-making—non-mechanical abilities of which only humans are capable, or the productivity of individual creativity.[41] The quantitative achievement of capital accumulation must cause a fundamental change in the quality of productivity, and

---

[40] "Nerve labor" is not a mental labor but a type of physical labor that frays the nerves, like "muscle labor." It also is the opposite of mental work.

[41] Some, but not all, of these abilities can be replaced by artificial intelligence (AI), but AI is also just a kind of machine (see Tomoyori, 2019). Nevertheless, the recent developments in internet and telecommunications technology give workers a high-capacity means of production at a small cost and have the potential to abolish capitalism. This is the significance of the post-industrial society that this book explains in subsection 1.1.6.

likewise, this transformation in the quality of productivity brings about a transformation in the mode of production.[42]

Moreover, the productivity of individual creativity is inseparable from individual workers. I emphasized this point in my example from the small retail stores in subsection 3.3.3, whether the important source of productivity, within or outside humans, directly affects whether command can be effective over labor. Thus, individual and creative productivity can be understood as workers' recovery of the productivity that was hitherto usurped by capitalism. Furthermore, at this time, labor will not be something to be avoided but will be truly self-realizing, and it will be supplied voluntarily as well. The true liberation of workers under post-capitalism must also include these changes in the nature of labor itself.

However, we should mention one implication of this change with reference to Bowles's contestable exchange theory. In subsection 3.1.4 of this book, I showed the incompleteness of labor contracts is a condition for contestable exchange, but the degree differs at each stage of the development of capitalism. For example, the great mechanized industrial stage of capitalism commands labor completely, and no incompleteness is possible. That is, absolute command over labor does not allow room for workers to choose their labor effort autonomously on site. However, the development of service industries such as administrative labor, sales work, and so on, strengthened the incompleteness of the labor contract and widened the possible area of contestable exchange.[43] Only this new situation gives workers room to choose the extent of the labor effort they employ, so that the optimally chosen labor intensity equal to the actual amount of labor supply becomes smaller than that under machine-controlled industry. Various types of automation in offices are the capitalists' countermeasure to this challenge by workers. However, if labor is developed to encompass the truly self-fulfilling features mentioned above, the labor supply of workers will be voluntary, and monitoring costs and unemployment will become useless (and therefore unnecessary) to force people to work. Thus, the true technological progress in the workplace consists of the change from machinery-controlled industries to the non-machinery type of service industries, and finally to a system based on the empowered individual creativity.[44]

---

[42] The author argued this point earlier. See Onishi (1991) and Part 2 of Onishi (1992). At that time, I called this new type of society a "software-based society" in which software was much important than hardware, and therefore in this the most important human ability becomes non-physical creative mental power.

[43] Contestable exchange theory should be understood as one that can explain the reality of the labor commanded by capital only when it includes and subsumes the non-mechanized portion of the labor process.

[44] Tazoe (2016) uses contestable exchange theory to analyze the change in labor from machine-based industry to a mature society.

4.4.5 Process of Latecomer Countries Catching Up to Early Starters

This zero-growth theory of the matured capitalism can explain the current situation in which the per-capita income of latecomer countries tends to catch up with that of the early starter countries. This is natural as long as any country reduces its rate of capital accumulation and then GDP growth as capital accumulates. **Figure 4.7** shows a typical case in which subjective discount rates and all technological conditions are assumed to be the same in both countries. Here, when a latecomer country cannot start to grow, its disparity with countries that began to grow earlier expands, but soon afterwards, the latecomer starts to grow, and begins to catch up. In the long run, its capital–labor ratio will reach the same level as the early starters. Indeed, developing countries in Asia, such as China, have followed this trend. This is identical to the Japanese experience about one hundred years ago, when its

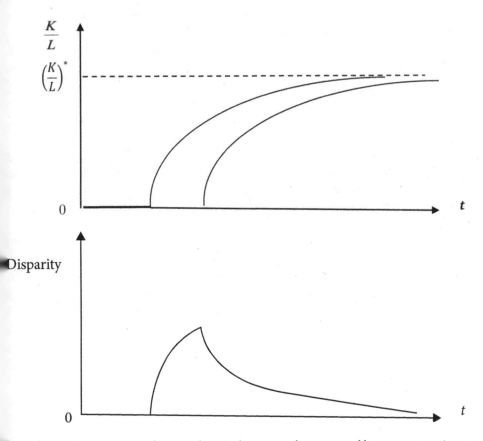

Figure 4.7 Catch-up process and income disparity between early starters and latecomer countries when both have the same subjective discount rate

backwardness was overcome and it became industrialized. The disparity between early and late starters arises at the early stage and then decreases later, as the well-known Kuznets curve projects.

However, in a textbook of Marxian economics we must acknowledge that this phenomenon was explained before Kuznets in Lenin's theory of the law of uneven development of world capitalism in his most important book, entitled *Imperialism, the Highest Stage of Capitalism*. In this theory, he focused on the international movement of capital during which early-starter countries invested in the least developed countries because their lack of accumulated capital leads to a higher profit rate in the latecomer countries.[45] Although dependency theory, as a leftist framework, states that the worldwide disparity must expand between the South and the North, the current circumstances have revealed that the reverse is true.[46]

However, this conclusion is true only under the special assumption that the countries share the same technological conditions and time preference, which is when there may be a disparity between them. For example, **Figure 4.8** shows a case where the rate of time preference of the early-starter is low and that of the latecomer is high, and **Figure 4.9** shows the opposite case. Both show that the disparity cannot disappear even at the final stationary state, and in fact, this is the reality. For example, Latin American countries had their own period of high growth, but now seem to have stopped catching up. On the other hand, China is overtaking some early starters, such as the Association of Southeast Asian Nations countries. In particular, the difference in time preference is critical in not only explaining international disparities among the nations composed of different ethnic groups, but also explaining the domestic disparities among ethnic groups inside a country. The disparity between white and black in the U.S.A. and the disparity between Han Chinese and ethnic minorities in China can basically be explained in this way. This also implies that if the cultural time preference of each ethnic groups does not change, then this disparity will never disappear.[47]

We now turn to an analysis of the important rate of time preference (subjective discount rate), in the possibility that it also comes under a specific historical law. One such change is the change in time preference to emphasize the future over the present (full of entrepreneurs) to cope with modern capitalism (Chinese ethnic minorities are now undergoing this process). However, the current advanced, therefore more mature, countries have

---

[45] See Onishi (1998b, 2003a), Ohnishi (2010).

[46] There are also two phases of expansion and reduction in disparities among regions, just as among countries. For example, provincial disparities in China have changed from their initial phase of expansion to the present phase of shrinking. See Mao (2003) and Onishi (2007).

[47] This type of conflict was a research field of mine for many years, and in Onishi (2012) I provide a compilation of these case studies.

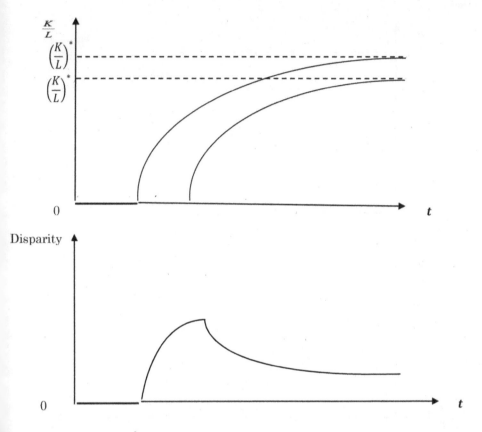

Figure 4.8 Catch-up process and income disparity between early starters that need to reach a higher target and latecomer countries

the opposite tendency to use their assets for their individual, selfish purposes, leaving no inheritance. This is modern individualism. Additionally, a zero-growth economy is now withering people's future-oriented and investment-oriented attitudes. Although social elites tend to dislike this tendency, this might shorten the gap of the target among different ethnic groups.

However, in any case, there is a tendency for this disparity to decrease. This is true at least in the cases of Figures 4.7, 4.8, and 4.9, where both countries accumulate capital to the respective target value. However, this is not true for the trend of disparity between the rich and poor within one country, even if it is true between rich and poor countries. For example, as shown in **Figure 4.10**, the disparity between the two groups continues to be twice same with the starting point. In this case, if both rich and poor groups

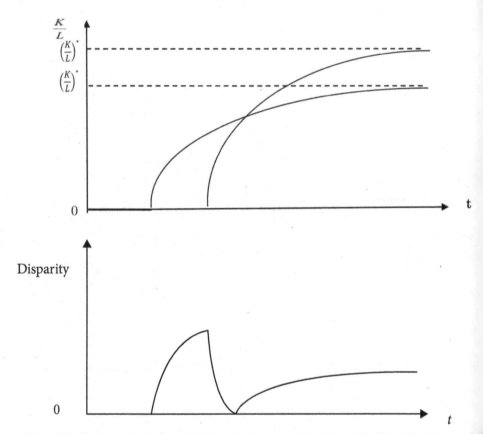

Figure 4.9 Catch-up process and income disparity between early starters with a lower target to reach and latecomer countries

have the same number of people, then the average accumulation level must be in the middle of both groups, shown as the dotted line $(K/L)^*$ in Figure 4.10. This is because the socially determined interest rate reflects the capital accumulation achieved and the marginal productivity of the whole society, and both subjects determine whether to invest or not based on the same interest rate in the same country. In other words, the situation differs only when both sides produce different goods and are unable to produce the other's goods because of particular social restrictions. In this case, the marginal productivity of the capital of one side can be higher than that of the other. However, this is not the case, and therefore the disparity between rich and poor will not disappear in reality.

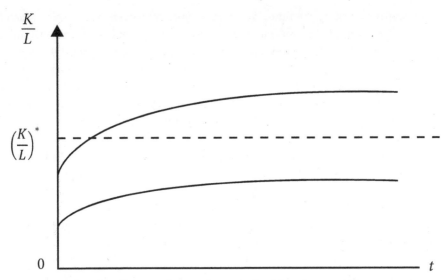

Figure 4.10 Growth path and income disparity between the rich and poor within one country

This thought experiment can clarify the cause of the generally narrowing process of disparities among countries. It is because the interest rates of different countries generally differ, which depends only on their own level of capital accumulation independently from that of other countries. In other words, if central banks are unified, as they are in the E.U., and financial markets are sufficiently internationalized, then the disparity tends to continue as it is in one country. A similar situation occurred under globalization, which creates strong connections between the financial markets of different countries.

4.4.6 Post-Capitalist Firms in the Marketplace: Shareholding Company Socialism

Therefore, we can imagine a post-capitalist society taking a form that differs considerably from previous forms, and the relationship to Marx's argument about socialism or communism through the *socialization* of the means of production arises as an issue that must be addressed anew. As this was identified first as the *nationalization* of private property, we need to consider (1) how the traditional nationalization theory should be understood now and (2) how we can imagine the typical organization of a joint-stock company system in the future.

Traditional nationalization theory was abandoned by Marxian economists in capitalist countries before the collapse of the Soviet Union and Eastern Europe. This is because the inadequacies of nationalized firms were apparent, not just in the Soviet Union, Eastern Europe, and China but also in capitalist countries. Furthermore, negative assessments were made not just of the business system but also of their politics and civil society. Thus, a trend started towards entrusting the future to different types of management systems, such as cooperatives and workers' control enterprises, together with a trend to identify methods of state control that did not depend on state ownership. Since it is unlikely that these forms of business organization will become the majority in the economy, the basic trend has been toward democratic regulation theory, which is an extension of workers' control. In fact, as a graduate student, I was a member of the Okishio and Nozawa (1982, 1983) study group that played a central role in democratic regulation theory in Japan.

However, what we identified as a result of these studies was the problem with this way of thinking itself. That is, even if it is very easy to restrict corporate activities using governmental regulations, in reality, it is very difficult to force people to open new businesses or to increase investment, employment, or production. In fact, as the leader of this democratic regulation theory, Okishio (1957) concluded that nationalization was unnecessary if the state is able to intervene in decisions on investment, employment, and production, but the theory itself did not discuss specifically how to actually bring about such an intervention. Even if laws prohibit the use of certain raw materials or revise investment, employment, and production downward, we cannot imagine concretely how boards of directors in each company could be forced to decide to produce or to make investments that they do not want to undertake, or how government officials would take part in such decisions. This also can be surmised from the fact that in the case of a severe economic depression, people tend to call for deregulation.

Thus, arguments in the direction not just of nationalization but for limiting or restricting market mechanisms continue to retreat, and even among Marxists, the center of debate is shifting to corporate social responsibility and improving corporate governance, or to issues of openness to civil society, such as corporate accountability or disclosure. In terms of peoples' movements as well, we now have a trend toward non-governmental methods of utilizing shareholder rights, such as the appointment of shareholder ombudsmen. This shows that the trend of corporate regulation has changed.[48]

We place particular importance on this trend because these changes of the social movement requiring listed firms to disclose various types of internal information and attempting to expand the scope of such disclosure can be understood as meaning that these firms have been deemed to be a type of social property. That is, this means that people have a right to know about the activities of these companies even if they themselves are not their owners. In fact, if this disclosure alone is sufficient, then companies cannot carry out bad actions such as corruption and fraudulent accounts, even without the presence of government directors. Additionally, to achieve progress in areas such as in employing

---

[48] The Shareholder Ombudsman movement is an example of this. See Morioka (2000) and Shareholder Ombudsman (2002). The recent suggestion "public interest capitalism" (Hara, 2017) can be also understand in this context, although we think this is a post-capitalistic movement.

people with disabilities, it is sufficient just to force companies to disclose the percentage of their employees that have disabilities rather than using the administrative guidance of government directors. In this way, the pressure for disclosure cannot be ignored, and it can ultimately be regarded as a method of *socializing* the corporation in the sense that they are forced to reflect the wishes of all constituent members of society by the tool of monitoring alone. That is, when using the term "*socialization*," we must consider what would enable the company to reflect the wishes of all constituent members of society instead of limiting ourselves to narrow definitions of ownership changes. Doing this leads to a completely different concept of socialization, which has the common root with the word socialism.

This concept of disclosure has deep ties to the joint-stock company system, particularly the stock market listing system. For the system of mass public shareholders trading in shares in listed corporations on stock markets to function effectively, the public needs accurate information on corporate performance. Only when this is the case will all constituent members of society be able to buy and sell stock in the corporations as potential shareholders. This is important because, as a result of such trading, stock prices rise if the company is performing well and fall if it is doing poorly. These trends sometimes directly reflect the ability of managers, and for this reason, they sometimes bring about changes in management personnel. This is why managers must work very hard to improve company performance. That is, management is under the strict daily supervision of all constituent members of society, meaning that the company is under the control of all these members. It is thus under the oversight of all these entities, not just government directors.

When considering the subject in this way, an issue of concern is the fact that disclosure or oversight is grounded in demands to protect the rights of shareholders. While this appears to have a different ideological origin from that of the socialist doctrine of workers' ownership of companies because it results from the expansion of the shareholder rights to the rights of all constituent members of society as potential shareholders, it is my view that the socialist doctrine does not focus strictly on the rights of direct workers in their respective companies to control a company itself but on the control of companies in general by society. I think the latter is more socialized than control by direct workers at a specific company; and that this is true socialization. Of course, the rights of workers at the company are important too, but these rights should be realized through negotiations with managers who represent the interests of all constituent members of society, or, more precisely, they will be realized only after a fundamental change takes place from a machine-based society to a human-based society in the sense of productivity. In this way, we think that the direct oversight of the managers' capabilities and efforts by all people is itself in conformity with traditional socialist doctrine.

However, it is also true that Marx did not share this perspective. While Marx characterized the new joint-stock company system soon after it had appeared as the "abolition of capital as private property," as "a necessary transitional phase towards the reconversion of capital into ... outright social property," and as "a transition toward the conversion of all functions in the reproduction process which still remain linked with capitalist property, into mere functions of associated producers, into social functions,"[49] Marx was critical of the stock exchange. However, in the Supplement to Volume 3 of *Capital*, Engels provided a slightly different commentary from that in these passages by Marx.

First, Engels says that "the stock exchange was still a secondary element in the capitalist system" before Marx describes it in 1885, and notes:

> At that time, the stock exchange was still a place where the capitalists took away each other's accumulated capital, and which directly concerned the workers only as new proof of the demoralising general effect of capitalist economy."

However, Engels also notes:

> But since 1885, ... a change has taken place which today assigns a considerably increased and constantly growing role to the stock exchange, and which, as it develops, tends to concentrate all production, industrial as well as agricultural, and all commerce, the means of communication as well as the functions of exchange, in the hands of stock exchange operators, so that the stock exchange becomes the most prominent representative of capitalist production itself."[50]

Certainly, even today, the stock exchanges are the sites of unceasing scandals and wrongdoing. One also must understand that shareholder rights at a deeper level include cases where they hinder the growth of a company. For example, when the Japanese internet provider, Livedoor, wanted to become a majority shareholder in Fuji Television by making Nippon Broadcasting a subsidiary, one of the reasons that Nippon Broadcasting and Fuji Television opposed its takeover was because it would be beneficial to Livedoor but detrimental to Fuji. This shows that the owners, or majority shareholders, of a company can use their rights as owners to make decisions that are detrimental to the company.

Let us look at another example from overseas; namely, the case in which Balkan Bulgarian Airlines was almost taken over by a foreign company. After the 1989 revolutions in Eastern Europe, the national airlines of those countries were privatized, and in Bulgaria a foreign company was preparing to take over the company's overseas assets, the value of which exceeded the purchase price offered, and to sell them to another company shortly afterwards. The reason for this was that even though Balkan Bulgarian Airlines was in financial ruin, the value of its overseas holdings, including its branches and routes, was considerable.

---

[49] Marx, 1907, 316.
[50] Marx, 1907, 644, 645. On the above points, Oguri (2005) is very instructive.

Fundamentally this is the nature of hostile takeovers, which differ from cases in which companies choose a merger or acquisition independently and for their own benefit. In this sence, too, hostile takeovers do not contribute to the growth of corporate society as a whole, making it clear that ownership rights should not be unlimited.

Shareholders own shares for two purposes: (1) to obtain dividends or increases in assets as a function of dividends and (2) to take over companies. However, above examples suggest that even if the former purpose is permitted, the latter must be restricted. Put another way, although the second type of shareholder rights has negative effects on economic development, this does not mean that the other type of shareholder rights is problematic. What is of particular importance here is that mass shareholding by the public in general is conducted for the first purpose but definitely not for the second.

Thus, while the development of the joint-stock company system presented new possibilities, it includes some problems to be solved. However, it is possible to improve the joint-stock company system while protecting the interests of mass public shareholders. This course of action makes it possible to envision a truly *socialized society*—a *socialist society*—premised on the market system.

Although descriptions of the "socialization of enterprises" are basically of big businesses, we also have small companies. Furthermore, we discussed the historical law of business size in the case of agriculture in Addendum VI and note 58 in subsection 4.5.2, and explained that company size, and therefore differences of scale, can shrink in certain conditions. Most Marxist economists, including Marx himself, did not doubt the increasing trend of the size of enterprises in terms of "advantages of scale," but this is only because these efficiencies could not be neglected in the past. In other words, our future will differ from the present because it will have a different technological base, such as non-mechanical productivity by creative individual human power. Of course, large companies will continue to exist in some industries, and for this purpose, the socialization of enterprises is required. The theory of shareholding company socialism is applicable to such big companies.

4.5 PRIMITIVE ACCUMULATION AND STATE CAPITALISM

4.5.1 Primitive Accumulation Theory and the Forced Formation of Wage Laborers

The discussion above presented an overview of reproduction after the formation of capitalism. However, this reproduction assumes the presence of accumulated funds, for which capitalist production is a precondition. Furthermore, conversely, this capitalist production presupposes the presence of large volumes of capital and labor. Since these preconditions form a vicious circle with capitalist production, to break free we need to envision a certain accumulation in advance of capitalist accumulation—that is, primitive accumulation.

The role of primitive accumulation in economics is the same as the role of the story of original sin in theology. This is because economists must explain why some capitalists are able to live in comfort, and so there is an attempt to explain that this is because they are a "diligent, intelligent, and, above all, frugal elite."[51] Alternatively, there must be an explanation why workers cannot live in comfort, and economists explain that they are "lazy rascals, spending their substance, and more, in riotous living."[52] While it cannot be denied that there are some such examples of industrious self-employed or other individuals who accumulate their initial capital using their own labor power, in most cases, the initial capital is formed through the coercive intervention of the state or through some kind of good fortune.

Marx says:

In actual history it is notorious that conquest, enslavement, robbery, murder, briefly force, play the great part. In the tender annals of Political Economy, the idyllic reigns from time immemorial. Right and "labor" were from all time the sole means of enrichment, ... [However,] as a matter of fact, the methods of primitive accumulation are anything but idyllic.[53]

One example is the fact that capital could employ producers after the latter were freed from their feudal bonds to the means of production. This is a precondition for the capital accumulation by capital itself, and it entails either that (1) craftsmen were given the ability to sell their own labor power as they like after having been freed from feudalistic labor regulations, such as the apprenticeship and trade restrictions of guilds or (2) peasants became freed from their bondage to the land under serfdom.[54,55] However, the use of the term "being freed" in this way paints too pretty a picture of the reality. This is clear from, for example, the tragedy that unfolded for peasants over the course of the second of these processes, especially the two enclosure movements that occurred in England.

---

[51] Marx, 1887, 507.

[52] Ibid.

[53] Ibid.

[54] In both cases, direct produces had become those who have no other commodity for sale except labor power, but can dispose their own labor power as their own commodity freely from any feudalistic bonds. This situation is called "dual sense of freedom," and was realized by the direct producers' losing the means of production.

[55] In Chapter 27 on primitive accumulation, Marx (1887) states: "In England, serfdom had practically disappeared in the last part of the 14th century. The immense majority of the population consisted then, and to a still larger extent, in the 15th century, of free peasant proprietors" (1887, 510). However, serfdom is defined in this citation narrowly to exclude serfdom with cash payments, and therefore, the enclosure movement kept serfdom, broadly defined, in place. This interpretation is based on Nakamura (1977).

The first enclosure movement unfolded over the final third of the fifteenth and the first decades of the sixteenth centuries. In response to the prosperity of wool manufacturing at the time and the resulting increase in wool prices, a new landed aristocracy with a strong desire for money promoted the conversion of cultivated land to sheep ranges and removed the peasants, who were no longer needed, from the land. Facing this situation, the English government at that time initially resisted this expansion of farming, which would lead to a decrease in the population, and enacted laws against it, but after the change in monarch, the government moved to reduce the power of the church and farmers from the lands belonging to the Catholic Church, which was the largest feudal landlord.

In addition, the former landed gentry that was restored with the 1660 restoration wanted modern land ownership rights rather than feudal rights, and the peasants thus lost the right to the land that they had originally held. That is, they lost their rights as serfs and were reduced to tenant farming. In the subsequent Glorious Revolution (1688–1689), the modernized former landed aristocracy were given state lands by the government or bought them at ridiculously low prices, as landlords or capitalist appropriators of surplus value.

On the other hand, since the end of the fifteenth century peasants in England had been losing access to the commons, which they had formerly used to farm. In the eighteenth century, this land was legally enclosed and became the private property of the landlords who owned it. After the first enclosure movement publicly available cultivated land was given over to privately owned sheep farms in the fifteenth and sixteenth centuries, and this conversion of the commons to private land by acts of parliament in the eighteenth century is referred to as the second enclosure movement. After the arable land that peasants formerly had access to was enclosed by various landlords, they were unable to make a living from farming. During most of this period, state violence and force were constant. This characteristic would come to fruition in the form of forced labor after the peasants were separated from the means of production.

Therefore, primitive accumulation is closely linked to the decline of the peasant mode of production. In this case, even on the domestic growth path of constant disparity shown in Figure 4.10, the fall of the peasant class and their transformation partly into wage laborers and partly into capitalists should show a fluctuating trend of the overall social income disparity. **Figure 4.11** shows these trends and tendencies.

This attack on the peasantry began with the enactment of several bloody laws against vagrants, or the unemployed, across Western Europe from the end of the fifteenth to the end of the sixteenth centuries. Despite the large-scale removal of peasants from the land, there was insufficient peasant labor to work in the newly developing manufacturing sectors. Several laws were enacted to force these unemployed people to enter the labor market to work in the new industries. For example, under a law enacted in England in 1530, apart from who were those granted a beggar's license because of

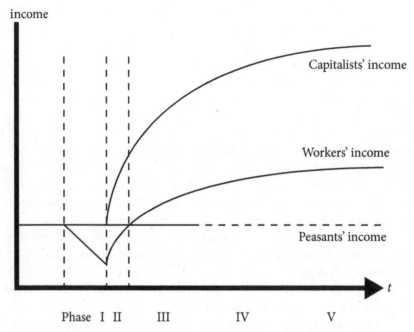

Figure 4.11 Income paths of the three classes during and after primitive accumulation[56]

---

[56] In this figure, the income paths of the capitalist and working classes are same as in Figure 4.10, but adding the peasant class gives several very interesting and realistic phases of income disparity because industrialization does not lead to any change in their income, but the incomes of the capitalist and working classes increase. There are five phases of income disparity:

Phase I) The process of primitive accumulation in a narrow sense, where some of the peasantry declines, typically as a result of the enclosure movements. The peasants are divided into two groups: those in decline and those not in decline, and so their disparity expands.

Phase II) The unsuccessful peasant farmers form the proletariat and sell the labor to the capitalists. Thus, capitalists gain profits and at the same time, capital accumulation also raises the incomes of both capitalists and workers. Thus, the income disparity between farmers and workers is reduced, and since both of these groups are in the majority in society, the income disparity of the whole society also decreases.

Phase II) With the development of the capitalist economic sector, the income of workers exceeds that of the farmers, leading to rising disparities between them.

Phase II) The expansion of the capitalist economic sector dramatically reduces the peasant population, so their income gradually becomes negligible in the long run. Therefore, income disparity shrinks in this new phase.

Phase V) If the farmer class disappears completely, then the subsequent income disparity becomes identical to that in Figure 4.10. In other words, it enters a state of constant income disparity.

their lack of work capacity due to old age, vagrants could be arrested and whipped or imprisoned. The law was later strengthened to require cutting off half of an ear on a second arrest and death for the third arrest. These punishments were legalized as "poor measures" by the Poor Laws.[57]

Furthermore, the 1547 Vagrancy Act in England required people who refused to work to become the slaves of those who reported them. While these slave owners were required to give them bread, water, weak broth, and such "refuse meat" as they thought fit, they had the right to whip and chain their slaves to force them to do any work they chose. If they took two weeks off work, these slaves would be reduced to the status of slaves for life and branded with an S on their foreheads or backs, and if they fled, the penalty was death. The slaves would be sentenced to death for any plotting against their owners. If an unemployed person who had no home was found to have been idle for three days, then they would be sent to their place of birth, branded with a V on their chest, and then chained to the land and forced to work. If they lied about their place of birth, they would be branded with an S and forced to be a slave to that land for life. Anybody could take their children and use them as apprentices to the age of 24 for a male or the age of 20 for a female. If they fled, then they would be made the slaves of their masters until they reached these ages. Their slave owners could place iron rings around their necks, wrists, and legs.

English laws during the Elizabethan period (the second half of the sixteenth century) were the same in their fundamentals. Some differences were the fact that the part to be branded was changed to the earlobe and the penalties became quicker, as even in the case of a second arrest an individual could get the death penalty if no one wanted to use them for two years. Marx cites a record saying:

> [R]ogues [this refers to homeless, unemployed people] were trussed up apace, and that there was not one year commonly wherein three or four hundred were not devoured and eaten up by the gallowes.[58]

Furthermore, under James I, the right to whip was given to justices of the peace in petty sessions and new measures were decided on for "incorrigible and dangerous rogues." Now, they would be branded, with an R and any "repeat offenses" would be punished with death.

---

[57] The U.S.A. presently has a similar law. According to Tsutsumi (2008), the Three Strikes Law in the U.S. mandates courts in twenty-eight states to impose life imprisonment on those who have been previously convicted of two prior serious criminal offenses. This should be understood as a kind of primitive accumulation of capital by creating cheap labor in prison.

[58] This description is in Marx. 1887, 526. Concerning these vagrancy laws, the Statute of Laborers and the two enclosure movements mentioned above, we rely on the descriptions in part 8 in Marx (1887).

These laws remained in effect until the beginning of the eighteenth century. Similar laws were enacted in France at roughly the same time.

Lastly, we touch on a variety of enforcement laws targeting not "vagrants" but workers who were employed. While portions of these were laws regulated minimum working hours, as discussed in subsection 3.2.1, these laws went further in constraining workers' freedoms. A tariff of wages was fixed for town and country, and employers paying higher wages than those recorded would be subject to ten days in prison, while their employees would be subject to 21 days in prison. A 1360 statute further strengthened this measure, giving employers the right to physically force people to work. Additionally, all attempts to organize masons and carpenters were declared null and void, and coalitions of laborers were treated as a heinous crime until 1825. These efforts reflected the view that while all wages could be as low as possible, the maximum amount should be dictated by the state. Legal prohibitions on trade unions remained in effect until 1859. In addition, while labor unions were finally legally authorized in 1871, an act passed in Parliament on the same day gave masters the right to treat strikes and lockouts as criminal offenses through their capacity as justices of the peace. This ordinance was enacted by Prime Minister Gladstone's Cabinet, which was said to be a relatively progressive administration. The same government also dug up earlier laws against conspiracy and applied them to coalitions of laborers. The situation was largely the same in France.

While real wages fell in the sixteenth century due to inflation, laws to keep wages down remained in effect. The Statute of Apprentices during the reign of Elizabeth I allowed justices of the peace to determine and change wages, and this was amended during later reign of James I to apply to all laborers. Additionally, a century later George II extended the law against coalitions of laborers to apply to all manufacturing sectors.

Throughout the manufacturing period, even though capitalism became strong enough to control laborers itself, the direct state control of workers remained in effect. For example, a law during the reign of George II prohibited silk-weavers from receiving a daily wage higher than two shillings, seven and one-half pence in and around London, except in cases of general mourning. A law during the reign of George III gave justices of the peace the authority to regulate the wage of silk-weavers. Even in 1799, the wages of mine workers in Scotland were determined by law as well. Although laws regulating wages were finally abolished in 1813, wages were controlled by the government until the middle of the industrial revolution.

4.5.2 Genesis of Industrial Capitalism

As seen above, the wage workers that compose the capitalist system were generated by "bloody legislation" (Marx, 1887, 522). Next, we see the formation of industrial capitalists (capitalists in the manufacturing sector), who are the other component of this system, was

THE GROWTH AND DEATH OF CAPITALISM

also as blood-soaked.[59] Starting as small guild masters, independent small industrialists, wage-laborers, and even usurer's capital and merchant's capital inherited from the period prior to the formation of capitalism, they would establish the first industrial capitalist class by overcoming various constraints and were helped by various state regulations. Specifically, they were assisted by the colonial system, the national debt, tax increases resulting from the national debt, the protectionist system devised prior to the industrial revolution, child labor, and the slave trade devised after the industrial revolution.

The Dutch slave trade is an example of the colonial system. This country invented a system of stealing people from Celebes Island to imprison slaves to use in Java. The captured boys were bound in secret prisons on Celebes until they were old enough, and then were loaded

---

[59] The argument in part 8 on primitive accumulation in the English version of *Capital* (1887) states that tenant farmers arose from serfs, and then became sharecroppers, and finally ended up in two classes: capitalist farmers and agricultural wage earners. However, they later changed again to become free peasant proprietors. This may imply that small-scale farming systems have a great ability to restore themselves. The capitalist agricultural system was not sustained for very long and reverted to small-scale farming, as stated by Nakamura (1977). This restoration can be explained by the special cost structure of agricultural production. Its initial cost, such as the cost of reclaiming farmland or training farmers to produce new crops, is very large, and therefore, economies of scale work well at the early stage of cultivation, say, for several years after reclamation. However, as shown in **Figure 4.12**, the importance of the initial costs become smaller or negligible after several decades. Thus, we can see that small farmers tend to be competitive with large farmers in the long run.

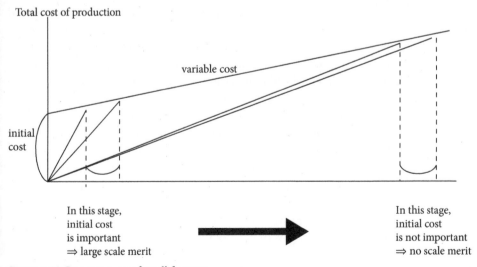

Figure 4.12 Cost structure of small farmers

onto slave ships. It was the thieves, interpreters, and slave traders involved in this trade, along with the native princes as chief sellers, who profited from it. In addition, England awarded monopoly rights to salt, opium, betel pepper, and other products to the viceroys of India and high-ranking officers of the East India Company, which was formed in 1600 and dominated trade with the East Indies. This monopoly made it possible for them to accumulate whatever wealth they wished. This caused a famine in 1769–1770 in India as a result of the purchase of all the rice in Bengal, to which the English were perfectly indifferent.

The indigenous peoples suffering under this colonial system were to be found not only in the West Indies, Mexico, and East India, which contained populations that possessed a certain degree of wealth. In 1712 devout Protestants offered rewards of 40 pounds for the capture or scalp of Native Americans in New England, North America, which was originally a colony, and later in 1720, they paid 100 pounds per scalp. Furthermore, in 1744, rewards of 100 pounds were offered in the new currency per scalp of a male aged 12 or above, 105 pounds for a captive man, and 50 pounds for a captive woman or child. Murder and scalping were proclaimed as "means that God and Nature had given into its hand" by the British Parliament.[60] Wealth captured from the colonies constituted the starting point of capital accumulation in European countries.

Next, the government bond system, first established in the Dutch Republic, also helped to make it easier to accumulate capital, and the promotion of investment in securities and modern banking controls also formed a starting point for accumulation. This system of government bonds was completed by the establishment of central banks in each country. These national banks earned massive amounts of interest on government bonds, easily securing initial funding from the public because they had a monopoly on the issue of banknotes and on investing these funds. The international credit system, which developed together with government bonds, was also important. The capitalist development of the Dutch Republic depended on the provision of funds from Venice, which repeatedly plundered worldwide, and in the eighteenth century, the Dutch Republic lent massive amounts of capital to England, which itself became the basic provider of funding to the U.S.A. in the nineteenth century. Today, the U.S.A. is the world's largest borrower, and its treasury bonds are purchased mainly by China and Japan. While the direction of today's financial transfers between mature countries and the emerging countries differs from that in the eighteenth and nineteenth centuries, it is important to recognize that it is only this international credit system that can extend the life of the U.S. economy in its current form.

Third, issuing government bonds led to the modern taxation system as an institutionalized system of tax increases, in which the issue of bonds systematizes later tax increases without taxpayers being aware of it. This is because, while it is difficult to obtain taxpayers' consent to new government expenditures involving tax increases, such

---

[60] Marx, 1887, 535.

expenditures can easily pass being unnoticed if they do not involve tax increases. However, this means issuing more government bonds, which eventually, imposes the taxpayers to repay these bonds. Furthermore, taxpayers cannot reject such increases because the money has already been spent. This cycle of tax increases and automatic increases in government bonds is a law of the modern fiscal system, and one that has come to fruition in today's world. Although Buchanan and Wagner, American advocates of small government, argued in the 1970s that this mechanism existed,[61] the true pioneer of this theory was Marx. They do not mention this fact.

Fourth, trade protectionism through methods such as protective tariffs and export subsidies is an important means for developing industrial capitalism to expropriate one's own citizens. Just as export subsidies are a form of expropriation, so are protective tariffs because they prevent the public from buying lower priced products from overseas. This system also broke down small independent industries in territories but outside the protective tariff zone. In the case of Great Britian, it completely destroyed Ireland's wool industry.

Fifth, the issue of child labor. The shortage of labor in provincial English cities after the industrial revolution was remedied by taking children from the poorhouses and elsewhere and forcing them to work, even at night. For example, in Lancashire, children aged between seven and 13 or 14 years old were procured from poorhouses in London, Birmingham, and elsewhere, and forced to work excessively, even to the point of death, at the whip.

Lastly, the continuation of the slave trade was a major issue in post-industrial revolution England. While England had rights to the slave trade first only in Africa and the British West Indies, the government extolled the trade as a victory for national policy when it secured rights to the trade in Spanish America through pressure on Spain. From 1743, England secured the rights to trade in 4,800 black slaves per year. There were only fifteen slave ships in Liverpool, which by 1730 was the center of the slave trade but this number would rise rapidly to 53 in 1751, 74 in 1760, 96 in 1770, and 132 in 1782.

In this way, industrial capitalists came into the world "dripping from head to foot, from every pore, with blood and dirt."[62]

4.5.3 The Necessity for High Levels of Accumulation at the Start of Capital Accumulation and State Capitalism

To summarize the above discussion, we have seen that the formation of wage-labor and the capitalist class were "as a matter of fact, ... anything but idyllic." Readers in countries such as Japan, Germany, Russia, and China are well able to understand this fact. Since

---

[61] Buchannan and Wagner (1977).
[62] Marx, 1887, 538.

readers in Japan and Germany know the severity of the systems that prevailed in their own countries in the prewar years, they tend to attend exclusively to their own case. However, to appreciate just what capitalism is we need to acknowledge that things were the same in the mother country of capitalism.[63] Readers in Russia and China, who know the faults of Stalin and Mao Zedong, can appreciate the true essence of the system by understanding that this is a general feature of the formative period of capitalism and not the special case of their own countries. This is the main reason why we have discussed the historical facts of this period in some detail.

However, having reached this point, we must review why this period needed to be so violent. Generally, Marx asks whether capital was strong enough to control and exploit workers in its own power. He explains that while it too weak to do so prior to the industrial revolution and the appearance of machinery, the industrial revolution gave rise to conditions under which it could control its own ability, describing this as "real subjection of labor to capital." This is what we refer to in the preceding chapter on the technological inevitability of the despotic command over labor that first appeared in modern industry. However, the problem here is the fact that while making such arguments Marx also argued that such violent "primitive accumulation" would grow stronger in the "infancy of modern industry."[64] How are we to understand this?

My answer is simple. Because the fundamental issue in *Capital* is to uncover the secret of capitalism, which is that exploitation can take place even under the principle of an equal exchange of commodities, the entire volume before the chapters on primitive accumulation did not recognize this violence. Therefore, Marx did not argue that there was no violence after the industrial revolution. What he explained was that the exploitation inherent in capitalism would occur with or without such violence. For this purpose, his argument advances without addressing violence, and he describes violence as a fundamental property of the pre-industrial revolution structure. However, despite this, Marx states that this violence developed in its most organized manner during the "infancy of modern industry"; that is, the initial stage of industrialization following the industrial revolution. In fact, this was the most important conclusion of the Marxian optimal growth model.

This is illustrated in **Figure 4.13**. This figure depicts the rate of allocation of labor to the sector producing the means of production, which is roughly equivalent to the ratio of investment, expressing the process of growth in the Marxian optimal growth model as $1-s$. It shows that in the infancy of modern industry following the industrial revolution, this ratio of investment was particularly high. That is, the production of the means of consumption was extremely limited. In terms of the levels shown in Figure 4.5, this means that a very high rate of capital accumulation was experienced in the infancy of modern industry. In modern

---

[63] See Ozaki (1990) on the importance of this perspective.
[64] Marx, 1887, 261.

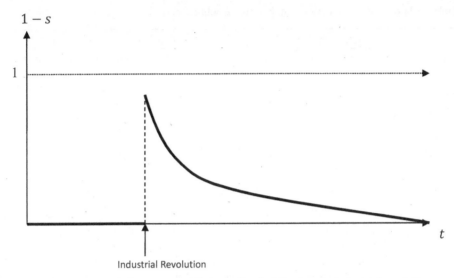

Figure 4.13 Investment ratio increases rapidly immediately following the industrial revolution

economics, this phenomenon is defined using the terms "big push" (Rosenstein-Rodan, 1961), "take-off" (Rostow, 1960), or "strong accumulation" (Minami, 1990). All these refer to the same thing. We cannot avoid considerable hardship in this period.

Nevertheless, we model the growth path that representative individuals *should choose* as representatives of society. Whether they do follow this course automatically is another matter. For example, it can be argued that social infrastructure such as roads and educational institutions that are required at the initial stage of capital accumulation involve a high degree of externalities, and therefore they tend to receive too little investment in a market economy. Alternatively, there are disparities between individuals in terms of their actual time preference, and this could bring about conflict between people with a lower time preference (i.e., those who prioritize the future) and those with higher time preference (i.e., those who prioritize the present).[65]

Furthermore, there may be a similar case in which investment becomes insufficient. It is because no matter how low one's time preference, and no matter how strong one's orientation toward investment, there are physical and biological limits restricted consumption even immediately after the industrial revolution because people were much poorer than they are now. If people lack the ability to make a rational choice, almost the identical situation will

---

[65] Inoue and Yamashita (2011) argue that this issue was a matter of effective differences in time preference between social planners and the general public. Thus, this type of conflict drags in the power of the state, which represents the interests of the bourgeoisie.

Table 4.5 Two stages of capitalism and their ruling political parties

	State capitalism	Turning point	Private capitalism
Japan	Imperial Rule Assistance Association	1945	Liberal Democratic Party
Germany	Nazi	1945	CDU
Indonesia	National Party (Sukarno)	1967	Golkar (Suharto)
Egypt	Nasser	1970	Sadat
China	Communist Party (Mao Zedong)	1978	Communist Party (Deng Xiaoping)
Vietnam, Laos	Communist Party	1986	Communist Party
Russia	Communist Party	1991	Yeltsin

Note: Although the difference between state capitalism and private capitalism is not as distinct in the cases of Indonesia and Egypt as in that of Japan and Germany or China and Russia, they have been added to the table with a focus on the changes in the nature of their ruling parties. This table also includes Vietnam and Laos, which advanced reforms and opening up eight years behind China.

occur. In all these cases, capital accumulation may be below the optimal level, and the state has to play a role in forced capital accumulation. The special stage at which the state plays a decisive role in capital accumulation is essential at the beginning of capitalism. It is natural to call this state capitalism. In contrast, the subsequent, non-state type of capitalism could be called private capitalism or market capitalism.

Above, we compared prewar Japan and Germany, the Soviet Union, and Mao Zedong's China with the violent formative stage of capitalism in England. **Table 4.5** summarizes these cases.

Table 4.5 requires some supplementary notes. One of these is the fact that in the creation of state capitalism the march to capitalism was progressive in terms of productivity and capitalism was superior to the feudal or serfdom system. In this recognition lies the essence of historical materialism. Thus, even if "[f]orce is the midwife of every old society pregnant with a new one. It is itself an economic power" in which all methods

> employ the power of the State, the concentrated and organized force of society, to hasten, hot-house fashion, the process of transformation of the feudalist mode of production into the capitalist mode, and to shorten the transition,[66]

then it can be said that, given time, progress toward capitalism is likely even in the absence of force. In an addendum to a study of the process of the emancipation of the Tibetan serfs

---

[66] Marx, 1887, 534.

in Onishi (2012), I estimated that the labor productivity of the serf immediately prior to emancipation had not reached the level at which they could earn wages. In this condition the old production system would therefore have broken up naturally over time. However, force, in the form of the central Chinese government's emancipation of the serfs, served as a midwife to speed up this historical transformation. Marx also recognized the existence of "purely economic causes" for primitive accumulation in agriculture, mentioning, "We leave on one side here the purely economic causes of the agricultural revolution. We deal only with the forcible means employed."[67] Even so, there is no case in which politics is uninterested and plays no part in the creation of a new social class.

However, if we acknowledge the possibility that the state may have been relatively uninvolved in the formation of capitalism, if it did intervene, there could be some variation in the degree to which it did so. In fact, as seen above, the differences are clear in the cases of England, Indonesia, and Egypt in the one hand and those of Japan and Germany, on the other, as well as between Stalin and Mao. I would position the second group of cases as central, while noting that there are also cases where there is little intervention, as in the first group, and others in which there is the highest level of state economic control, as in the third group. These depend on whether the circumstances urgently need industrialization,[68] or on the degree to which it is needed. To expound on this, these countries depended on circumstances such as the threat of war or invasion, such as Japan's urgent need to prevent colonization after opening up to the outside world, the joint Allied intervention and the Cold War in the Soviet Union, as well as the Korean War and the Sino-Soviet split in China. This point cannot be forgotten. In the Soviet Union, Eastern Europe, Maoist China, North Korea, Cuba, Vietnam, and Laos a special type of state capitalism was born from their particular situation.

4.5.4 Ways of Transition from State Capitalism to Private Capitalism: The Wisdom of Deng Xiaoping

Incidentally, we notice that when comparing countries in Table 4.5, even if these countries were essentially undergoing the same systemic transition, there are two different types of cases: those in which this transition encountered considerable chaos and those in which it went smoothly. Examples of the former cases include those of Japan and Germany, in which the transition could not occur without a terrible war that entailed the slaughter of large numbers of Asians and Jews; that of the 1967 Suharto coup d'état in Indonesia, in

---

[67] Marx, 1887, 510.
[68] Generally, this is the case in latecomer countries. Obata (2009) states that the intervention by the mercantilist state in England resulted from its status as a latecomer that needed to catch up with the Dutch Republic.

which hundreds of thousands of members of the former ruling Communist party were slaughtered; and the upheaval in the former Soviet Union in 1991, which resulted in an economic collapse that decreased total output dramatically by about one-half. While these can be divided into hardships resulting from catastrophic upheaval (in the former Soviet Union and Indonesia, in which the hardships occurred during or after reforms) and those resulting from the need to implement reforms in a catastrophic form (as in Japan and Germany, in which the hardships occurred mainly prior to reform), both cases are referred to generally as radical reform.

However, the transitions in China, Vietnam, and Laos stand in sharp contrast to the other countries in the sense that nobody was killed except for four political leaders in China and they were undertaken amid accelerating economic growth. Their type of transition may result from the need to avoid missing the opportunity to reform at a comfortable speed due to continued resistance to reform, and that this made the reform so gradual that all constituent members of society could respond to the transformation properly. In sum, this is an example of gradualism. In subsection 1.3.3, we discussed the fact that the human mentality related to each mode of production differs. The same can be said in the case of state capitalism and private capitalism. Because this will not change overnight, gradual change is desirable. This is one reason that gradual reform is most successful.

However, this leads to the question why the former group of countries did not employ this method and why only a very small portion of countries could adapt this way. This can be explained as the difficulty to change the ruling parties from directing state capitalism to directing private capitalism or, more generally, the difficulty to make the party of the previous administration responsible for the change to the next administration in the transition of the modes of production. In fact, Table 4.5 shows that only China, Vietnam, and Laos had the same governing party before and after reform. In the absence of very special circumstances, different modes of production need to be led by different parties (political groups).[69]

For example, unlike the ordinary non-Marxian understanding of the event, China's Cultural Revolution was not just a struggle for power but a struggle between classes. This is because the free market was beneficial to the entrepreneurs but not to the ordinary people, and therefore as a social class the ordinary people supported Mao, at least at the beginning of this revolution. In other words, the 1978 transition too was a transition of leadership from one class to another. If this transition had occurred in another country

---

[69] In Mexico, the same gradual transformation process was in progress in the 1980s and 1990s in the form of the large-scale privatization of state-owned enterprises, the demolition of the commons (*Ejid*), and the joining of NAFTA under the Salinas administration. During this process, civil war broke out with the Zapatista National Liberation Front (see Marcos and Le Bot, 1997).

besides China, then it is possible that there would have been different political parties representing the different interests of the former ruling class and the latter, respectively. A change in the ruling class would have necessarily involved a change in ruling parties. Thus, even if the former ruling party sensed there would be a transition in the ruling class at the base and tried to switch class, the new ruling class would already have its own party that it trusted more than the former ruling party that suddenly promised to represent its interests. Because the former ruling party had until then treated them as enemies, asking them to align with their party immediately would not succeed. Even worse, such a proposal from the former ruling party would invite distrust and disaffection from the class that it had represented until then, and in the end, the party would lose the support of its old class while being unable to gain the support of the new one. This dilemma presents itself in nearly all former ruling parties, and it explains quite well why such parties fail to change the classes they represent. Thus, the changes in ruling class necessarily involve changes in ruling party.

The question that then arises is why China, Vietnam, and Laos are the only exceptions to this rule. Because Vietnam and Laos (instead of learning from *perestroika*, which began in the Soviet Union in 1985) largely imported the Chinese method unchanged, the true exception to the rule is Deng Xiaoping's reforms in China, and we must consider how he was able to effect this special, exceptional achievement. To do so we cannot ignore the fact that conditions were favorable for reform due to the public's exhaustion with the immediately preceding Cultural Revolution and the death of the nation's founding leader (Mao Zedong).[70] However, we know that an exquisite distribution of benefits took place to benefit the class that was disadvantaged by this change in policy. For example, public officials who lost their posts were provided with opportunities to become managers of newly privatized companies. Around 2040, China will be forced into conditions of zero growth, and will need a new transition in its ruling class to transform itself to true socialism or communism. For this reason, the issue is whether it will be able to apply Deng's wisdom to prevent a change in the ruling party or not—in other words, whether it can carry out this transition while avoiding the hardships of revolution or war.

We conclude by touching on the unique nature of the Communist Party itself. It is because all ruling parties in China, Vietnam and Laos are communist parties. Thus, it is conceivable that there could be some grounds for deciding to aim for a total transformation under Marxism, their guiding ideology. For example, as argued in subsection 1.2.2, Marxism is objective because it is independent of any social classes that may conflict with the turning point, and therefore, the Communist Party of China, as an ideological Marxist

---

[70] North Korea and Cuba also have an intentionality to adopt reforms in the Chinese way, and in both countries the fact that Kim Il-sung and Fidel Castro have gone is important condition for these reforms. The same is true of Vietnam.

party, might bring about balanced policy management in China.[71] Furthermore, a historical understanding of historical materialism, which argues that social systems must change with the times, may be also very important. In the mid-1980s, Deng Xiaoping said that the target year for Chinese economic development would be 2150. I sense the influence of Marxism—or historical materialism—in this projection of the many years needed for perceiving, recalling, and planning policy management, and in the thought oriented to historical laws that can relativize the present and look ahead to a distant future. In the preface to the first edition of *Capital*, Marx said:

> (E)ven when a society has got upon the right track for the discovery of the natural laws of its movement—it can neither clear by bold leaps, nor remove by legal enactments, the obstacles offered by the successive phases of its normal development. But it can shorten and lessen the birth-pangs.[72]

However, what must not be misunderstood is the fact that this does not mean that only a Marxist administration (or Marxist thinkers) can make progress toward new modes of production. History moves forward, and has in fact moved forward, regardless of whether or not such social forces are present. This is the thesis of historical materialism. However, without keeping this historical progress under control, it can only advance in a chaotic way or accompanied by economic ruin in the form of revolutions or wars. Marxism is not essential to historical progress. However, if we want to move history forward without chaos, it is preferable to put this wisdom to use. While the term social science refers simply to an objective knowledge and understanding, its role in social transformation is massive.

---

[71] Strictly speaking, this Chinese wisdom could have been realized by Prime Minister Hua Guofeng's balancing policy based on Marxist ideology. In addition, while some are of the opinion that having a single-party system deters changes in administration, even under the Soviet one-party system changes in administration occurred. This cannot be deemed a fundamental condition for choosing the path of gradual reform. At most, it can be said to be merely a condition that favors it.
[72] Marx, 1887, 7.

5

The Distribution of Surplus Value Among Industries and to Non-Productive Sectors

5.1 THE SUBJECTS AND STRUCTURE OF *CAPITAL*

So far, Chapters 2, 3, and 4 discussed the various economic laws in the historical stage called capitalism. As mentioned at the opening of Chapter 2, this roughly corresponds to what Marx argued in Volume 1 of *Capital*. Marx's argument is that the exploitation of labor exists under capitalism, which is a society based on commodity production that is supposed to be a world of equal exchange based on absolute equality; and that this system will not last forever.

There was visible exploitation in pre-capitalist class society. For example, serfs who had to spend three days a week tending the lord's land, leaving them with only four days a week to work their own land, did not need a special economic theory to comprehend what proportion of their labor was exploited. However, conditions are different in a capitalist society premised upon the exchange of equal value. In this world everyone is equal and it seems there is no social irrationality; or, if anything is irrational, it appears as a divergence from the law of commodity production societies; that is, the law of exchange of equal value. Thus, non-mainstream and non-Marxist economists who oppose neoclassical economics try to discuss irrationality, externalities, and imperfect competition in a society by attributing them to a failure in the market mechanism. These non-mainstream and non-Marxist economists believe that there is nothing absurd in the "perfect market mechanism"; thus, they conclude that any social absurdity should be explained by reference to a divergence from the principle of commodity production societies (the principle of the exchange of equivalents).

However, Marx's approach was different. He saw that exploitation, the fundamental irrationality in capitalism, emerges under the principle of exchange of equivalents, and that it will emerge even if the market mechanism works perfectly. His approach to the question is completely different from the one taken by the non-mainstream modern economics, and *Capital* is written from that perspective. I would urge the reader to confirm that *Capital* contains no reference to externalities, negation of the rational Homo economicus, or asymmetric information.

However, while surplus value, which *Capital* explains was produced by industrial capital in the production process, commercial capital and financiers have, in reality, also acquired profit, and landowners received rent. Consequently, the task in *Capital*, which aims to explain the mechanism of exploitation and surplus value, cannot be completed without explaining these additional processes. Volumes 2 and 3 of *Capital* were written to achieve this aim.

In fact, these types of surplus value are explained in the latter half of Volume 3 of *Capital* as commercial profits, interest, and rent. Furthermore that, the problem of the distribution of profit to various sectors within industrial capital is examined in the first half of Volume 3. To expand the argument developed in Volume 1, which abstracted the relationships with other forms of capital and focused on direct production processes, Volume 2 of *Capital* analyzes the circulation process that mediates the production processes. Without this circulation; that is, without its social entanglement with other forms of capital, capital in the original form of money cannot be transformed into production capital (the means of production and labor force). Furthermore, without circulation, it is impossible to transform the commodity thus produced back into the form of money. This point permits the analysis of commercial capital developed in Volume 3. In addition, the shift in focus from the analysis of processes to the analysis of time taken prepares the ground for an analysis of the profit gained by "advances"; that is, interest. In addition, because circulation includes the intersectoral arrangements underlying total social production, which consists of the consumption goods and the means of production sectors, the conditions for social reproduction that must be satisfied among these sectors are discussed in the form of the reproduction scheme described in Volume 2 of *Capital*.

As we have already discussed this reproduction scheme in Chapter 4, this chapter examines other issues addressed in Volumes 2 and 3 of *Capital*.

5.2 THE CIRCULATION, TURNOVER, AND SOCIAL REPRODUCTION OF CAPITAL: THE CIRCULATION OF CAPITAL

5.2.1 The Circulation of Capital and the Circulation Process

The simplest formula for defining capital as self-valorizing value, which we first discussed in Chapter 2, is $G - G'$, but as **Figure 5.1** shows, this cycle represents only one turnover in a process that is repeated infinitely. Marx referred to this as "form I" of the circulation of capital. Similarly, Figure 5.1 also contains the circulation of $P - P'$ and that of $W - W'$, which Marx referred to as "forms II and III," respectively, providing detailed discussions of the characteristics of each. In sum, form I illustrates clearly that the purpose of the movement of capital lies with the increase in value, while form II in contrast, represents money merely as something that mediates production activity. In other words, he emphasis the continuous reproductive activities of this movement, which in turn shows that the production activity in any society is the essence of its continuation and argues that there is a social rationality behind the whole process. On the other hand, form III shows clearly on that the circulation process is a premise of production activity in that it starts from the transformation of

THE DISTRIBUTION OF SURPLUS VALUE AMONG INDUSTRIES

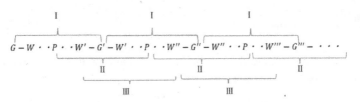

Figure 5.1 Infinitely repeated circulation of capital

commodity *W* into money *G*. In other words, this shows that this process is premised on the existence of a buyer, whose presence is required by the commodity produced. In turn, the means of production and the seller of labor are necessary for the next production. In this way, the form liberates our analysis from a perspective that is stuck on individual capital and suggests that it is necessary to move to an analysis of social interaction. Marx's reproduction scheme, as discussed in Chapter 4, is developed on the basis of this understanding.

What is important in the analysis is that this circulation process constitutes time lost to capital. Merchants who deal with the process independently can reduce this loss. However, if industrial capital concentrates on production activity by commissioning a merchant (or money dealer) to undertake this process of circulation, then it has to pay part of the benefit to the merchant (or money dealer) as its contribution. This is the source of commercial profit (or the money dealers' profit). Commerce itself does not produce value or profit, but can received his share from surplus value generated by industrial capital. To illustrate this, we examine the distribution of profit to commercial capital in two steps.

The first step is to identify the contribution of commercial capital to industrial capital by reducing the time of circulation. Suppose Δp is the original length of time of production, Δc the original length of time of circulation, and $\Delta c'$ the reduced length of time of circulation due to the activities of commercial capital, with the surplus value m originally produced by industrial capital. In this case, industrial capital can lengthen its time of production to $(\Delta c - \Delta c')/\Delta p$ times longer and additionally produces $\{(\Delta c - \Delta c')/\Delta p\}m$ amount of surplus value. In this case, it also needs a certain additional production cost for its additional production, but this additional cost can be covered by a faster turnover.

The second step is to identify the conditions in which this additional new surplus value can cover the whole cost of the commercial sector and its profit. Suppose c_c, v_c, and m_c are the transaction cost, wage, and profit of commercial capital.[1] In this case, this condition can be shown as

---

[1] Marx identifies storage and transport costs as expenses related to the original activity of production, even though they are part of the distribution process. Consequently, even if these are paid by commercial capital, in this context, they need to be understood as capital investments in the productive sectors that produce value, not in the commercial sector.

$$\frac{\Delta c - \Delta c'}{\Delta p} m > c_c + v_c + m_c \tag{5.1}$$

Furthermore, representing the constant and variable capital of industrial capital after introducing the commercial capital as c_p and v_p, respectively, and assuming an equalized rate of profit r between the two types of capital, the above inequality can be transformed into

$$\frac{\Delta c - \Delta c'}{\Delta p}(c_p + v_p)r > (c_c + v_c)(1+r) \tag{5.2}$$

and again into

$$\frac{\Delta c - \Delta c'}{\Delta p} \cdot \frac{r}{1+r} > \frac{c_c + v_c}{c_p + v_p} \tag{5.3}$$

Until now we have discussed individual industrial and commercial capital. However, this inequality can be changed into an equal sign if the scope of consideration is not limited to individual commercial capital but applied to commercial capital in general. This is because when commercial technology becomes sufficiently widespread, less productive commercial capital will enter the field. Therefore, replacing this inequality sign with an equal sign, we have

$$\frac{\Delta c - \Delta c'}{\Delta p} \cdot \frac{r}{1+r} = \frac{c_c + v_c}{c_p + v_p} \tag{5.4}$$

The left side tends to rise under a given rate of profit because Δp tends to decrease as a result of technological progress in the productive process and $\Delta c - \Delta c'$ tends to rise as a result of technological progress in the circulation process. This tendency shows the rising trend of the ratio of the commercial sector under the given rate of profit. Although it is true that the Marxian optimal growth theory proves the falling rate of profit, we can understand that technological progress has the effect of increasing the ratio of the commercial sector. Marx said that the relative reduction of the weight of the commercial sector is led by technological progress, such as a reduced turnover time and innovations in transportation that expand the size of $\Delta c - \Delta c'$[2]. Although our conclusion is the opposite of Marx's, note that modern Marxian economics can explain the expanding trend of the commercial sector as a part of the service industry that has become the major sector of the modern economy.

5.2.2 The Turnover of Capital

So far, we have examined the whole process of capital circulation by focusing on circulation time. However, something more needs to be abstracted in relation to "time" in the

---

[2] Marx, 1907, 87.

THE DISTRIBUTION OF SURPLUS VALUE AMONG INDUSTRIES

analysis above, in that invested capital contains both the capital that is constantly being recovered and the capital that can be recovered only after a long time. This issue has not yet been taken into account. The former refers to variable and circulating capital such as raw materials and the latter refers to fixed capital such as machinery, equipment, and the buildings that house them; namely, the type of capital that has featured as the main protagonist after the industrial revolution. Marx deals with this issue as one of the discrepancies in turnover time, which represents the sum of the times of production and circulation.

Let us consider the following case. If we suppose that a factory building will last for twenty years after its construction, then its value depreciates over these 20 years. In this scenario, the annual depreciation rate is one-twentieth, which means the turnover time is 20 years. Similarly, the turnover time of a machine that lasts for ten years is ten years. On the other hand, if we suppose materials are brought in every month, then there are twelve shipments (i.e., sales) per year. Lastly, if we suppose that workers are paid monthly, this also means twelve investments are made per year. In these cases, depreciation turnover time is one-twelfth. Consequently, the following relationship is established between the annual number of turnovers, length of the year, and turnover time:

$$\text{annual number of turnovers} = \frac{\text{length of year}}{\text{turnover time}} \tag{5.5}$$

Although this shows the annual number of turnovers of individual items of goods and the labor force, this is not important to capitalists. Instead, what is important to them is the number of turnovers as a whole, which is determined by the proportion of capital invested in, for instance, the building, machinery, materials, and labor. **Table 5.1** illustrates this. C_1, C_2, C_3, and V represents the amount of invested capital in each of the four kinds of input factors (the amount of payment invested in each factor that is functional at a given point of time).

As can be seen, because C_1, C_2, C_3, and V have their own turnover time in which the capitalist makes investments, the annual turnover of each factor (equivalent to the annual depreciation for buildings and machinery) is shown in the farthest right column. Consequently, the overall annual number of turnovers is

$$\frac{\frac{C_1}{20} + \frac{C_2}{10} + 12C_3 + 12V}{C_1 + C_2 + C_3 + V} \tag{5.6}$$

This can also be expressed as

$$\frac{C_1}{C_1 + C_2 + C_3 + V} \cdot \frac{1}{20} + \frac{C_2}{C_1 + C_2 + C_3 + V} \cdot \frac{1}{10} + \frac{C_3}{C_1 + C_2 + C_3 + V} \cdot 12 \\ + \frac{V}{C_1 + C_2 + C_3 + V} \cdot 12 \tag{5.7}$$

Table 5.1 Relationship between the amount of invested capital by input factors and the annual turnover

Input factor	Invested capital	Turnover time	Annual number of turnovers	Annual amount of turnover (annual amount of recovery/depreciation)
Building	C_1	20 years	1/20	$C_1/20$
Machinery	C_2	10 years	1/10	$C_2/10$
Materials	C_3	1 month	12	$12C_3$
Labor	V	1 month	12	$12V$
Total	$C_1 + C_2 + C_3 + V$			$C_1/20 + C_2/10 + 12C_3 + 12V$

The overall annual number of turnovers is equal to the weighted average per input factor of the annual number of turnovers. Consequently, if we denote the overall invested capital of all input factors as C_i and each annual number of turnovers as N_i, the annual number of turnovers as a whole can be expressed as

$$\sum_{i=1}^{the\,number\,of\,input\,factors} \frac{C_i}{\sum_{i=1}^{the\,number\,of\,input\,factors} C_i} N_i \qquad (5.8)$$

When the proportion of fixed capital in total capital increases, the annual number of turnovers as a whole decrease.[3] This formula thus displays one of the basic principles of the capitalist economic system, in which fixed capital has an expanded role. Hereafter, the reader is asked to bear in mind that c, v, and m are calculated in this manner.

Before examining the number of turnovers, we must take into account the fact that C_3 and V circulate twelve times a year to calculate the rate of surplus value. Only the depreciation of C_1 and C_2 are considered by multiplying them by 1/20 and 1/10, respectively. However, when examining the annual number of turnovers, the rate of surplus value is worked out over a defined period so that the annual rate of surplus value is calculated by multiplying the rate of surplus value by the annual number of turnovers, as above. In other words, the faster the speed of turnover (i.e., the more turnovers per unit time), the higher the true rate of surplus value when turnover is

---

[3] However, the time of turnover can also be reduced when the time of production is shortened due to the introduction or reinforcement of machinery and cooperation. This is because this machinery can rise the number of turnover of circulating capital by shortening the production time, while it reduces the annual number of turnovers by increasing the proportion of constant capital in the total capital.

THE DISTRIBUTION OF SURPLUS VALUE AMONG INDUSTRIES

taken into account. This is why the capitalist undertakes a variety of tricks to increase the speed of turnover.

Volume 2 of *Capital* then undertakes its famous contrivance of the reproduction scheme under the title, "The reproduction and circulation of the aggregate social capital." This book dealt with this in the previous chapter, because it understands the problem of exploitation as an issue of consumption and investment. Marx discusses this at the end of Volume 2 of *Capital* (Part 3) because the reproduction scheme describes the relationship among multiple industrial sectors. An exploration of the intersectoral relationship in industrial capital is the precondition for identifying the relationship between commercial and industrial capital, and Marx's reproduction scheme deals with this as the relationship between industrial capital as the producer of the means of consumption and industrial capital as the producer of the means of production, respectively. The reader is asked to review the previous chapter from this point of view.

5.3 THE CONVERSION OF SURPLUS VALUE INTO PROFIT, INTEREST, AND RENT: THE PROCESS OF CAPITALIST PRODUCTION AS A WHOLE

5.3.1 Equalizing the Rate of Profit Among Industrial Sectors and Price of Production

Marx next discusses the distribution of surplus value among industrial capital in *Capital*, Volume 3. However, the distribution of surplus value has been already discussed in subsection 5.2.1 in the example of the industrial and commercial sectors as the equalization of the rate of profit. That is, each sector has the same right to receive same proportion of surplus value to the amount of invested capital or rate of profit). Now, we first examine this problemas a problem among industrial capital.

In truth, Marx's numerical example in the simple reproduction scheme was carefully set to have same rate of profit. That is,

$$\left.\begin{array}{l}6000W_1 = 4000c_1 + 1000v_1 + 1000m_1 \\ 3000W_2 = 2000c_2 + 500v_2 + 500m_2\end{array}\right\} \quad (5.9)$$

In this case, the first sector's rate of profit ($m/(c + v)$) is 1000/5000 =1/5 and the second sector's rate of profit is also 500/2500 = 1/5. The rates of profit were set equally. However, because these are special cases, we need to introduce a much general case, and for this purpose, first, we suppose the following reproduction scheme of the two sectors:

$$\left.\begin{array}{l}W_1 = c_1 + v_1 + m_1 \\ W_2 = c_2 + v_2 + m_2\end{array}\right\} \quad (5.10)$$

Note that this system does not have to be a simple reproduction, although it appears the same as system (4.18) in subsection 4.2.1. Under this assumption, then, let us further suppose that the rate of surplus value in both sectors, ε, is equal and that $m_1 = \varepsilon v_1$ and $m_2 = \varepsilon v_2$. Then, the rate of profit of the respective sectors are

$$\left. \begin{array}{l} \dfrac{m_1}{c_1+v_1} = \dfrac{\varepsilon}{\dfrac{c_1}{v_1}+1} \\[2em] \dfrac{m_2}{c_2+v_2} = \dfrac{\varepsilon}{\dfrac{c_2}{v_2}+1} \end{array} \right\} \tag{5.11}$$

In this case, when the organic composition of the two sectors (c/v) differs, the rate of profit naturally differs. Consequently, to achieve equality among the capital in both sectors, a part of the surplus value of one sector must be transferred (redistributed) to the other. We now designate the new surplus value after redistribution as m_1' and m_2' and the two sectors' sales as W_1' and W_2', respectively. Because the rate of profit is equalized,[4] we can write:

$$r^0 = \frac{m_1'}{c_1+v_1} = \frac{m_2'}{c_2+v_2} \tag{5.12}$$

On the other hand, because redistribution does not increase the total value, the redistributed total value should be the same as the original. That is,

$$W_1 + W_2 = c_1+v_1+m_1+c_2+v_2+m_2 = c_1+v_1+m_1'+c_2+v_2+m_2' = W_1'+W_2' \tag{5.13}$$

When substituting the simplified formula $m_1+m_2 = m_1'+m_2'$ (meaning that redistribution does not increase surplus value) in the equal rate of profit formula, we obtain

$$m_1+m_2 = r^0(c_1+v_1+c_2+v_2) \tag{5.14}$$

Consequently, the equal rate of profit that is applicable to the entire society (the general rate of profit) is

$$r^0 = \frac{m_1+m_2}{c_1+v_1+c_2+v_2} = \frac{\varepsilon}{\dfrac{c_1+c_2}{v_1+v_2}+1} \tag{5.15}$$

---

[4] In Chapter 2, Section 3 Okishio (1978) proves that these rates of profit can be equalized and that this equalized rate of profit is stable.

THE DISTRIBUTION OF SURPLUS VALUE AMONG INDUSTRIES

The new transformed values become

$$\left.\begin{array}{l}W_1' = (c_1 + v_1)(1+r^0) \\ W_2' = (c_2 + v_2)(1+r^0)\end{array}\right\} \quad (5.16)$$

Marx designated the newly determined sales of the two sectors W_1' and W_2', divided by the amount of each produced goods, as the price of production. In this manner, Marx's writing becomes more concrete from the general level of value to the level of price. With regard to the rate of profit, what is important here is that there must be consistent equality among capital and an intersectoral redistribution of surplus value to achieve this.

Marx knew all too well that the redistribution described above was not sufficient, which is also clearly stated in *Capital*. This is because when the value of one unit of product in each sector is transformed in the price of production, the amount of payment for the means of production borne by the capitalists in both sectors, c_1 and c_2, also change. In addition, if the real wage the worker receives is held constant, its amount, v_1 and v_2, also has to change. Shibata (1935) shows that a continuous calculation of the redistribution of surplus value, the re-redistribution of the redistributed surplus value, and so on, will lead to a stable mathematical solution. More concretely, due to changes in the cost component, the formulas for both sectors change. Thus, now, we must find the new average rate of profit, r^1:

$$\left.\begin{array}{l}W_1'' = \left(c_1 \dfrac{W_1'}{W_1} + v_1 \dfrac{W_2'}{W_2}\right)(1+r^1) \\ \\ W_2'' = \left(c_2 \dfrac{W_1'}{W_1} + v_2 \dfrac{W_2'}{W_2}\right)(1+r^1)\end{array}\right\} \quad (5.17)$$

However, this re-redistribution is not yet complete because this formula does not consider the new W_1' and W_2' for the re-calculation of c_1, c_2, v_1, and v_2. Thus, we still need to find another average rate of profit, r^2, expressed as

$$\left.\begin{array}{l}W_1''' = \left(c_1 \dfrac{W_1''}{W_1} + v_1 \dfrac{W_2''}{W_2}\right)(1+r^2) \\ \\ W_2''' = \left(c_2 \dfrac{W_1''}{W_1} + v_2 \dfrac{W_2''}{W_2}\right)(1+r^2)\end{array}\right\} \quad (5.18)$$

MARXIAN ECONOMICS

Table 5.2 Reproduction scheme in the level of price of production

	c^p	v^p	m^p	Total
Sector 1	$c_1 \dfrac{W_1^*}{W_1}$	$v_1 \dfrac{W_2^*}{W_2}$	$\left(c_1 \dfrac{W_1^*}{W_1} + v_1 \dfrac{W_2^*}{W_2}\right) r^*$	W_1^*
Sector 2	$c_2 \dfrac{W_1^*}{W_1}$	$v_2 \dfrac{W_2^*}{W_2}$	$\left(c_2 \dfrac{W_1^*}{W_1} + v_2 \dfrac{W_2^*}{W_2}\right) r^*$	W_2^*
Whole society	$(c_1+c_2)\dfrac{W_1^*}{W_1}$	$(v_1+v_2)\dfrac{W_2^*}{W_2}$	$\left\{(c_1+c_2)\dfrac{W_1^*}{W_1} + (v_1+v_2)\dfrac{W_2^*}{W_2}\right\} r^*$	$W_1^* + W_2^*$ $= W_1 + W_2$

Finally, we obtain the equal rate of profit, r^*, and each sector's sales, W_1^* and W_2^*:

$$\left. \begin{array}{l} W_1^* = \left(c_1 \dfrac{W_1^*}{W_1} + v_1 \dfrac{W_2^*}{W_2}\right)(1+r^*) \\[1em] W_2^* = \left(c_2 \dfrac{W_1^*}{W_1} + v_2 \dfrac{W_2^*}{W_2}\right)(1+r^*) \end{array} \right\} \quad (5.19)$$

In this example, the repetitive calculation will converge (Shibata, 1935).[5]

Note that the structure of c, v, and m at the price (of production) level derived above can be translated into the reproduction scheme shown in **Table 5.2**. In this table, (1) the costs (c, v) are multiplied by W_1^*/W_1 and W_2^*/W_2 by this transformation, (2) the sales prices (W_1, W_2) also change, and (3) the amount of profit obtained as a result are different from the original surplus value. In summary, all the value, that is, the c, v, and m originally in terms of embodied labor become different amounts through this transformation. For this reason, c, v, and m are here expressed as c^p, v^p, and m^p, respectively, on the level of the price of production.

5.3.2 Aggregate Equalities Propositions and "New Interpretations" in Western Marxism

Because this new reproduction scheme, translated from the original value term in Table 5.2 (which is called the transformation from value to the price of production or abbreviated

---

[5] This passage is drawn from Okishio (1977) Chapter 4, Section 3. For Shibata's proof, see Okishio (1977), Chapter 4, Section 2.

THE DISTRIBUTION OF SURPLUS VALUE AMONG INDUSTRIES

as "transformation"), has two equations and three unknown variables (W_1^*, W_2^*, and r^*), it needs one more equation to determine the unknowns. However, because Marx insisted on the aggregate equality proposition, which consist of two equations, (1) the total value = the total price of production condition and (2) the total surplus value = the total profit condition, this became a problem of overdetermination. In short, both these conditions cannot hold simultaneously, and thus only one of them can hold. This point was revealed by Bortkiewicz (1906) and has become a major problem in Marxian economics for many years. This controversy is called the transformation problem. As for the above two formulas, condition (1)

$$W_1 + W_2 = W_1^* + W_2^* \tag{5.20}$$

and condition (2)

$$m_1 + m_2 = \left\{ c_1 \frac{W_1^*}{W_1} + c_2 \frac{W_1^*}{W_1} + v_1 \frac{W_2^*}{W_2} + v_2 \frac{W_2^*}{W_2} \right\} r^* \tag{5.21}$$

cannot hold at the same time generally.[6]

---

[6] Special cases in which both conditions hold can be obtained as follows. First, because the profits of both sectors after the transformation from value to price are $W_1^* - \left(c_1 \frac{W_1^*}{W_1} + v_1 \frac{W_2^*}{W_2} \right)$ and $W_2^* - \left(c_2 \frac{W_1^*}{W_1} + v_2 \frac{W_2^*}{W_2} \right)$, respectively, and the surplus values are $W_1 - c_1 - v_1$ and $W_2 - c_2 - v_2$, respectively, the condition total profit = total surplus value becomes

$$W_1^* - \left(c_1 \frac{W_1^*}{W_1} + v_1 \frac{W_2^*}{W_2} \right) + W_2^* - \left(c_2 \frac{W_1^*}{W_1} + v_2 \frac{W_2^*}{W_2} \right) = W_1 - c_1 - v_1 + W_2 - c_2 - v_2$$

This can be transformed into

$$(W_1^* - W_1)\left(1 - \frac{c_1}{W_1} - \frac{c_2}{W_1}\right) + (W_2^* - W_2)\left(1 - \frac{v_1}{W_2} - \frac{v_2}{W_2}\right) = 0$$

However, because our assumption that total value = total price ($W_1^* + W_2^* = W_1 + W_2$) can be transformed into $W_1^* - W_1 = -(W_2^* - W_2)$, substituting this equation yields

$$\left(\frac{c_1}{W_1} + \frac{c_2}{W_1} - \frac{v_1}{W_2} - \frac{v_2}{W_2} \right)(W_2^* - W_2) = 0$$

Although this equation means that $W_2^* = W_2$ or $\frac{c_1}{W_1} + \frac{c_2}{W_1} = \frac{v_1}{W_2} + \frac{v_2}{W_2}$, only the latter condition has meaning because $W_2^* = W_2$ is the condition in which there is no transformation from value to price. Therefore, our task is to identify the meaning of $\frac{c_1}{W_1} + \frac{c_2}{W_1} = \frac{v_1}{W_2} + \frac{v_2}{W_2}$, that is $\frac{c_1 + c_2}{W_1} = \frac{v_1 + v_2}{W_2}$.

However, this overdetermination problem can be solved by using an additional unknown valuable. A group of scholars led by Foley and Duménil, called New Interpretations, tried to do so. In their opinion, v_1 and v_2 are wages given to workers and are not the means of consumption itself, so we do not need to reassess them by multiplying W_2^*/W_2 in this by wage. In this case, what we need is only the ratio obtained by dividing the total added value in money terms by the total amount of labor (e.g., yen/hour), which can translate the money term wage into the value term. They call this ratio the monetary expression of labor time (MELT). Therefore, with this idea, using the ratio M, the above two formulas (5.19) are changed to

$$\left.\begin{aligned} W_1^* &= \left(c_1 \frac{W_1^*}{W_1} + v_1 M\right)(1+r^*) \\ W_2^* &= \left(c_2 \frac{W_1^*}{W_1} + v_2 M\right)(1+r^*) \end{aligned}\right\} \quad (5.22)$$

In this case, we now have four unknown variables (W_1^*, W_2^*, r^*, and M), and therefore, in order to match the number of conditions, we can add the two aggregate equalities propositions. However, because condition (1) counts $c_2 = v_1 + m_1$ twice, condition (1) is substituted with condition (1)' "value of net product = total added value." Therefore,

---

By using the symbols used in Section 4.3 to explain the extended reproduction scheme, this condition can be rewritten as

$$\frac{v_1+v_2}{c_1+c_2} = \frac{W_2}{W_1} = \frac{W_2 - v_1 - v_2}{W_1 - c_1 - c_2} = \frac{m_1(v) + m_2(v) + m_1(k) + m_2(k)}{m_1(c) + m_2(c)}$$

It can also be rewritten again as

$$\frac{m_1(c) + m_2(c)}{c_1 + c_2} = \frac{m_1(v) + m_2(v)}{v_1 + v_2} + \frac{m_1(k) + m_2(k)}{v_1 + v_2}$$

This left side is the accumulation rate of constant capital and the right side is the accumulation rate of variable capital plus the ratio between total capitalist consumption and total variable capital input. There is no guarantee that both sides match. For example, if we substitute the results of the calculation shown in Table 4.2 in Chapter 4, the above equation becomes

$$\frac{\delta \dot{K}/B}{\delta K/B} = \frac{\beta \delta \dot{K}/B + (1-(1-\beta)\dot{s})L}{\beta \delta K/B + (1-(1-\beta)s)L}$$

showing that this equation cannot hold at the steady state $\dot{K} = \dot{s} = 0$ because the right side cannot be zero when the left side is zero.

$$v_1+v_2+m_1+m_2=(v_1M+v_2M)+\left\{c_1\frac{W_1^*}{W_1}+c_2\frac{W_1^*}{W_1}+v_1M+v_2M\right\}r^* \quad (1)'$$

$$m_1+m_2=\left\{c_1\frac{W_1^*}{W_1}+c_2\frac{W_1^*}{W_1}+v_1M+v_2M\right\}r^* \quad (2).$$

(5.23)

However, as a kind of natural extension of this way of thinking, other economists have proposed translating not only the wage part ($v_1 + v_2$) but also the constant capital part ($c_1 + c_2$) by M. That is,

$$\begin{aligned}W_1^* &= (c_1M+v_1M)(1+r^*)\\ W_2^* &= (c_2M+v_2M)(1+r^*)\end{aligned} \quad (5.24)$$

This new idea was originally called the macro-monetary interpretation by its advocates to explain that all costs are translated by MELT as a kind of macro-level variable, but after the appearance of a much newer research group using a concept called the temporally single system interpretation (TSSI), it changed its name to simultaneous single-system interpretation (SSSI). Here, both the SSSI and TSSI are regarded as a single system because both use MELT to characterize practically everything in terms of the price level only. On the other hand, in Fundamental Marxian Theorem (FMT), as explained in subsection 3.1.2 in this book, there are both the value term (embodied-labor term) system of equations and the price term system of inequalities, and therefore, we should face the transformation problem of transforming the former to the latter. It is natural to understand the Marxian system as a double system, but SSSI and TSSI erase this problem by noting that the two systems can be reduced into one system.[7]

---

[7] Here we discuss the translation from the value term to price term by MELT, but we can also use MELT to translate the price term into the value term (the embodied-labor term) conversely. Recall the value equations of the two sectors (3.6) shown in subsection 3.1.2 of this book. That is,

$$t_1 = a_1t_1 + \tau_1$$
$$t_2 = a_2t_1 + \tau_2$$

This system of equations indicates the relationship of the actually embodied labor per unit of product in both sectors. If the amounts of product of each sector are X_1 and X_2, respectively, then the input-output structure of the total labor in this society must be

$$X_1t_1 = X_1a_1t_1 + X_1\tau_1$$
$$X_2t_2 = X_2a_2t_1 + X_2\tau_2$$

An additional point of issue in TSSI is that it not only reinterprets the Marxian system as a single system, but also explained that it is unnecessary to recalculate c and v at the point of the transformation from value to price of production because they are paid at different times. Mathematically speaking, this proposal implies that economic process should be expressed in the form of a differential or a difference system of equations, not in the form of simultaneous equations. As a matter of fact, our Marxian optimal growth model also has the time structure of input and output as differential or difference equations, so we can communicate with them. All the single system schools including New Interpretations, SSSI and TSSI emphasize that the reason for using MELT is to describe the actual process of capitalistic competition between capitals (see Morimoto, 2014, 59), and this also our intention with our Marxian optimal growth model. Since our model has two distinct dimensions: price and value (embodied labor) as discussed shortly, our standpoint is to confront the single system, but in the sense described above, our intention is like that of single systems schools.

This is a labor input structure, which exists objectively not only in capitalism but in all human societies as well (see Izumi, 2014, 23–24). However, there is also the price level in capitalism. That is,

$$X_1 p_1 = X_1 a_1 p_1 + X_1 w_1 \tau_1 + X_1 \pi_1$$

$$X_2 p_2 = X_2 a_2 p_1 + X_2 w_2 \tau_2 + X_2 \pi_2$$

Here, p_1 and p_2 are prices per unit of both goods, w_1 and w_2 are wages per unit of labor of both sectors, and π_1 and π_2 are the profits earned by the capitalists per unit of both goods. In this case, using MELT $M = \dfrac{X_1 w_1 \tau_1 + X_1 \pi_1 + X_2 w_2 \tau_2 + X_2 \pi_2}{X_1 \tau_1 + X_2 \tau_2}$, this labor input structure can be written as

$$\frac{X_1 p_1}{M} = \frac{X_1 a_1 p_1}{M} + \frac{X_1 w_1 \tau_1}{M} + \frac{X_1 \pi_1}{M}$$

$$\frac{X_2 p_2}{M} = \frac{X_2 a_2 p_1}{M} + \frac{X_2 w_2 \tau_2}{M} + \frac{X_2 \pi_2}{M}$$

It is obvious that these values are different from those of the original labor input equations $t_1 = a_1 t_1 + \tau_1$ and $t_2 = a_2 t_1 + \tau_2$, which are the actual labor inputs, not the ideal ones. While the above two equations ($t_1 = a_1 t_1 + \tau_1$ and $t_2 = a_2 t_1 + \tau_2$) of the labor input express the actual input structure of labor, the former, converted system of equations divided by MELT is just an imaginary system of equations calculated using MELT as a kind of macro variable. In this case, unpaid labor such as domestic labor or volunteer work are ignored because they are not evaluated by their price, and even if a certain labor is paid, measured labor using MELT counts only the paid money. That is, if that labor is not paid enough, then that measured labor must be underestimated. This criticism is also made by Izumi (2014), 23 and 306. Shaikh and Tonak (1994) raise another criticism, arguing that the single system school is only a labor commanded theory of value, which is Adam Smith's labor theory of value.

THE DISTRIBUTION OF SURPLUS VALUE AMONG INDUSTRIES

Returning to the problem of transformation, thus, we have shown a way to solve it, but the single system school itself also argue that it is not a solution, just an interpretation. Thus, still we have a question whether two conditions (1) and (2) (or (1)' and (2)) hold or not, and my answer is to take only (1) and abandon (2). That is, abandoning (2) is not crucial, because we already have a mathematical proof in the FMT which proves that profit is premised on surplus value. The late Professor Okishio also took this position. When profits exist in all sectors of society (in this case, the means of consumption sector and the means of production sector), labor exploitation is the condition of profit. In other words, there is no fact other than that the gross profit is still just a redistributed total surplus value.[8]

5.3.3 Division of Profit into Interest and the Enterprise's Profit: Interest-Bearing Capital

Like commercial profit, interest is also part of the surplus value created and distributed from the industrial capital. It is received by financial capitalists such as banks and investors, who gain from industrial capital. For example, banks play an important role in raising funds from many potential investors who are scattered across society to invest it in industry. As far as it is a socially required cost, the bank reserves the right to obtain an equal rate of profit proportional to its cost ($c + v$). This constitutes interest. On the other hand, functioning capitalists who are not investors; that is, those who use the money that they have in hand, receive the business's net profit, which is the profit minus the interest. Thus, part of the surplus value that is created by functioning capitalists' own business activities belongs to these functioning capitalists. This can be shown by working out the difference between the rental price of industrial capital and the interest rate shown in subsection 4.3.3. This difference constitutes the functioning capitalists' share. Therefore, the distribution to both the financial capitalists and the functioning capitalists constitutes the whole distribution to capital.

To show this point clearly, let us now revisit the rental price of capital in terms of the means of production, r_k, and the interest rate, $\tilde{r}$ introduced in subsection 4.3.3. If we place the rental price of capital in terms of the means of production, r_k, and the interest rate, $\tilde{r}$, respectively, on the left-hand side of the equation, as in the earlier case, and substitute the shadow price of the means of production, μ, with the price of the means of production, p_k, these two variables can be expressed as

---

[8] Strictly speaking, among these two equality conditions, condition (1) is absolutely necessary, but the total profit in condition (2) can be higher or lower than the amount of total surplus value because it is superficial and is indirectly determined by the basic exploitation system of the real economy, which depends crucially on market conditions. If there is no surplus value, then no profit can be realized.

$$r_k = \delta + \rho - \frac{\dot{p}_k}{p_k} \tag{5.25}$$

$$\tilde{r} = \rho + \frac{\dot{Y}}{Y} \tag{5.26}$$

The second formula can be replaced by equation (5.27) when instantaneous utility is expressed by $\log Y$ if the price of the means of consumption in terms of utility is expressed as p_c.

$$\tilde{r} = \rho - \frac{\dot{p}_c}{p_c} \tag{5.27}$$

This is because $p_c = \partial \log Y / \partial Y = 1/Y$ and is generally applicable to when instantaneous utility is of the constant relative risk aversion (CRRA) type. In this case, substituting (5.27) from (5.25) makes:[9]

$$r_k - \tilde{r} = \delta + \frac{\dot{p}_c}{p_c} - \frac{\dot{p}_k}{p_k} \tag{5.28}$$

Because equations (5.25) and (5.28) include depreciation in the right sides, pure profit and the share of functioning capitalists become

$$r_k - \delta = \rho - \frac{\dot{p}_k}{p_k} \tag{5.29}$$

$$r_k - \delta - \tilde{r} = \frac{\dot{p}_c}{p_c} - \frac{\dot{p}_k}{p_k} \tag{5.30}$$

The division of profit into interest and the functioning capitalists' share in the Marxian optimal growth model is clear. Note that:

(1) The share of interest shown in equation (5.27) is the sum of the time preference rate and the rate of the fall in price (according to the Marxian optimal growth theory because in terms of utility prices will fall due to the improvement of productivity, which means $\dot{p}_c/p_c$ is negative). In other words, this is the real time preference rate.

---

[9] Equation (5.27) and the right side of equation (5.28) can be expressed accurately by the real rental price of capital in terms of means of consumption, as follows:

$$r_c = \frac{p_k}{p_c}\left(\delta + \rho - \frac{\dot{p}_k}{p_k}\right)$$

$$r_c - \tilde{r} = \frac{p_k}{p_c}\delta + \frac{p_k - p_c}{p_c}\rho - \frac{\dot{p}_k - \dot{p}_c}{p_c}$$

These are shown in Onishi and Kanae (2015).

(2) The functioning capitalists' share of profit, shown in (5.30), represents the falling rate of the relative price of the two goods acquired through business activities. When we look more closely, business activities raise the average labor productivity (though no proof is provided here), making p_k fall faster than p_c and making $\dot{p}_c/p_c - \dot{p}_k/p_k$ positive. Consequently, the functioning capitalist who has brought this about gains what has been achieved.[10]

Because the functioning capitalist gains this achievement by hiring workers and using them efficiently, it appears to each individual functioning capitalist as remuneration for the work of supervision (wages of superintendence) or remuneration for administrative improvement,[11] but it is impossible to calculate each amount accurately. Thus, a conflict of interest emerges between financial and functioning capitalists. In any case, financial capitalists receive a reward for deferring consumption for a certain period by lending money, and functioning capitalists gain from the achievement of their function.

In Marxian optimal growth theory discussed in this book, the rise in labor productivity, which is the capitalists' ultimate function in this case, can be achieved only by approaching the optimal capital-labor ratio by promoting capital accumulation (this is the historical role of the capitalist in capitalism). Therefore, the effect of the function—which appears as a fall in the price of both goods—stops when the optimal capital labor ratio is achieved, which is equal to the end of growth. Consequently, at this point, the share of profit for functioning capitalists becomes zero (and consequently, the rate of profit shown in (5.29) will also fall). Still, since the acquisition of profit in (5.27) remains at this point, achieving the complete abolition of exploitation requires some external influence in one way or another. For example, changes in time preference in the society as a whole, as seen in subsection 4.4.5 and the emergence of a social norm[12] opposing unearned income as the price of deferring consumption might

---

[10] When we include intangible production technology in capital and thus classify research and development as a kind of investment, then the reward for the entrepreneur's investment in technological innovation must also be included in this portion. In this case, the physical price of the means of production, p_k, does not fall, but the overall price of the investment goods, including both tangible and intangible goods, diminishes, so that $-\dot{p}_k/p_k$ is still positive.

[11] Marx notes that the "economy in the employment of constant capital" has become a function of capitalists but does not concern the workers. This is because the outcome of administrative improvements does not benefit workers directly. Here, the capitalist system works only in the interests of the capitalist (furthermore, Marx says, generally in the interest of "the most worthless and miserable sort of money-capitalists" (in Marx, 1907, 74).

[12] This norm is historically natural and not limited to the Koran. It also appears in the New Testament and the Old Testament. However, attacks on usury by the Church in medieval Europe also had the character of a class struggle against the usury class that was competing to take the social surplus value. We cannot ignore the Church's role as the major feudal lord in medieval Europe, or the role of

be the most important conditions. If we had such norms, governments could easily suppress financial capitalists' demand for interest, and it would be easy to raise the wage share of workers even within the enterprise. In this case, financial capitalists, who are traditionally the ultimate providers of finance, but who could not endure such an absurdity, may tend to exit their profession. However, this would not leave the field empty. On the contrary, there will be an increasing number of donor-type funders who do not demand interest. One of these could be not-for-profit organizations established by quasi-entrepreneurs who do not demand profit. In fact, in this situation, a zero share of profit for functioning capitalists would mean that all modern enterprises became not-for-profit organizations. To express this situation positively, functioning capitalists can become non-exploiters. This might be an image of post-capitalistic enterprises, while rentiers will be redefined as a class of exploiters that continue to pursue income from nonproductive monetary transactions.

The banks that emerged as lenders do not invest their own money, instead, they raise funds from others and invest or lend them. The banking system has its origin in the pre-modern commercial credit system, in which merchants and producers used credit in dealing with each other, and this played a large role in the development of capitalism. With the establishment of the banking system, small funds are collected to form one large fund that is then put to use by a business (the functioning capitalist). This promotes capital accumulation as well. Other developments such as joint-stock companies and the investment trust system followed this pattern. Funds are quickly raised in fields with a high rate of profit and the fields with lower rates of profit will lose. As competition among capital intensifies, the gap in the rate of profit naturally closes and the "general rate of profit" becomes more realistic.[13]

5.3.4 Land Monopoly and Rent

The problem of equality among different types of capital leads to a relationship in which special profit due to preferential conditions on production assigned to a business is acquired by its provider. Let us now consider the case illustrated in **Figure 5.2**, which is similar to Figure 3.2 in Chapter 3. Unlike Figure 3.2, however, Figure 5.2 presents the case of an agricultural business (1) that is provided with agricultural land by the landowner (2) that includes land with different fertility levels, not workers with different levels of productivity; (3) a, b, c, d, e, f, g, h ...represents each agricultural business; and (4) assumes

---

the usury class to overthrow the power of the Church and the small producer class controlled by the Church. Indeed, the fundamental change in social consciousness to accept the acquisition interest was indispensable for subsequent capitalist development (see Marx, 1907, Chapter 36).

[13] Banks are also managed in order to make a profit, and they therefore also require a general rate of profit. However, the total surplus value that banks can take is restricted by the interest rate, shown in equation (5.31). The total amount of banks' investment should be smaller than (total surplus value/ general rate of profit.

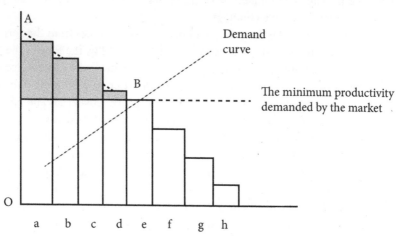

Figure 5.2 Differential rent

that the horizontal line D–B represents the minimum level of productivity demanded by the market for a particular product (if it falls below the line, then there is no profit). In this case, only agricultural businesses a, b, c, d, and e can continue with their business activities. What is at issue here is that the source of gain in a situation in which a earns more than b, c, d, and e; b earns more than c, d, and e; c earns more than d and e; and d earns more than e, depends on the fertility of the land. In this case, the landowner who provides the land would demand a certain distribution according to each tenant's contribution. Marx called the rent that thus emerges (the grey area on Figure 5.2) "differential rent," because each pays a different amount. At this point, the rate of profit among agricultural capitalists levels off and the portion of equality that does not depend on the land is secured.

However, fertility can be improved by tenants' efforts to make additional investments, for example, by bringing in better quality soil and developing irrigation in the area. The result of the improvement in fertility in this way also constitutes a kind of differential rent. This is called differential rent II,[14] and is distinct from the differential rent gained through no additional investment by tenants, which is called differential rent I. The effect of an additional investment by tenants on differential rent II is examined by Onishi (2021a) which clarifies that an additional investment with constant total production and constant

---

[14] When land improvement is carried out by tenants, not landowners, the excess profit is taken by the tenants as profit instead of land rent. However, after the time stipulated in their contract with the landowners, this profit will belong to the landowners, and if the land improvement has a permanent character, the result of this capital investment will be incorporated into the land itself and will eventually be taken as land rent by the landowners.

price has the effect of maintaining the total land rent and raising land rent per area under a Cobb-Douglas type of technology.

In addition, although the example under discussion comes from the agricultural sector and the productivity of the land is expressed as its fertility, the idea of the fertility of the land can also be transposed to situations where the issue is the existence or non-existence of a vein of ore and its quality in mining, the location of a business (whether it attracts customers or not) in commerce, and the ease of securing materials and labor in manufacturing. The rent emerges according to the same logic in all these cases.

During the course of the above discussion, the notion of equality among capitals has been extended to the right to equality between the landowner and the capitalist, which enables us to understand what rent is. However, some fundamental differences are also important between Figure 5.2 and the case illustrated in Figure 3.2. In the latter case, it is assumed that workers with a high level of productivity (or workers who make a lot of effort) do not obtain profit in the grey area, as this is obtained by the capitalist. In Figure 5.2, the assumption is different and the profit from that area goes to the landowner, who provides the factor of production; that is, land. Consequently, this assumes that landowners have a special power that enables them to do so. The source of this power is found in the finiteness of the land. Under capitalism, even highly productive workers have become unskilled and basically easily replaceable. In contrast, agricultural lands are finite and even if there is land with the same level of fertility elsewhere, reaching that location incurs transport costs. In this way, the land is fundamentally finite and this determines the power of negotiation of the owner of this factor of production.[15]

Consequently, the problem of the land is at the same time the problem of monopolies, and for this reason, we can imagine a situation in which even the landownere in Figure 5.2 can obtains rent. **Figure 5.3** shows such a case in which the availability of the land is restricted because the portions of land named f, g, and h do not exist even when demand for agricultural products is higher than before. Where demand is higher, the demand curve (which rises to the right since the vertical axis is productivity, unlike normal demand curves) moves to the right compared with Figure 5.2, which leads to the emergence of rent even for the worst land, e. Simultaneously, the same amount of rent is added to other lands. Marx refers to this as absolute rent.[16]

---

[15] Put differently, when land is infinite, it is very difficult for rent to emerge. This was observed when the Europeans colonized the New Continent. The price of land in the New Continent, which had a large quantity of land, was far lower than that in Europe (Piketty, 2013, Chapter 6).

[16] Marx questions whether the source of the absolute rent that thus emerged was found in the surplus value of what was produced by the respective sector (in this case, agriculture) or whether it was found in the transfer of surplus value produced by other sectors. He then uses the term "monopoly rent" to describe the situation in which strict restrictions on land and a higher demand escalated the latter condition. However, the reason why this happens basically remains the same with the absolute rent.

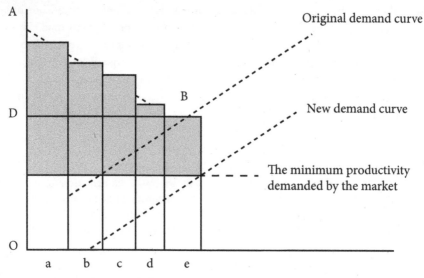

Figure 5.3 Differential rent and absolute rent

Although rent can be discussed using a similar framework to the demand-supply theory of modern economics, Marxian economics can provide its own argument based on the labor theory of value (LTV). Marx's *Capital* discussed rent in the last part of its theory of value (income theory), which was the concluding part of Volume 3 of *Capital*, where the logical level of abstraction is the lowest to explain a superficial phenomenon. However, because it is also a part of *Capital*, the source has to be explained in terms of his LTV. Marx uses the following example to explain that the source of the rent lies in the surplus value created in the productive sector.

Let us suppose that there is a mill by a waterfall. In this case, because the milling machine can be operated without a steam engine, the cost of production is low. Therefore, the provider of the land in which the waterfall is located demands a certain rent. This scenario supposes that both milling factories with steam engines and those with waterfalls exist. Now, if the demand for flour decreases to such an extent that only milling factories with waterfalls with a fundamentally low cost survive and those with steam engines with a fundamentally high cost disappear, or if the amount of land with easy access to a waterfall increases to the extent that it can meet all demand for flour (in other words, a land monopoly no longer exists), the use of the waterfall becomes free to everyone and the landowner can no longer take rent. In other words, rent disappears. This is the same as in the case in which air and water is plentiful enough and available to all; the providers of air and water cannot charge. To put it differently, only when the unlimited use of those resources is impossible and a monopoly therefore exists, a labor input from outside nature (in this case, the production of steam engines, human-made air or water) becomes necessary. This part emerges as rent. Thus, the amount of rent is determined by how much labor has been reduced compared with the marginal land.[17] This is why the value in terms of labor input is decisive in determining rent. Marx understood rent as the redistribution of surplus value that like the cases of commercial profit and interest discussed earlier.

---

[17] This conclusion about the importance of its difference from marginal land, such as land that cannot use the power of falling water, can also be shown by the mathematical formulation of this differential land rent. In general, if we denote production, total factor productivity, land, and factors of production other than land as Y, A, N, and K, respectively, and if we set the production function, assuming constant return to scale, as $Y = AK^\alpha N^{1-\alpha}$, the marginal productivity of land becomes

$$\frac{\partial Y}{\partial N} = (1-\alpha)AK^\alpha N^{-\alpha}$$

In this case, when we denote the marginal land as N^*, subtracting the marginal productivity of the marginal land from the marginal productivity of each piece of land, we can calculate the differential rent in each piece of land as

$$(1-\alpha)AK^\alpha N^{-\alpha} - (1-\alpha)AK^\alpha N^{*-\alpha} = (1-\alpha)AK^\alpha (N^{-\alpha} - N^{*-\alpha})$$

Therefore, the total differential rent becomes

$$\int_0^{N^*} (1-\alpha)AK^\alpha (N^{-\alpha} - N^{*-\alpha}) dN = (1-\alpha)AK^\alpha \left[\frac{1}{1-\alpha} N^{1-\alpha} - N \cdot N^{*-\alpha}\right]_0^{N^*}$$

$$= (1-\alpha)AK^\alpha \left(\frac{N^{*1-\alpha}}{1-\alpha} - N^{*1-\alpha}\right) = \alpha AK^\alpha N^{*1-\alpha}$$

which shows a completely opposite result from the product-exhaustion theorem in mainstream economics explained in subsection 4.3.3 of this book. This is because $AK^\alpha N^{1-\alpha}$ is the amount of production (Y) itself, and therefore landowners and entrepreneurs share the total production in the ratio of α to $1 - \alpha$. This means, conversely, that it is the degree of contribution to production by the factors of production other than land, indicated by α (the power of K), that is the source of land rent. In the example in this book, this is the effect of the non-natural labor input that substitutes for falling water must be calculated as land rent. Thus, Marx's emphasizes the amount of labor invested in his land rent theory. Note that it is only when natural forces are monopolized that we can include natural forces in Marx's theory of value and price.

A comparable situation in which a monopoly on natural forces leads to rent can occur in a situation that excellent capital facilities and a labor force cannot be accumulated or put to use immediately. Although excellent capital facilities and labor create excessive profit corresponding to their quality, this will diminish in the long run, because capital facilities and labor of the same quality can be formed over time. However, it still takes time and therefore excessive profit is produced in the meantime. This is called quasi-rent in modern economics and as the theory of special surplus value in Marxist economics.

In addition, beautiful natural environments have the same essence as the products of non-labor and are evaluated by their scarcity, and are thus also assigned a monetary value. For example, we pay money for eco-tourism and to save these environments. Ecologists say these expenses show their own value, and then deny the LTV, which asserts that only labor can create value. However, these expences are only rent; that is, it is a kind of rent that has a price without labor value.

5.3.5 Rising Asset Prices as a Law of Capitalism

In the twentieth century, capitalism evolved into a new phase called monopoly capitalism, characterized by high levels of capital accumulation and concentrated industries, especially heavy industries such as the steel industry, and shipbuilding, electricity, and railroad industries that make use of economies of scale. Over the course of this development, banks, which played a crucial role in connecting these monopoly capitalist enterprises and developing them, changed their characteristics to become bank capital and industrial capital, which is known as financial capital. Furthermore, in the twenty-first century, the role of the financial sector expanded, creating a new phenomenon called financialization or financial capitalism. However, the rise in asset prices as the core characteristic of financial capitalism occurs not only in modern capitalism but also throughout capitalism.

To show this point, we first show the revaluated asset:

$$\frac{\text{income produced by the asset in question}}{\tilde{r}} \tag{5.31}$$

Here, $\tilde{r}$ is the market rate of interest. For example, a piece of land that yields one million yen per year will be valued at twenty million yen if the market rate of interest is 5 percent. It is not because this land is prepared by 20 million dollar valued labor, but because this amount of interest which could be created by 20 million dollar bank deposit if interest rate is 5 percent. In other words, what matters is not the amount of labor expense to produce that that land, but the amount of income that land yields. This type of asset pricing is called "capital reduction."

Using this formula (5.31), we show that the price of stock should be calculated via capital reduction, comparing the interest rate with the return on risk-bearing stocks. Of course, the latter is higher than the former because of risks; therefore, the return from stocks is shown as $\tilde{r} + r_p$, and the stock price becomes

$$\frac{\tilde{r} + r_p}{r_p} \tag{5.32}$$

times larger than the original price. Here, $\tilde{r}$ and r_p are the interest rate and risk premium,[18] respectively, which are created by risk-taking investors, and this rise in stock prices creates the founder's profit. This is how the securitized total price of stocks becomes larger than the original.

Another reason that stock prices rise is that the denominator $\tilde{r}$ of the above fraction tends to fall according to the law of the falling rate of profit.[19] Marx explains this mechanism in his discussion of price of land in *Capital* Volume 3, Chapter 37. Therefore, we understand that the present trend for rising asset prices is caused by the inevitable law of capitalistic development.

On the other hand, the long-term tendency for rent to rise can also be shown by Marxian rent theory. This is because the total differential rent becomes $\alpha A K^\alpha N^{1-\alpha}$, as shown in the note 17 in subsection 5.3.4, which shows a rising trend in the course of capital accumulation (a rise in K). Furthermore, the rent per land $\alpha A K^\alpha N^{1-\alpha}$ also rises over the course of capital accumulation.[20]

Although N is not the product of labor and therefore has no value, landowners claim more rent in the price level. We should note here that this rise is caused by the scarcity of land N, which does not expand as K or L do. Furthermore, the falling rate of interest also accelerates this tendency; that is, there are two reasons for the price of land has to rise.[21]

---

[18] Although Marx and Engels do not use this word, they knew this category. See Marx, 1907, 445. Furthermore, they paid attention the specially high risks in the shipping business. See Marx, 1907, 151 and Engels's supplement on p. 618 in Marx (1907).

[19] However, the interest rate has a bottom to fall at the rate of time preference ρ shown in equation (5.28). Marx acknowledges this in Marx, 1907, 316.

[20] Marx also states that capital accumulation leads to a rise in rent (see Marx, 1907, 511).

[21] Because of the absolute shrinking of the agricultural sector in modern times, the power of the owners of agricultural land diminishes. Instead, the numbers of landowners' leasing apartment houses and land for residences and commercial shops has grown (see Figures 6–5 in Piketty, 2013). Differently from Marx, Piketty defines anything that produces interest as "capital" and thus, emphasizes the rising "capital" share including land rents. Therefore, it is increasingly important to re-examine the role of such landowners in contemporary capitalism. Many economic bubbles have also been related to land. This is an important theme that requires further examination.

5.3.6 The Bubble Economy in the Marxian Optimal Growth Model

In addition, under a given population size and technological conditions, the Marxian optimal growth model shows that capital accumulation stops at a certain level, after which economic growth (an increase in Y) also stops. This indicates that the $\tilde{r} + r_p/\tilde{r}$ and rent also become stable at that time. Therefore, if various asset prices do not stop increasing in that stable state in reality and if technology does not change, then there might be a bubble. In this way, Marxian economics can distinguish between a normal rise in the asset price and a bubble.

Since the bubble that will burst later deviates from the rational equilibrium path explained above, it is shown as an unstable path that deviates from the normal accumulation path in the Marxian optimal growth model. For this reason, I now show in **Figure 5.4** how bubbles come about by adding two types of unstable path to Figure 4.3 in the previous chapter.

Three basic arrows have been added to this figure. The thick dashed line from the upper right via point G to the final equilibrium point E in the center shows the saddle path from the right to equilibrium. This line is not shown in Figure 4.3 because capital accumulation usually starts at a point lower than K^*. Here we add it to deal with the consequences of capital accumulation beyond K^* due to a bubble. The other two thin arrows go from point D' and point D'' to the right or upward. These arrows show the transition paths when the transversality condition shown in p. 000 in the Mathematical Appendix at the end of this

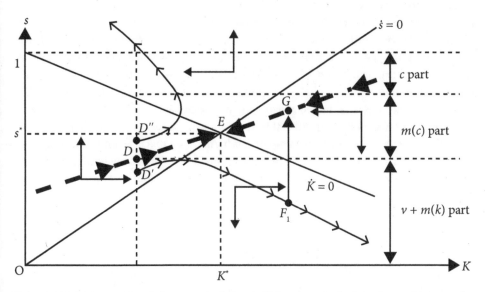

Figure 5.4 Multiple transitional dynamics including bubble process in the Marxian optimal growth model

book is not satisfied. The transversality condition is one that guarantees that the present value of the asset price of a capital good in the distant future will be neither positive nor negative, that is, that a rational investor in the present, considering the future, will neither hold much of it infinitely nor immediately sell off all of it. This means, conversely, that when this condition is absent, the rational investor's infinite desire to accumulate or sell out rises. The arrow starting from point D' represents the former path (the bubble process), and the arrow starting from point D'' represents the latter path.

As shown in this figure, the bubble process from point D' will enter the $\dot{s} < 0$, $\dot{K} > 0$ region in the middle of the process, so from there on, s is reduced, and the path is changed in the direction of increasing the ratio of total labor allocated to the sector of the means of production. This raises the shadow price, μ, of the means of production, as shown by the equation

$$\frac{\dot{\mu}}{\mu} = -\frac{\dot{s}}{s} \tag{5.33}$$

(4.35) in subsection 4.3.2.

That is, $\dot{s} < 0$ leads to $\dot{\mu} > 0$, and in this form there will be an infinite rise in asset prices: a bubble economy. This will exceed the normal rise in asset prices described in the latter half of subsection 5.3.5, and in this process unnecessary restrictions on consumption will continue in the form of a decline in s. Therefore, in Marxian optimal growth theory, a bubble economy is explained as an entirely useless economic activity that accumulates unnecessary capital and restricts consumption more than necessary.

However, these irrational increases in asset prices cannot continue forever. That is why at a certain point, people realize that they are in a bubble and will correctly identify the expected rate of return. This means a shift to a path that satisfies the transversality condition, which is expressed as a jump from point F to point G in Figure 5.4. Note here that a sudden jump in s means a sudden collapse in asset prices, as described above. This is what we mean by a bubble bursting. After this, the economy passes on the saddle path toward point E, but since the s-coordinate at point G is higher than that at point E asset prices will overshoot and collapse. In any case, that the economy will be heavily burdened and this situation can be represented by the Marxian optimal growth model.

5.3.7 The Conflict Between the Landowners and Other Classes

Above discussions revealed a very special characteristic of the landowner class, and then landowners are in conflict with both the capitalist and working classes. They served as an obstacle to the early development of capitalism and the capitalist and working classes united to fight against them. This relationship can be expressed as follows. Let us first recall the price base formula, the simplest form of which was introduced in the explanation of

THE DISTRIBUTION OF SURPLUS VALUE AMONG INDUSTRIES

Okishio's Fundamental Marxian Theorem in subsection 3.1.2. If the general rate of profit, r, is applicable to both sectors, then the two formulas can be rewritten as

$$\left. \begin{array}{l} p_1 = (a_1 p_1 + \tau_1 R p_2)(1+r) \\ p_2 = (a_2 p_1 + \tau_2 R p_2)(1+r) \end{array} \right\} \quad (5.34)$$

Now, let us suppose that rent is produced in the second sector because agriculture—in which the typical landowner appears—mainly produces the means of consumption. Marx adopts a similar assumption. When Ω represents the real term rent measured in terms of price of the means of consumption, the above formulas can be reformulated as

$$\left. \begin{array}{l} p_1 = (a_1 p_1 + \tau_1 R p_2)(1+r) \\ p_2 = (a_2 p_1 + \tau_2 R p_2)(1+r) + \Omega p_2 \end{array} \right\} \quad (5.35)^{22}$$

Let us now investigate the ways in which Ω relates to the capitalists' rate of profit, r, and the workers' real wage rate, R. If we focus on p_1 and p_2, the two formulas are expressed as follows:

$$\left. \begin{array}{l} \left(\dfrac{1}{1+r} - a_1 \right) p_1 - \tau_1 R p_2 = 0 \\ \dfrac{1}{1+r} p_2 = a_2 p_1 + \tau_2 R p_2 + \dfrac{\Omega}{1+r} p_2 \end{array} \right\} \quad (5.36)$$

The relative price p_1/p_2 is then

$$\left. \begin{array}{l} \dfrac{p_1}{p_2} = \dfrac{\tau_1 R}{\dfrac{1}{1+r} - a_1} \\ \dfrac{p_1}{p_2} = \dfrac{\dfrac{1-\Omega}{1+r} - \tau_2 R}{a_2} \end{array} \right\} \quad (5.37)$$

The latter formula shows that an increase in real rent, Ω, leads to a fall in the relative price of the first goods over the second goods, p_1/p_2. In other words, the price in the sector where rent is taken goes up and the price in the sector in which there is no rent falls, and that sector is disadvantaged.

---

[22] Chapter 5 in Negishi (1985) incorporates a similar formation in the value equation and denies the proportionality between value and embodied labor. However, such a formation can be applied only in price equations, because land input cannot create any value.

In addition, because the two formulas have to match,

$$\frac{\tau_1 R}{\frac{1}{1+r} - a_1} = \frac{\frac{1-\Omega}{1+r} - \tau_2 R}{a_2} \tag{5.38}$$

This can be expressed as:

$$a_2 \tau_1 R = \left(\frac{1}{1+r} - a_1\right)\left(\frac{1-\Omega}{1+r} - \tau_2 R\right) \tag{5.39}$$

Here, as shown in the first equation in (5.38), $\frac{1}{1+r} - a_1$ is positive, which means that if Ω increases, r has to decrease if the other conditions are held constant. This shows that the interests of the landowners and those of the capitalist class are contradictory.

What about the relationship between the interests of the landowner and those of the working class? Let us reformulate the above formula as follows:

$$\left(\frac{1}{1+r} - a_1\right)\left(\frac{1-\Omega}{1+r}\right) = a_2 \tau_1 R + \left(\frac{1}{1+r} - a_1\right)\tau_2 R \tag{5.40}$$

Since $\frac{1}{1+r} - a_1$ is positive, if Ω increases, R has to decrease. In other words, the interests of the landowners contradict those of the working class.

In this context, under capitalism, the interests of the landowners are in opposition to those of the newly emerging capitalist and working classes, and thus the latter two classes united to fight against the landowners. The basic outline of the early phase of history of capitalism is a history of class struggle, as described here.[23]

---

[23] This passage is based on Chapter 1, Section 12 of Okishio (1977).

6
Precapitalistic Economic Formations

6.1 AGRICULTURE AS ROUNDABOUT PRODUCTION

6.1.1 The Leap in Production Resulting from the Agricultural Revolution

Chapters 2, 3, and 4 discussed the basic content of the capitalist system and its birth, growth, and death. Chapter 5 discussed the re-distribution of the surplus value produced among various sectors. These chapters constitute a modernized interpretation of the three volumes of Marx's *Capital*, his most important achievement and publication. However, Marx also analyzed precapitalistic modes of production. His notes of these studies were found in 1939 in the Soviet Union and subsequently published under the title, *Pre-Capitalist Economic Formations*. In these notes, we can find numerous points that were new (at that time). This implies that Marx devoted himself to studying the most recent archaeological and historical findings. He believed that these studies, as well as the content of Volumes 2 and 3 of *Capital*, were very important.

In light of this, Marxian economics also needs to be updated in light of the current archaeological and historical scientific findings of the twenty-first century, and this means we must update our understanding of precapitalistic economic systems, as is consistent with our understanding of capitalism. For this purpose, we therefore first provide a clear definition of agricultural society to replace previous explanations found in Marxian textbooks.

Agricultural society includes stages such as the last phase of the primitive communism, the slavery system, and serfdom. All these systems have their own special characteristics, but first we explain the essential identity that is common to these systems as parts of agricultural society. The birth of agriculture approximately 10,000 years ago was a huge leap for human society. In this book, we call this the agricultural revolution, and its significance was much larger than the influence and importance of the industrial revolution in terms of productivity and civilization.[1]

---

[1] However, it was not a single event, and, like the industrial revolution, the agricultural revolution developed over a very long time. This is because there are many aspects to growing plants: in addition to sowing seeds, they need to be weeded, fertilized, watered, and transplanted, and understanding these processes may precede their implementation. As this can also apply to the domestication of animals, Scott (2017) widened the concept of domestication to include plant growing in its widest sense.

This significance can be understood by imagining the very low productivity and living standards in hunting-gathering societies, which were succeeded by the agricultural revolution. For example, in North Borneo there was a hunting-gathering society with very low productivity, which by the middle of the twentieth century, had developed a special population-control system in the form of a head-hunting culture. In that society, grooms in hunting-gathering tribes like the Iban hunted the male heads of other villages before they could marry. This ritual signified that only the strong males would have their own descendants, thereby restricting population growth. In other words, low productivity led to this special culture. Thus, the superstructure as well as the population size were determined by productivity, which was rather limited in hunting-gathering societies.[2] For example, consider how many rabbits needed to be hunted in a year to keep one family alive, and how large an area one family needed to occupy by to catch enough rabbits to live. One small mountain may not have been sufficient for this, and this condition determined the population density.

However, the invention of agriculture changed the entire society. For example, in the holy land of the Iban, North Borneo, in the late nineteenth century English colonialists brought Chinese farmers from southern China to create a balance among the natives, Chinese, and the colonialists. This policy was successful due to the rapid population growth of the Chinese farmers and to agricultural development. Even if the number of immigrant farmers was small and the land they occupied was marginal, these farmers were able to raise many children owing to their agricultural productivity. This was the basic difference between Chinese farmers and the hunting-gathering society, and for this reason, the Chinese population was able to catch up with that of the natives. In 2000, in the Sarawak Province of North Borneo, the population balance between the Malay, Chinese, Iban, and Bidayuh was 22:26:29:8. In numerical terms the Chinese farmers had caught up with the native Iban, although the latter have a longer history in this area. Such is the power of agriculture.[3]

In fact, human history is replete with cases in which agricultural people overtook the former hunting-gathering inhabitants. In the third century BC in Japan, the Yayoi people came to the islands from the Chinese continent and brought with them developed

---

[2] Each primitive society had its own special culture for restricting population growth. For example, the Australian Aborigines cut a part of the penis for birth control, and the Japanese relegated the elderly to the mountains. The population of nomadic societies did not increase as they lacked the ability to care for infants as the elders of the settlement society did.

[3] The Bidayuh were also hunter-gatherers, but they are fewer than the Iban. The population size in Malaya was maintained at a certain size because they enjoyed relative wealth as merchants. Therefore, we focus on the population balance between the Chinese and Iban or Bidayuh, because our main interest is in the difference in the productivity of agricultural and hunting-gathering societies.

agriculture; subsequently occupying these islands by pushing out the natives known as the Jomon people or occasionally by uniting with the natives. Although the Jomon people also practiced an early type of agriculture, they were basically dependent on hunting and gathering at least before its early period.

Another example is the Thai people who lived in the southern part of China in ancient times. They brought rice cultivation to the Thai plains, where the former inhabitants were engaged mainly in primitive fishing. Therefore, the ancestors of those living in this plain are these latecomers, who managed to occupy the plain due to the higher productivity by their agricultural practices. We find a similar history in Vietnam, which was inhabited by a different people before the present dominant population arrived there.

6.1.2 Agricultural Society as a Land-Accumulating Society

The cultivation of plants—agriculture—began approximately 10,000 years ago in the Near East and lower Yangtze River area, and thus, we must understand that almost all societies throughout human history were hunter-gatherers. If we take human history as starting from the time of the evolution of *Homo erectus* one million and several hundred thousand years ago, humans were agriculturalists for over less than one percent of that time; and even taking it as starting from the emergence of *Homo sapiens*, they were agriculturalists for less than one-tenth of that time. Thus, from a historical perspective, it is evident how slow the development of agriculture was.

For example, let us consider the first crops that were planted in the ancient Near East. They were wheat and beans, which could be eaten only by boiling or roasting, and they were modified so that the ears of the grain would not be scattered by the wind. The ability to manipulate fire was a precondition for agriculture, and humans needed to be farsighted and have a great imaginative ability to gather the very primitive crops whose seeds were limited, and then to modify plants from them using a so long process. Furthermore, the agricultural revolution required a sedentary society as its precondition. If people are on the move all year round, not only will they not be able to take care of their crops, but the crops they have planted will be stolen by ithers.[4] The Japanese Jomon period is a typical example of this.

In fact, shell middens existed even before the establishment of agriculture. These were generally found in the mid-latitudes, as in the Japanese Jomon period, where people had settled because the conditions for gathering food were stable all year round, not only in the springtime, when they could harvest shellfish. For example, the Ainu people in Japan and the indigenous people of the northwestern coast of North America were engaged in fishing,

---

[4] Before the sedentary period, people were able to escape from the enemies, but now they had to fight against the invaders. Therefore, sedentary societies created their own defense system, such as moats around the settlement.

mainly salmon. Scott (2017) notes that the conditions were suitable for sedentarization in the big and damp lower river areas where the ecosystem changes throughout the seasons, because an adequate diet depended on their ability to fish, hunt, and collect wild food. In general, these facts indicate that there was a revolution in the mode of production that enabled the mass to survive by efficient acquisition of aquatic products before agricultural goods started to be produced. Although a rival view holds that sedantarization was the result of animal husbandry rather than fishing, here I follow Nishida (1986) and assume that it was due to the cultivation of wild plants and advances in fishing technology during the period of vegetation changes in the mid-latitudes in the era of global warming about 10,000 years ago. Obata (2016) proposes that the time-lag between the sedantarization and the cultivation of plants was relatively short. because legume-based farming, though not full-scale grain cultivation, began in the early Jomon period. In addition, Fagan (2004) and Miyamoto (2005) assert that global climate changes forced the people to plant crops to maintain the population that had grown before climate change occurred.[5] At that time, maintaining the population had become a central preoccupation, not only due to climate change but also due to seasonal shortages of food in winter.

Special products of nature, like marine products, were gifts of fortune, and this fact reminds us that primitive accumulation in European countries had the additional advantage of the gold and silver inflows from the New Continent. Furthermore, the role of climate change in forcing people to adopt agricultural improvements is also interesting, because the industrial revolution was also accelerated in some late industrializing countries by certain exigencies such as wars or invasions in many cases. Finally, livestock farming is similar in the sense that the food (here, meat and milk) can be taken only by growing or breeding livestock. That is, livestock farming and agriculture are both roundabout production systems. We thus include the invention of livestock farming in the invention of agriculture in this book.

In fact, as roundabout systems of production, agriculture and livestock products can be used as a means of production as well as a means of consumption. Seeds can be invested in as well as consumed, and livestock can either be raised and eaten or fed so the animals give birth to calves.[6] Furthermore, one more important characteristics of both industries is that they requires a specific means of production; namely, land for cultivating or serving for livestock, just as modern industry requires machines for production; therefore, at least in the case of agriculture, the agricultural revolution had started "land accumulation," like the

---

[5] See Yasuda (2004).

[6] The most important seed, grain, has special characteristics as it can be stored, accumulated, and measured. Those characteristics became the fundamental base for private property, monogamous societies and enslaving debtors. Furthermore, because grain is harvested at the same time, taxation became easy and this was a prerequisite for state formation (see Scott, 2017).

industrial revolution started capital accumulation. Therefore, to explain this agricultural revolution, we can replace the terms used in Figure 4.5—the industrial revolution, feudalism, capitalism, and K/L—can be replaced with the agricultural revolution, primitive communism, slavery or serfdom system, and arable land/L, respectively.[7]

However, there is an opinion that land can be a means of production derived from nature without the need for land improvement and, therefore, "land accumulation" is not an appropriate term. Furthermore, Marx's concepts of absolute and differential rent I abstract land improvement. Of course, Marx knew that land improvement was important,[8] but his perspective of land cultivation might be different from that of Asians.

In my opinion, this difference has come from the difference between the topography of England and Asia. English farmland was easily transformed into grazing land by the two enclosure movements, while Asian land could be made cultivable only by cutting or burning trees and digging up tree roots. This difference arises from the difference in the density of trees, but in a much more important way, from the difference in the amount of rain in the two regions.

If we focus on Asian agriculture, we can understand the difficulties farmers faced. Creating paddy fields was laborious. Paddy fields need to be completely level and have a good irrigation and drainage system. Thus, it is difficult to make and maintain terraced rice fields in mountainous areas. In southern China and some plains in Japan, a good drainage system is also needed. For this purpose, people first made ponds or creeks in order to keep small pieces of lands free from excessive water. In other cases, constructing tall levees by rivers is necessary for creating arable lands by protecting from floods. While low levees could protect only small pieces of land, tall levees could secure large areas for cultivation. In this way, the construction of levees can be considered a concrete form of land accumulation and cultivable land is to be thought of as a man-made means of production, at least in Asia. Cultivable land is the most important factor of production in agriculture,[9] and agricultural societies have always prioritized land accumulation over any other activity. Some nations took over cultivable land from other nations through wars, while others made land cultivable by their own efforts. In both cases this necessitated a certain amount of political power, a

---

[7] We cannot find a system which corresponds to "socialism or communism" in Figure 4.5, because attaining the optimal arable land/L does not mean attaining capitalism.

[8] Form II differential rent is created by improving the land. According to Marx, "extractive industries, ... such as mining, hunting, fishing (and agriculture, but only in so far as it begins by breaking up virgin soil), where the material for labor is provided directly by nature" (Marx, 1887, 129). This indicates that Marx understood the importance of land improvement in agricultural production after breaking up virgin soil.

[9] While the most important factor of production in capitalism is machines, the most important factor of production in agricultural society is cultivable land.

typical example of which was oriental despotism in large-scale irrigation systems theorized by Wittfogel (1957).[10] These special efforts were aimed at the violent accumulation of land, just like the violent accumulation of capital soon after the industrial revolution.[11]

6.1.3 Evolution from Primitive Agriculture to Intensive Farming Caused by Population Pressures

Here we examine the difference between primitive and developed forms of agriculture. For example, paddy-field farming is different from primitive farming[12] in terms of intensity and it created its own superstructure. Our next task is to analyze the different stages of agricultural societies and discuss what led to the evolution in agriculture. Based on Boserup (1965), we explain that the cause was population pressure.

Just as an agricultural revolution was necessary to maintain the population level before global cooling and drying about ten thousand years ago, Boserup (1965) maintains that the evolution of agriculture occurred as a result of population pressure. The author, Boserup, spent many years instructing agricultural technologies to African countries where advanced, intensive agricultural technologies were not adopted by the local people, no matter how many times she instructed them. This experience made her believe that this is not because

---

[10] The argument that the public regulation of water utilization requires a tyrannical state supports the argument that the need for state-led capital accumulation in primitive accumulation required state capitalism.

[11] Generally speaking, after accumulating a certain amount of land, each society underwent their own industrial revolution and began to accumulate capital with a speed that was determined by the level of land they had accumulated. This problem was studied by Roxiangul and Kanae (2009) who built a model with three factors of production: labor, capital-1, and capital-2. In this case, we can regard agricultural land as capital-1 and machinery as capital-2. The result is shown in **Figure 6.1**, indicating that after the industrial revolution there should have been a period with no consumption for a certain length of time (from m to $m + i$). Of course, no consumption is impossible, but consumption should be close to this ideal situation. In terms of the calculation, this is because the model seeks to balance processes of accumulation between the two types of capital; therefore, matured agricultural economies need a much longer time for capital-2 to catch up to the level of land accumulation that has already been achieved by capital-1. However, this also implies that after catching up, more matured agricultural economies can build highly developed industrial economies. That is, the degree of agricultural development determines the degree of industrial development; thus, it is difficult for less developed countries to catch up to highly developed ones. Each country must develop step by step.

[12] This image can be seen in the cultivation of soybeans and other crops, as identified by Obata (2016) using the Jomon period as an example, or in the slash-and-burn agriculture that is still practiced in tropical forests today. The Jomon period is characterized by the fact that not much labor was devoted to the cultivation of arable land.

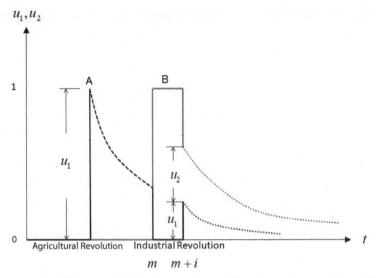

Figure 6.1 Ratio of total labor used for land accumulation (A) and capital accumulation (B)

they are unaware of the new technology, but because the old technology is somehow more suitable. More specifically, they thought that at low population densities, productivity per capita was higher using the old technology, which was more unrestrained, and that it was lower with the new technology. Such an agricultural technology can be represented by a model.

First, we introduce an agricultural production function under the assumption that land area is constant and only the population is growing.[13]

$$Y = (1 - c\beta)L^\beta \tag{6.1}$$

Here, Y and L represent agricultural production and labor input (the population), respectively, and we assume diminishing return to labor; that is, $0 < \beta < 1$. β expresses labor intensity. A high β implies higher labor intensity, and a low β implies lower labor intensity. $c\beta$ expresses the fact that the production cost goes up in accordance with the rise in β. Therefore, $c > 0$, and under these assumptions, we can show that the optimal labor intensity increases in accordance with the population growth. First, we differentiate the above production function with respect to β:

$$\frac{\partial Y}{\partial \beta} = L^\beta \log L - cL^\beta - c\beta L^\beta \log L \tag{6.2}$$

---

[13] Boserup's idea was first modeled by Robinson and Schutjer (1984), and Tazoe and Liu (2011) improved on their model.

Therefore, the optimal β can be obtained by making equation (6.2) equal to zero, with the result that

$$\frac{1}{c} - \frac{1}{\log L} = \beta \qquad (6.3)$$

This shows that the labor intensity of agricultural production should rise in accordance with population growth, which was what Boserup found. In addition, we can introduce per capita production as

$$\left(\frac{Y}{L}\right)^* = c(\log L)^{-1} L^{\frac{1}{c} - \frac{1}{\log L} - 1} \qquad (6.4)$$

and again, we differentiate it with respect to L in order to check the effect of population growth. Therefore:

$$\frac{d(Y/L)^*}{dL} = (\log L)^{-1} L^{\frac{1}{c} - \frac{1}{\log L} - 2} \{(1-c) - c(\log L)^{-1}\} \qquad (6.5)$$

Whether this result is negative or positive depends on the value of $(1 - c) - c(\log L)^{-1}$. However, remember our assumption—$0 < \beta < 1$, which implies:

$$0 < \frac{1}{c} - \frac{1}{\log L} = \beta < 1 \qquad (6.6)$$

This condition can be transformed into $(1 - c) - c(\log L)^{-1} < 0$. Therefore:

$$\frac{d(Y/L)^*}{dL} < 0 \qquad (6.7)$$

That is, population growth leads to a decrease in per capita production if other conditions are constant. According to Boserup, the intensive agriculture that is brought about by population growth leads to a fall in per capita income. This is the situation that Malthus (1798) discusses in his *An Essay on the Principle of Population*.

6.1.4 Productivity Growth by Land Accumulation: Slavery System and Serfdom

However, this decrease in income has not occurred homogenously even if African primitive agriculture suffered, and it must be noted that our original standpoint is that arable land is not constant, but land is accumulated. Therefore, we here introduce the concept of land accumulation, and show the course of income increase in accordance with the intensification of agricultural production.[14]

---

[14] This model was also first provided by Tazoe and Liu (2011).

For this purpose, first, we assume that land accumulation occurs in accordance with population growth, but that additional land creation diminishes with this growth because new land can be assumed to be less productive than old land. Therefore, arable land area N can be expressed as

$$N = L^k \qquad 0 < k < 1 \tag{6.8}$$

Here, it must be noted that N expresses not only quantity but also quality, for example, the quality of soil. In this case, the former production function (6.1) can be transformed into

$$Y = (1 - c\beta) N L^\beta \tag{6.9}$$

Therefore, substituting (6.8) into (6.9) leads

$$Y = L^{\beta+k} - c\beta L^{\beta+k} \tag{6.10}$$

Then, we repeat the calculations we did earlier. The first step is to differentiate the new production function with respect to β, and making the result equal to zero yields

$$\frac{1}{c} - \frac{1}{\log L} = \beta \tag{6.11}$$

Since this result is the same as the former result (6.3) in the previous subsection 6.1.3, we can assume that $0 < \beta < 1$, and therefore, $c < \log L < \frac{c}{1-c}$ again. However, Y/L now becomes a little different:

$$\left(\frac{Y}{L}\right)^* = c(\log L)^{-1} L^{\frac{1}{c} - \frac{1}{\log L} + k - 1} \tag{6.12}$$

and then,

$$\frac{d(Y/L)^*}{dL} = c(\log L)^{-1} L^{\frac{1}{c} - \frac{1}{\log L} + k - 2} \{(1 + ck - c) - c(\log L)^{-1}\} \tag{6.13}$$

Therefore, depending on whether $(1 + ck - c) - c(\log L)^{-1}$ is negative, zero, or positive,

$$\text{if } \frac{c}{1 - c(1-k)} < \log L < \frac{c}{1-c}, \quad 0 < \frac{d(Y/L)^*}{dL} \tag{6.14}$$

$$\text{if } \log L \div \frac{c}{1 - c(1-k)}, \quad 0 \div \frac{d(Y/L)^*}{dL} \tag{6.15}$$

$$\text{if } c < \log L < \frac{c}{1 - c(1-k)}, \quad \frac{d(Y/L)^*}{dL} < 0 \tag{6.16}$$

This result shows that the effects of population growth on per capita production depend on the size of population. If the population is small, population growth leads to a decrease in per capita production. After the population size reaches a certain level, population growth leads to an increase per capita production with an increase in labor intensity (β^*). Therefore, we can identify the following three stages of an agricultural society:

Stage (1): very extensive primitive agriculture with an extremely low population density.
Stage (2): transition between extensive and intensive agriculture.
Stage (3): intensive agriculture.[15]

These correspond to the following three stages given by Marx:

Stage (1)': the clan or agricultural community where class and state are still in the process of formation.
Stage (2)': the slavery system.
Stage (3)': serfdom.

When classes and a state are formed, a society needs surplus products to pay for them. However, in the first stage this society could not produce sufficient surplus; therefore, no class and state could be formed. For example, even if such a community took slaves from other communities by means of war, such slaves could not produce any surplus because they would consume all the products that they produce in that stage, and therefore, useless slaves were killed or eaten. Thus, there could be no slavery system in this a low stage of development.

The difference between the second and third stages is based on the degree of agricultural intensity. While the slavery system requires direct command over slaves, serfdom needs only rules governing the land (via landlords) if the arable land is fruitful. This is because serfs do not want to excape from their landlords and they pay tax if the land allocated to them is sufficiently fruitful. Before these conditions could be fulfilled in the slavery period, slave owners needed to have absolute power to control their slaves. However, this control did not necessarily take the form of the large-scale management that is conventionally associated with Ancient Greek and Roman slavery or the slavery of black people in eighteenth and nineteenth-century America. In fact, the main agricultural product—grain—in Ancient Greek and Roman agriculture did not need large-scale slavery to produce it, and Nakamura (1977) suggests that the general form of slavery at that time was land-occupying slavery and patriarchal slavery. In the case of Japan, patriarchal slavery, which was listed in the family registers of Mino Province, which was an advanced region at that time, and was

---

[15] This differs slightly from Boserup's method of dividing into various stages. While Boserup provides five stages, we provide only three.

found in the Shosoin archives of 702, corresponds to this form of organization. The point is that in slavery, peasants were directly subjected to personal control.[16]

In summary, we have three criteria for identifying the slavery system and serfdom:

(1) degree of agricultural intensity,
(2) degree of (personal) freedom,
(3) land ownership system.

The third criterion was introduced by Nakamura (1977), who identified the special land ownership system in serfdom, where both lords and serfs had double land property rights: an upper property right and a lower property right. Nakamura (1977) identifies this double ownership system in specific slavery systems, but the pure double ownership system was established only in serfdom. In Chapter 4, we explained the process of two enclosure movement and the development of a class of almost-independent farmers in England. In reality, these farmers had become independent as a result of the existence of much stronger type of property right than that available to typical serfs in previous eras.

To understand slavery and serfdom in this way allows us to free our conceptions of slavery as being limited to the large-scale slavery of Greece and Rome, and our conceptions of feudal serfdom[17] as limited to Europe, Japan, and Tibet. Indeed, ancient Rome practiced large-scale slavery in orchards and mines, and feudal serfdom was established after the Great Migration of Peoples. This is true. But there was also state slavery,[18] which Marx called total slavery, and

---

[16] Although Tibetan serfs were not chained by landlords, Tibetan slaves were confined by leg-chains (leg irons) before liberation in 1959. This is a typical slave system. Also in ancient Rome, slavery was controlled by offering a price for the capture of escaped slaves, notices in market places, hiring professionals to capture runaways, obtaining support from government officials, and sometimes, by casting spells. As long as the slaves believed in the effect of spells, spells functioned as a major psychological barrier to attempting escape. As for the slaves who had once escaped, they were made to wear metal collars so it was impossible for them to try again. A more moderate way to avoid runaways was to make the slaves have children (see Toner 2014, Chapter 5). In the Edo era in Japan the Satsuma domain harbored all their ships in the Amami Islands so that other islanders could not escape. This can be understood as a kind of state slavery system.

[17] Europe in the Middle Ages, Japan, and Tibet (occasionally including the Indian Zamindari system) are referred to as feudal serfdoms, which comprise a system in which serfs were ruled over in private manors. Marx called this pure feudalism. The term feudalism expresses a form of personal dependency and is therefore used to characterize the apprenticeship in small manufacturing crafts before the industrial revolution. That is, apprenticeships involved personal dependency, but one that was completely different from human relationships under slavery. A detailed history of human relations would show the development from strong dependency through less freedom to pure freedom. This is the field of Marxian historical materialism known as the history of human dependency, which was first described in Marx's *Foundations of the Critique of Political Economy* (Marx, 1973).

[18] In my opinion, the Spartan Helot slave was a typical example which was subordinate to the state directly, like slaves under the Asiatic mode of production.

there was also state serfdom during the serfdom period. This is the same as the existence of state and private capitalism in capitalism. Since the intensification of agriculture proceeded not only in certain regions but throughout the world, the resulting shift from slavery to serfdom; that is, changes in personal freedom and land ownership, had to occur. This is a very important concept in generalizing the historical stages in Marxian historical materialism.

If this is so, when did serfdom system exist in China? In my opinion, the great leap from extensive to intensive agriculture was realized by the introduction of iron-made farm implements and ox-drawn plows. As plows and oxen are types of capital, this leap can be regarded as a change in the factors of production from land and labor to land, labor, and capital. In north China, where the soil is very hard, ox-drawn plows are needed to cultivate the land effectively. This system was invented in the Spring and Autumn Period in China and then popularized during the Han dynasty. Therefore, this great leap in productivity was expressed through the growth in China's population, which numbered sixty million in 4 AD at the end of the former Han dynasty. Before this time, China had a population of only ten million, and even if it fluctuated repeatedly due to wars and famines, the figure of sixty million was able to grow at most to only 100 million (Lee and Wang, 1999) by around 1600 AD when China introduced corn from the New Continent. This implies that the same stage of productivity continued throughout that period. Therefore, to determine whether population growth or productivity led to the other, we must acknowledge there was a big difference in the periods before and after the Han dynasty.

In the serfdom system, there was a system of double land ownership, but this duplication developed in two stages. In the first stage, the Han dynasty played the role of the senior owner of land belong to the state. This was a result of the state-led reorganization of land division project called the *qianmo* land system, which was introduced during the Warring States period (see Yoneda, 1968). Even after this land system was dismantled, the despotic Chinese government continued to hold onto this land, according to Nakamura (1993, 2013). However, from the Song to the Qing period, this system of duplicate land ownership developed from being state-led to a more private type in which both senior and lower ownership were private. In other words, the stronger class that had developed in the private sector had become in essence the senior owners under a new landownership system called "one piece of land, two owners' system" (see Terada, 1983). In this process, the status of direct producers started to rise, and during the Ming and Qing dynasties they became more independent, like the yeoman in England up to the seventeenth century. Fang (2000) describes this change as a kind of leveling of the farm size. and Niida (1962) describes it as the first emancipation of serfs. The Chinese serfdom era can be divided into state serfdom in the previous period and land-owning serfdom in the latter period.[19]

---

[19] Similar historical processes can be found in Slavic and West European regions, ranging from state to private feudalism at the *uklad* level. See Fukutomi (1972).

This is the same in the capitalist era, which also has two periods; state capitalism in the former and private capitalism in the latter.[20]

This division can be applied to Japanese serfdom, which was copied from China. Based on the rise in agricultural productivity led by the spread of iron farm implements after the sixth century, an ancient Japanese land reformation started by Taika Reformation in the seventh century initiated the allotment system of paddies for cultivation during the cultivator's lifetime and the permanent private ownership law of developed lands in the eighth century completed the nationalization of the landholding system both on nationalized and privately owned farmland (Kitamura, 2015). The former constituted state serfdom, but after several centuries it was gradually dismantled and became feudal serfdom ruled by the samurai class. This historical process was basically same as that in China.

This analysis characterizes the Yayoi and Tumulus period before the Taika Reformation as a slavery system period, while most Japanese studies of ancient history characterize as a chiefdom (Sho Ishimoda), an early state (Hiroshi Tsude) and a pre-state (Shozo Iwanaga). In our analysis we designate the oldest class society as a slavery system in so far as it consisted of various social classes, aside from the question of the maturity of the state, which is discussed later.[21] Most anthropologists agree that a chiefdom is a class society. (e.g., Diamond 1997).

Another important related question is whether North America underwent serfdom; because the New Continent did not have iron before Columbus. However, we know that the Aztec civilization was very sophisticated and created many big constructions based on its economic power; therefore, our question can be translated into the question of how that civilization enjoyed such high productivity without iron. The answer is that it was based on a special agricultural technology known as *chinampa* where farmers scooped up mud from the bottom of lakes using baskets and used the mud for intensive agriculture. That civilization was located on a big lake that was reclaimed after the invasion, and this technology did not need iron only but baskets. This was the secret why it achieved particularly high productivity without iron. Furthermore, what should be noted here is that such fruitful arable lands were developed by the state, and the state delivered them for the farmers. In this sense, this system can be defined as a kind of "state serfdom" which was the early stage of serfdom.[22]

---

[20] One of the origins of double ownership was the agglomeration of agricultural land. In practice, the government created farmland directly by using the *qianmo* land system to reorganize parcels of land and in this way the state system of one-land two-owners was established. On the other hand, the private system of one-land two-owners was partially developed by farmers' creation of private land (see Kusano, 1970). The latter succeeded as a form II type of differential rent in capitalist land ownership.
[21] Following Nakamura (1977), we regard the Asiatic mode of production as a state slavery stage.
[22] On Aztec and Teotihuacan civilizations, see Onishi (2005). Also, related to this issue, on the economic history of the North-east American Indian, see Onishi (2003b, 2003c, 2004).

6.1.5 Land Accumulation in Intensive Agriculture by Ox-Drawn Plows

Thus, intensive agriculture opened a new era that was completely different from extensive agricultural methods, and it was based on the improvement of soil productivity. We explain this improvement by introducing ox-drawn plows as a new factor of production. Before this, soil productivity was maintained with a hand plow, but afterwards, its productivity was typically maintained by plowing the land in early spring to replenish it after the previous year's cultivation. Therefore, the ox-drawn plow can be understood as an investment to compensate for the depletion of the productivity of the land, and thus we need two additional production functions: one is to maintain land quality and the other is to cultivate using the ox-drawn plow. This production system as a whole can be expressed as follows;

$$\left.\begin{array}{l} Y = AN^\alpha (s_2 L)^{\beta_2} \\ \dot{N} = B(s_1 L)^{\beta_1} K^\gamma - \delta_N N \\ \dot{K} = C(1 - s_1 - s_2)L - \delta_K K \end{array}\right\} \quad (6.17)$$

Here, the ox-drawn plow (including the ox itself) is expressed as K as a kind of capital that does not function directly in agricultural activities but is used to create and maintain land productivity. The labor productivity or the total factor productivity $[(1 - c\beta)$ in both cases] in the previous agricultural production functions (6.1) and (6.9) in subsection 6.1.3 and 6.1.4 is simplified as A, the effect on agricultural production by land input is set as the αth power, and a constant return to scale ($\alpha + \beta_2 = 1$) is assumed. This is a standard assumption for the agricultural production function. In this case, the target of land accumulation $(N/L)^*$ and ox-plow accumulation becomes

$$\left.\begin{array}{l} \left(\dfrac{N}{L}\right)^* = \dfrac{BC^\gamma \gamma^\gamma}{\delta_N}\{(\delta_K + \rho)\beta_1\}^{\beta_1} \left[\dfrac{\alpha\delta_N}{\{(\alpha\gamma + \alpha\beta_1 + \beta_2)\delta_K + (\alpha\beta_1 + \beta_2)\rho\}\delta_N + \beta_2(\delta_K + \rho)\rho}\right]^{\beta_1 + \gamma} L^{\beta_1 + \gamma - 1} \\[2em] \left(\dfrac{K}{L}\right)^* = \dfrac{C\alpha\gamma\delta_N}{\{(\alpha\gamma + \alpha\beta_1 + \beta_2)\delta_K + (\alpha\beta_1 + \beta_2)\rho\}\delta_N + \beta_2(\delta_N + \rho)\rho} \end{array}\right\}$$

(6.18)

according to Yoshii (2020). From this result, we see that the target of accumulation rises sharply after the introduction of the ox-drawn plow. As for $(K/L)^*$ we can see easily that it was zero before the introduction of the ox-drawn plow's system because γ was zero at that time, and therefore, it was lower than that after its introduction. However, as for $(N/L)^*$ first we need to calculate the optimal final ratio of labor allocation to land accumulation under the serfdom system. This result is as follows:

$$s_2^* = \frac{\beta_2(\delta_N+\rho)(\delta_K+\rho)}{\{(\alpha\gamma+\alpha\beta_1+\beta_2)\delta_K+(\alpha\beta_1+\beta_2)\rho\}\delta_N+\beta_2(\delta_K+\rho)\rho} \tag{6.19}$$

Because γ was zero before the introduction of ox-drawn plow, the optimal final ratio of labor allocation to land accumulation in the slavery system can be written as:

$$s_2^*|_{\gamma=0} = \frac{\beta_2(\delta_N+\rho)}{(\alpha\beta_1+\beta_2)\delta_N+\beta_2\rho} \tag{6.20}$$

On the other hand, we assume that total production after the introduction of the ox-drawn plow must be larger than it was before. This assumption of reasonable technological selection is realistic. In this case:

$$AN^{*\alpha}(s_2^*L)^{\beta_2} > AN^{*\alpha}|_{\gamma=0}(s_2^*|_{\gamma=0}L)^{\beta_2} \tag{6.21}$$

where N^* and $N^*|_{\gamma=0}$ are the target of land accumulation in the slavery and serfdom systems, respectively, and can be transformed into:

$$\left(\frac{N^*}{N^*|_{\gamma=0}}\right)^\alpha > \left(\frac{s_2^*|_{\gamma=0}}{s_2^*}\right)^{\beta_2} \tag{6.22}$$

Therefore, we can use the result of the above calculation of s_2^* and $s_2^*|_{\gamma=0}$, and this leads to:

$$\left(\frac{N^*}{N^*|_{\gamma=0}}\right)^\alpha > \left[1+\frac{\alpha\gamma\delta_K\delta_N}{\{(\alpha\beta_1+\beta_2)\delta_N+\beta_2\rho\}(\delta_K+\rho)}\right]^{\beta_2} > 1 \tag{6.23}$$

where we also use the condition $\gamma > 0$. This inequality shows $N^* > N^*|_{\gamma=0}$, and therefore $(N/L)^* > (N/L)^*|_{\gamma=0}$, showing that use of the ox-drawn plow changed agricultural society discontinuously, as shown in **Figure 6.2**.

6.1.6 From Symbolic Memorials to the Birth of Class and State

We now turn to the development of symbolic systems that developed in addition to language in the history of different forms of production. This probably began at the first stage, identified in subsection 6.1.4, when there was no class division, and its agriculture was very primitive. This is because in the first stage, there was a tendency for per capita production to fall in line with population growth, even if primitive agricultural productivity is different from that in hunting-gathering societies. Human required a certain amount of time before they formulated classes and states.

In this context, Terasawa (2000) provides important information on the formation of primitive states in Japan. According to him the formation of Japanese classes in the

MARXIAN ECONOMICS

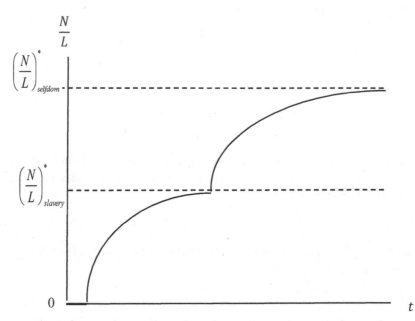

Figure 6.2 Changed target of accumulation by replacing non-intensive agriculture with intensive agriculture

northern Kyushu area occurred around 400 BC, after rice cultivation had become established. However, because the cultivation of primitive dry rice and other cereals can be traced back for thousands of years, we know that there must have been a long time lag between the origin of primitive agriculture and that of class divisions.

According to Terasawa, small communities divided by class started appearing in this area in the fourth century BC, followed by primitive states with primitive kings in the third century BC, small states with kings in the second century BC and finally, states clustered into several federations such as the Ito and the Na federations. Again, it took some time before societies developed from having a class structure to having a state.

In world history, we can say that the first crucial change in society was sedentarization (when nomads settled on the land), which subsequently became the basis of agriculture. Later, people built symbolic memorials like Stonehenge in England and Native American mounds in North America. These symbolic memorials are closely related to agriculture, because cultivation needed a much deeper understanding on the law of the movement of heavenly bodies that indicates changes in seasons. This was why all ancient agricultural societies had their own way of worshipping heavenly bodies. Stonehenge's central line faces the direction from which the sun rises on the summer solstice, and the direction in which it sets on the winter solstice.

Renfrew (2007) notes that these symbolic memorials played a crucial role in enlarging communities that consisted at most only of a small number of families and moved around frequently. People began staying in one place as the building of symbolic memorials required people to gather, build their memorials, and fix the special points that were to be worshipped. Simply as a result of people gathering together in one place changed societies substantially. Once they were living together they needed leaders to make rules for people to obey. Thereafter, human societies needed to take only one step forward to form social classes.

Furthermore, small communities that utilized laborers from other communities to build memorials must be understood as the leaders of other communities, and this was a typical type of formation of ruler-ruled relations among societies. While occasionally these communities might obtain the cooperation of other communities peacefully, sometimes this cooperation was coerced.[23] This is because during that period productivity was directly dependent on natural conditions; thus, there was a large gap in productivity among communities. Furthermore, some primitive states, for example, ancient Egypt, Mesopotamia, Meso-America, and Peru left pictures and reliefs that depict the humiliation of conquered people and slaves.[24] This suggests that the most important issue at that time was the balance of power among these communities. In this manner, class societies were generated not only by internal social stratification but also through the ruler-ruled relations among communities.[25]

In addition, when such coercion took the form of a war, the status of soldiers in the communities was elevated.[26] This is because wars against other communities would not be successful otherwise. Shamans who deified nature were also given elevated status because their successful interpretation of natural forces was decisive in the communities' productive activities. In many cases, shamans or soldiers subsequently became the ruling classes. However, bear in mind that both building symbolic memorials and exploiting other communities needed a certain level of productive power as a decisive precondition of surplus production. This condition was met only when agriculture had developed to at least the second stage, which I showed in subsection 6.1.4.

Therefore, it is evident that occasionally the most successful communities became rulers of other communities, and occasionally these leaders became rulers of the whole society that consisted of the subordinate communities. This suggests that the lands and members

---

[23] During the Yayoi period Japan witnessed numerous wars among communities, and occasionally the winners became the rulers of the losers. This is an example of coerced ruler-ruled relations among communities.
[24] See Renfrew (2007).
[25] Shiozawa and Kondo (1989) make much of the latter coerced relations among communities as the tributary mode of production. However, it is not clear whether this mode was universal.
[26] See Renfrew (2007).

of the ruled community became the possessions of the rulers, and this social mode is termed the "state slavery" system by Nakamura (1977). In such a system, all members and the entire land of the communities were generally controlled by certain ruling groups. They can be regarded as states because each community had just one ruler and each constituent community had just one ruling community. This ruling system was not decentralized.

Based on this analysis, we must remember that internal stratification in communities creates a small set of wealthy members. The concept of wealth requires the existence of the family[27] and private property. If a certain community member has accumulated wealth, we can assume that they owned the wealth and had the right to use or dispose it. Furthermore, if this accumulated wealth is inherited by their descendants after their death, the concept of the inheritance of property comes into being and the family becomes the unit of inheritance. In fact, succession to the throne is also a kind of inheritance. This is also based on the interlinking of the concept of inheritance with that of the family. According to Engels (1902), the concept of private property was created by the domestication and breeding of livestock. Furthermore, we can imagine other type of origin for the concept of private property. For example, skillful archers might believe that the game they hunt is their private property and farmers might believe that their harvest is their private property.

However, this inequality was not sufficiently marked to be called class division. Class division requires a special human relation, like that between a lender and a borrower, which can occasionally lead to bankruptcy. For example, when the production system became more roundabout and substantial funds were needed to make investments, producers tended to borrow money and occasionally defaulted on this debt, leading to debt servitude. Certain societies institutionalized donations to avoid this risk, but the lender-borrower relationship, in general, became a class relationship.[28]

Furthermore, solving such complex civil problems in the lender-borrower relationship and the inheritance system needed intervention from the court system and the chiefs of communities were given jurisdiction over such disputes. True rulers have the right to control property rights. That is, rulers' control of jurisdiction enabled them to form states, which rule over society. In combination, these features characterize the class state.[29]

In summary, the concept of class developed via the progression of certain steps in human history after the birth of agriculture. All these phenomena resulted from the invention of agriculture, particularly the part it played in the phase of increased labor productivity. Thus, the development of the slavery system and serfdom can be explained in the context of agricultural development.

---

[27] Here, the concept of family includes a joint family such as a compound family.

[28] This analysis is based on Renfrew (2007). Graeber (2011) claims that debt slaves are the crucial factor of class formation.

[29] See Nakamura (1977), 215.

6.2 SLAVERY AND SERFDOM IN CRAFT PRODUCTION, ANIMAL HUSBANDRY, AND FISHING

6.2.1 Craft Production Before and During the Feudal Period

The above discussed the long pre-capitalist period that is collectively referred to as the "agricultural era," but of course, industries other than agriculture also existed during this period as well. The hunting and forestry industries did not die out, and certain types of fisheries existed as extractive industries. In this sense, the term "agricultural society" or "agricultural period" indicates that agriculture had come to stand at the center of the society, for example, in the Han Chinese areas in China and Japan. In other regions, "pastoral societies" (one form of which was "nomadic societies") and "fishing societies" were established as pre-capitalist societies, all of which should be called "industrial societies" with the establishment of capitalism. This is because industry has replaced agriculture, animal husbandry, and fisheries as the center of society.

However, strictly speaking, manufacture was relatively primitive before the feudal stage of manufacture, where learning skills was not required so much. This is because there were no sophisticated tools in existence before the iron and bronze era that required special high skills in use. Only handmade stone tools were used in that period, and such primitive tools could not reflect the level of skill on the quality of products. In other words, only sophisticated tools could reflect skill, and therefore, only sophisticated tools required sophisticated skills. Because this requirement did not exist in the Stone Age, the feudal apprenticeship relationship was not required.

Let us imagine the difficulty of building houses using only stone tools. Without iron planes, it was impossible to shave off thin pieces of wood and make completely flat boards.[30] As a result, manufacture did not require special training processes led by masters; therefore, people made their necessity goods without assistance. They built their own houses using self-made bricks or knitted their own straw sandals, and this indicates that manufacture was part of everyday life and not an independent industry.

The existence of iron and bronze tools fundamentally changed the nature of craft production. Iron was used in farming to make ox-drawn plows. Although iron rusts easily, it is stronger and more suitable than bronze for use in sophisticated craft. Furthermore, iron implements could be used for ox-draw plows, whereas wooden implements could not. That is, iron intensified agricultural production, and made a similar contribution to craft

---

[30] Wooden plates could be made during the Stone Age but it was very time-consuming, and their surfaces were not completely flat because people could use only stone knives to shape them. Furthermore, completely flat surfaces provided a new possibility in different fields of production. For example, Japanese buckwheat noodles are made by cutting the dough on completely flat boards with a kitchen knife; thus, buckwheat noodles could not be made before the Iron Age.

Table 6.1 Historical development of nomadic societies

Circa 9000 BC (circa 21000 BC by another report)	Emergence of agriculture in the Fertile Crescent
7600 BC	Domestication of sheep and goat in the Fertile Crescent
7000–6500 BC	Hunter-gatherers supplement livestock rearing in the grasslands of the Fertile Crescent. Later, the livestock was enclosed with a fence for protection during the winter or at night only. Grazing commenced
6000–5000 BC	A settled, complex economy including agriculture, animal husbandry, and hunting on the Eurasian grasslands
5500 BC	Prevailing nomadic culture in western Asia resulting from climate warming, and the complete domestication of animals (prepared for by large-scale settled community and agriculture)
350 BC	Invention of the wheel in Mesopotamia, which spread to the Eurasian grasslands 200–300 years later
3000–2200 BC	A settled, complex economy including agriculture and animal husbandry in the present Ukraine area
circa 2000 BC	Emergence of primitive horse-riding in Mesopotamia
2600–2000 or 1900 BC	A settled, complex economy of agriculture and animal husbandry in the eastern Urals
2000 or 1900 BC	Bronze culture in the Eurasian grasslands
Latter half of 14th century BC	Emergence of horse-riding in Egypt
13–12th century BC	Emergence of horse-riding in the Aegean Sea area
1300 century BC	Late bronze culture in the east Eurasian grasslands (the Karasuk culture). Bridles invented to control horses
10–9th century BC	Prevalence of horse-riding in the Eurasian grasslands (the late Karasuk culture). Emergence of a nomadic horse-riding culture and its prevalence in north China
10–9th century BC	Emergence of royal authority in the Mongolian plateau (Uushgiin Ovor ruins in Mongolia. No grave goods but deer stones with curved moon and sun)
9th–mid-7th century BC	From the Cimmerian (pre-Scythian) age to the early Scythian age. (In the latter half of this period, various nomad groups moved west)
Latter half of 9th century–first half of 8th century BC	Emergence of royal authority in southern Siberia (north of Mongolian plateau) (bronze harnesses, armor and accessories were excavated in Arjan burial mounds)
7–6th century BC	Early Scythian age
BC 5–4 (or early 3rd century)	Classical Scythian age (in some cases, divided into a middle period and a late period) From the Cimmerians (pre-Scythian) to this period, burial mounds continued to be built
BC 4th century	Emergence of the Hsiung-Nu Huns (Modu Chanyu's reign from 209 to 174 BC)

production. While iron transformed agriculture from semi-primitive farming to intensive farming, it also transformed craft into the feudal type of production. This is how agriculture developed into serfdom societies and craft production developed into feudal ones.

6.2.2 The Pastoral Revolution and Serfdom in Nomadic Tribes

The development of iron and metal objects also led to a decisive transformation of the animal husbandry in the sense of mode of production. Here we summarize the development of a nomadic society with a developed pastoralism system from the viewpoint of historical materialism (Hayashi, 2007, 2009; Yukijima, 2008).

This chronology shows: (1) Periods of nomadic cultures, which differed by region, started in Egypt, moving from west Asia to the present Ukraine area and afterwards became established in the central Asian grasslands. (2) Settled agriculture preceded animal husbandry in all these areas. (3) Animal husbandry emerged as a type consisting of a half-farming half pastoral economy. (4) To develop into nomadism, the invention of wheels and bronze harnesses were necessary. (5) The nomadic economy was accompanied by royal leaders.

The fourth point also includes the importance of the iron in agriculture, because horse-riding was decisive for the development of a nomad culture and metal bridles were decisive tools for horse-riding. A bridle is the headgear inserted between the front and back teeth of the horse, and is used to control a horse and to move it to the right or to the left. In addition, a saddle was necessary for riding a horse whose back vertebrae protrudes, but the bridle was the most important piece of equipment. Equipping horses with adequate reins was crucial for expanding the livestock's grazing range and increasing the grass available to them. All this equipment crucially raised the nomads' productivity. The speed with which horses can move in relation to cattle gave the nomadic people a decisive new ability not only to respond to sudden movements of livestock but also to enter into combat as a cavalry. Bronze inventions also developed weaponry, expanding the grasslands that each nomad tribe could occupy, and increased the number of livestock. This is why nomadic people often went to war. Tribes and nations with strong armies could successfully occupy and rule very large areas, such as Scythia in the north of Persian empire and later the Hsiung-Nu Huns in the north of China. Furthermore, the development of this combat capability and long-distance mobility enlarged their ability to acquire agricultural products from other peoples by trade and through warfare, which allowed them to specialize in animal husbandry. In conclusion, the establishment of a nomadic economy as a result of the invention of the bronze harness led to a big leap of productivity. Since the harness was small, compared to agricultural tools such as the hoe and plow, bronze implements met the needs of these societies without needing the invention of iron objects. Therefore, the crucial means of production in animal husbandry was the horse, just as the ox was in agriculture.

We now examine whether the mode of production of Scythia and the Hsiung-Nu Huns corresponds to slavery or serfdom in agriculture. Traditionally, Marxist historians, such as Ito (1995) and Gao (1980), determined that the advanced nomadic system in the Mongolian empire was a feudal system and that those of Scythia and the Hsiung-Nu were ancient modes of production, or slavery. However, development from a primitive communism economy that depends solely on hunting and gathering to the half-farming, half-pastoral society resulted from a great leap in productivity, unlike the leap from a half-farming and half-pastoral economy to a nomadic one. The former leap, which took place around BC 7600, represented progress to the first type of class society; namely, slavery. It is true that agriculture at that time was not yet intensive and the number of pastoral livestock societies was also limited to the use of grasslands around their settlements. However, these societies had developed a roundabout production system by reclaiming land for agriculture and saving a portion of their livestock as the means of production, instead of eating it. Thus, we characterize the thousands years from 7600 BC to the tenth or the nineth century BC as a period of slavery. While the previous section explained that agriculture gave way to slavery, we now claim that the half-farming and half-pastoral economy gave rise to the first type of class society in the grasslands.

However, since the existence of pure slaves at this time has not been recorded, we must again clarify why it should be called slavery. This is because slavery in this book is defined in contrast with the personal freedom of the serfs under serfdom systems that are based on intensive agriculture. Agricultural land at the stage of serfdom started to show high levels of productivity as a result of the use of iron tools, and farmers who were given lower ownership rights did not want to escape from heavy exploitation. This is because high productivity was secured from this land, so in the serfdom system there was no need for direct violent control over the serfs by the exploiting class. Even in the case of the nomadic economy discussed here, individual nomads who belonged to a group with combat capabilities owned their livestock privately, but they were permitted to use the land as the main means of production only under the dominant control of the lords.[31] Thus we can see that, as in serfdom in agriculture, rights to the ownership of the means of production were multilayered.

It is only after the emergence of the nomadic economy that individual pastoralists started to acquire the same personal freedom as serfs. In the half-farming half-pastoral grasslands, since they had been settled by small, separate groups that were little more than clans, and surplus products were taken by clan rulers or other the clans that dominated the clan in question. This is why these relations of production can be understood as a slavery system in the sense that a certain amount of compulsion was present at this time.

In addition, there were slaves in the Mongolian plateau in the Hsiung-Nu Hun era called *bogol*, who were prisoners of war from China and other tribes and who were owned

---

[31] See Wuritaoketaohu (2006), 100–101.

by individual pastoralists and engaged in handicrafts and agriculture.[32] However, as these were only peripheral industries in this society and small in terms of number, their status can be compared with the handicraft slaves that existed under Tibetan serfdom. If they could make a large contribution in time of war, they could raise their social position and in some cases, become aristocrats with a family and possessing a small amount of private property. This status is understood as serfs by modern Chinese historians such as Wuritaoketaohu (2006).

We have shown here the basic difference between slavery and serfdom in the pastoral grassland, but this type of serfdom can be divided into the former age, such as in Scythia and among the Hsiung-Nut Huns, and the latter age, such as the Mongolian empire. In the former period, when clans had a big presence, an influential clan became the aristocratic one, and some of their military leaders, such as Modu Chanyu, became kings. Therefore, here the administrative organization became a military organization,[33] and the fact that the pastoralists belonged directly to the clan means that they belonged to the administration and military organization of the state and were also warriors. This is almost same as in the state system during the Chinese Warring States period. In this sense, we can understand this nomadic society as a form of "state serfdom."

However, after this period, the Mongolian empire developed a professional army, and established the strong political administration pyramidally organized down from Hahn, the supreme leader to the feudal lords and their retainers and vassals. This is why Ito (1996), Chinese historian Wuritaoketaohu (2006) and Gao (1980) describe this era as a feudal lord system or feudalism. Since individual pastoralists were directly affiliated to the overall ruler who dominated the land as the landlord, this was similar to the feudal system in Europe, Japan, and Tibet. It is a type of private serfdom different from state serfdom. Therefore, we claim that there were also two stages of pastoral serfdom: state serfdom and land occupation serfdom (private serfdom), identical to the serfdom system in agriculture.[34]

6.2.3 The Development of Modes of Production in the Fisheries Industry

This book emphasizes the technological changes brought about by the industrial revolution in the manufacturing sector, while mentioning other industries such as commerce, construction, hospitals, schools, and finally, agriculture and pastoralism. To complete this

---

[32] See Sawada (1996), 144–45.

[33] Ibid., 127–31.

[34] However, since the Qing dynasty, the range of nomads gradually narrowed due to population increases caused by the improvement of their security and productivity, so that the nomad system has changed finally to transhumance, or a system of grazing in which pastoralists use determined areas seasonally. This also corresponds to transformation of systems by the population pressure, discussed earlier.

section, I now extend the discussion to another important industry, fishing. This is because it is particularly well developed in Japan and because in this industry the development of the means of production directly determines the mode of production.

However, it is not possible to say in general that the fishery industry as a whole had an "α" mode of production from the xth to the yth century. This is because the means of production in the fishery industry varies depending on the species of fish and there are also large regional differences in this industry due to ocean currents and topography.

I start by describing the most primitive form of fishing for shellfish on the beach and note that shell middens were not formed until after people moved into settlements, and so they were not characteristic of hunter-gather societies. The first form of settlement was brought about by fixed-net fishing (Nishida, 1986). Nishida focuses on the Jomon period in Japan (roughly corresponding to the Mesolithic period in world history), which was settled before the fully fledged agricultural age started, and notes that fishing at the time was based on the newly invented, non-transportable fixed net (and dugout canoe). In addition, the invention of the *yana*, which was a fixed net made from bamboo and was more effective than fixed-net fishing on a large scale, seems to have had the same effect on sedantarization. And, of course, settling down as a precondition for agriculture contributed greatly to the development of these societies.

However, at this stage, the fishers' main occupation depended on the season, and they fished for shellfish in the spring and hunted in the summer. During the Jomon period societies produced goods by hunting, fishing, and gathering (as well as using early agriculture from the early period onward), while the Yayoi period witnessed a complex society of farming and fishing.[35] By the late Yayoi period the invention and improvement of the octopus pot and other earthenware technologies led to the specialization of octopus fishing and salt harvesting.[36] However, specialization in terms of family fisheries engaged in coastal fishing with the use of small-scale fishing nets (small fisheries) had to wait for the general use of constructed or semi-constructed wooden boats, that were developed by the use of ironmongery (i.e., by bolting together plates without leaks), and the accompanying increase in the size of the nets. This also raised the issue of the ownership of the means of production, and the need for an apprentice system of skill formation to pass on the complex skills of ocean-going craft from generation to generation. In other words, this is like the history of the feudal type of craft production system discussed above. In both cases, the use of iron played a decisive role at the start.

---

[35] Yasuda (2009) calls this a rice-farming and fishing civilization derived from the Yangtze River and contrasts it with the field-cultivating and pastoral civilization derived from northern China. It is important that both types of early agricultural civilization were accompanied by their own special sectors which acquired animal protein.

[36] For more information on this point, see Hamasaki (2012), Chapter 2.

Furthermore, during this period, competition for fishing grounds led to the development of the right to use them, like the right to possess land in agriculture. The general use of constructed wooden boats and the increase in the size of nets by using strong linen led to competition between types of fishing methods,[37] and this resulted in the superior ownership of the fishing grounds going to the lord and the inferior ownership going to the fishers or their community.[38]

Under these conditions, there was a differentiation between small-scale family fisheries and organized fisheries (group fisheries) in the Edo period. The former include single-line fishing for skipjack tuna, which expanded beyond inland waters and coastal areas to offshore areas, and the latter include trawling, purse seining, and whaling. In the latter case, the domination of the industry by influential people with financial power (single or multiple) allowed it to grow even larger. For example, in the case of the large sardine purse seine fishing in Kujukuri beach, each influential fisherperson (*amimoto*) owned not only a net but also two boats, and even set up a surveillance cabin to oversee the entire fishing ground and direct the fishing. This cabin was also found in net fishing industries in the Uchiura Bay of Nishi-Izu, where individuals (*tsumoto*) also owned fishing rights in the Meiji period. The net fishers in the sardine trawling fishery in Kujukuri beach borrowed huge amounts of money from landowners to carry out their business,[39] indicating that the former were beginning to function as an entrepreneurial or capitalist class.[40] In this case, in addition to the fishers called *okabataraki* who made and repaired the nets, the *amimoto* hired forty to fifty fishers called *daisengata*, and 200 to 300 net pullers called *okamono*, paying money to the former in advance and to the latter in kind with sardines after the catch.

These changes in the relation of production were reflected in changes in the inferior ownership of fishing grounds permitted by feudal lords. Originally it was the fishers who were the inferior owners, but the feudal lords gradually started to sell the right to inferior ownership to influential people. This is how the *amimoto* system was established, in which the users of the fishing grounds, which were originally the fisher's commons, were allowed to choose their own representative (*senushi*), who would then take over the fishing grounds in exchange for tribute and money.[41] As a result of this system, the *amimoto* were subject to the feudal patronage of their lords, who imposed non-economic forms of compulsion on the

---

[37] Ibid., 335.

[38] See Kawaoka (1987), 332–3 and Taguchi (2018), 201.

[39] In small-scale net fishing, where the cost was extremely low, a few fishers who worked together took turns to be in charge of the work, preventing the formation of domination and subordination like the Amimoto system.

[40] See chapters 6 and 7 of Yamaguchi (2007) for the circumstances of Kujukuri beach ground towing net fishing and Nishiizu Uchiura's open-cut net fishing.

[41] See Takahashi (1995), Chapter 1, Section 4 and Chapter 4.

general fishers (for example, by prohibiting them from operating freely and preventing them from adopting new technologies). The modern ownership system was not established before the Meiji period, when the contracting system turned into system where you had to bid for a fishing ground and the grounds themselves became privately owned.[42] This corresponds to the establishment of the landowner-tenant system in agriculture via the land tax amendment.[43]

Besides those highly developed large-scale organized fisheries, the majority were small family-run fisheries. However, since capitalist relations of production begin with large-scale fishing, if we want to follow the changes, we need to mention whaling. In Japan, whaling with harpoons started in the Jomon Period and developed later in recent times to become large-scale fishing, which used several boats and employed more than 200 people, including those on land, using harpoons and nets to catch dolphins, toothed whales, bowhead whales, and right whales. Here, the division of labor was similar to that of large-scale craft production into tasks such as rapidly observing the whales' movements, throwing the harpoon with accuracy, enclosing the whales with nets, and finally performing various processes on land.[44] This stage of productivity corresponds to the wholesaler-based domestic industry system in the manufacturing sector in that (1) each fisher used their own boat and harpoon, and (2) unlike in purse seine fishing, the risk of not catching the whale was so great that only the wealthy wholesalers who sold the whale could provide the funds.[45] Each fisher was contracted by the wholesaler to participate in the cooperative work. In other words, this mode of production corresponds to the early stage of capitalism, just before, but not quite at, the machine-based industrial stage.

In this way, the fully developed type of capitalistic fishing industry necessitated new technologies such as Norwegian whaling technology in the whaling industry, and power vessels, refrigeration equipment, and fish finders in other type of the modern and large-scale organized fishing industry. A typical example is the modern pelagic offshore fishery with large and medium-sized purse seine fisheries for tuna longline, mackerel, and horse mackerel. Among these new technologies, the invention and spread of automatic net hoisting devices and fish finders expanded the operating space for small boats owned by individual fishers, and fishing industry is still divided into a large capitalistic fishing

---

[42] See Kamiya (1967), Section 3.

[43] This was a tax reform in 1873 in Japan which clarified who owned the land and established a system to collect taxes from the owners.

[44] This assessment is based on Kawaoka (1987), 251. The harpoon fishery described here later changed to net fishing, but it was still a large-scale craft production type of division of labor.

[45] Similar characteristics are seen in the sardine seine fishing in Kujukuri beach. The merchants who bought the sardines were responsible for processing them into fish fertilizer (dried sardines), but they also provided the net owners with funds and acted as guarantor, borrowing money from the landowners on behalf of the net owners. Therefore, their role was more than mere providers of funds. See Yamaguchi (2007), Chapter 7, Section 2.

industry and small family-run fisheries. However, at least in the case of Japan, fishing cooperatives, which organize small fishermen in fishing villages, function in practice as commercial capitalists to provide fuel for fishing boats and other necessities to fishermen and buy fish from them.

Finally, it should be added that a different kind of capitalist fishery is now growing, constrained by the overfishing that accompanied the rapid increase in productive capacity and the restriction of fishing grounds overseas by setting exclusive economic zones. These are cultivated fisheries and aquaculture fisheries, some of which are based on employed labor. In fact, aquaculture fisheries now account for the majority of Japan's fishery production tonnage, so it is no exaggeration to say that the way they are managed will define the future of fisheries. In any case, the development of the means of production has led to the same type of development in fisheries as in agriculture and manufacturing.

6.3 THE PART PLAYED BY HUNTING IN THE TRANSITION FROM APE TO MAN

6.3.1 Human History After *Sahelanthropus*

We now address the birth of primitive communist society well before the institution of slavery. This section addresses the boundary between humans and apes and the evolutionary process leading from anthropoids to *Homo sapiens* via *Sahelanthropus* or *Homo erectus*. This necessitates grasping the nature of human beings. In the nineteenth century, Engels studied the novel academic findings in this field and authored a chapter titled "The Part Played by Labor in the Transition from Ape to Man" in his book *Dialectics of Nature*. He notes in this study that labor and the development of labor caused the development of highly sophisticated organs of the human body, including the brain.

In the nineteenth century, knowledge of paleoanthropology was very limited. For example, scientists in the field had just discovered organic evolution and the Neanderthals, and they believed that the development of the brain preceded bipedal locomotion. However, the reverse was true. Bipedal locomotion and free hands helped the brain develop, a causal relationship that was first proposed by Engels. The brain capacity of the first human being, *Sahelanthropus*, was only 320–380 cc, approximately the same as that of chimpanzee. The brain capacity of *Homo erectus* was also small, at 871 cc, which is just under 60 percent of the brain capacity of *Homo sapiens* (1500 cc).

I claim that the beginning of meat-eating corresponds to the beginning of hunting (and fishing), as well as tool-making. This is because I place importance on the means of production and the profound imaginative ability deepened by its development, as noted in subsection 1.1.4 of this book.

The importance of meat-eating and the means of production has been proposed many times by modern paleoanthropologists after Engels. The first prominent supporter of this idea after Engels was Raymond Dart, who discovered *Sahelanthropus* in Africa in 1924, and argued for the killer ape hypothesis or the hunting hypothesis (Dart, 1953). Later, Ardrey (1961, 1976) supports this idea with the argument that humans are to be seen as killers and that males are superior. However, public opinion opposed Ardrey's gender bias and his hypothesis needed to be revised. Stanford (1999) insists on the decisive role played by hunting in the formation of human beings, and this has become the accepted theory in modern U.S. paleoanthropology.

We accept this hypothesis because we emphasize means of production in human history. When humans started hunting, they may have made and used tools because they lost the claws and fangs with which other animals catch preys. Therefore, tools were invented at that time, and humans generated the first industry using this invention, that is, hunting industry. Just as the industrial revolution gave birth to capitalism and the agricultural revolution gave birth to slave societies, the "hunting revolution" formed human beings itself. Since our early beginnings, humans have been continually creating industries.

However, among the approximately 200 species of primates besides *Homo sapiens*, I could show only three exceptions—the baboon, the Brazilian longtail monkey, and the chimpanzee—that partially hunt,[46] and these are, strictly speaking, our relatives. We do not know whether these apes began to hunt before or after human did; however, we do know that hunting by chimpanzees is not based on their DNA but on their culture, which is passed on from one generation to another.[47] This may be why the Brazilian longtail monkey and chimpanzee have the highest encephalization quotient of body weight and only these primates and humans have culture and industry. In other words, industry implies culture, even if the hunting by these primates is very limited.

Therefore, based on Stanford (1999), the similarities and differences between chimpanzee and the earliest human beings are crucial and should be listed here. They are similar in these ways:

(1) Their dependency on meat for food is very limited, and they hunt or scavenge[48] in a haphazard way. Only in the case of food shortages do they eat flesh.

---

[46] Because bonobos and orangutan seldom eat meat, we omit them here.

[47] See Stanford (1999).

[48] An argument against this hypothesis is the scavenger hypothesis, which accepts the importance of meat-eating but denies that the earliest human beings hunted. However, Stanford (1999) criticizes the dichotomy between hunting and scavenging. According to Henry Van, lions may scavenge, and hyenas may also hunt. That is, hunting does not exist without scavenging and vice versa. Another criticism of the scavenger hypothesis that of Boyd and Silk (2009), who say that chimpanzees do not eat flesh from a dead animal they have come across by accident, and that the earliest humans were able to hunt. Thus, modern paleoanthropologists place more importance on hunting than previously.

(2) They hunt only within their territories.
(3) They hunt in groups.
(4) Males share the game, females, and a few exceptional individuals of their community to secure their own interests or for propagation.
(5) They preferred eating marrow and brains to other parts of the kill.
(6) Their prey is usually small, under 40 kilograms in weight.

By contrast, there are two differences between them:

(1) Humans used stones to cut up the bodies of their prey 2.5 million years ago.
(2) These stone tools gave humans an ability to hunt much larger game, and meat thus became significantly more important to humans.

However, human history did not start 2.5 million years ago, and *Sahelanthropus* had a much older history (i.e., seven million years ago). The earliest human beings used wooden and bone tools, not stone tools to hunt. This is the same among chimpanzees, which are also capable of making primitive tools.

Using tools for hunting allowed humans to be clearly differentiated from all other animals,[49] and the historical development of these tools determined the development of their methods and productivity. The development of cooking tools was also crucial because it widened the variety of food that could be eaten, for example, roasted nuts or smoked fish. Later, earthenware was invented, and its use led to boiling and cooking.

The development from *Sahelanthropus* through *Homo erectus* to *Homo sapiens* can be identified in the development of human society. Human were able to become true hunters only after *Homo erectus* had entered on the scene in that they were able to catch large adult beasts, and only *Homo sapiens* could hunt large animals by ambushing them (Stanford 1991). Thus, we can describe the historical development of primitive society from *Sahelanthropus* to modern society. Donald (1991) identifies four stages in the development of human cultural and cognitive faculties, as follows:

(1) Primitive cognition: episodic culture. Humans had the cognitive level of anthropoids, which merely reflects external stimuli.
(2) Human cognition without language: mimetic culture. Cognition without language, from *Sahelanthropus* to the appearance of *Homo erectus*. However, unlike apes, these early human beings could mimic others' motions and understand differences in tones

---

[49] Nishida (1986), Chapter 9, notes that it was our ancestors' stones and clubs that made their weapons superior to the fangs and claws of the macaque monkeys who competed with them as medium-sized primates several million years ago. He argues that such weapons were useful in stopping them from invading the territories preferred by our ancestors. This was not hunting in the narrow sense, but was essentially similar to it.

of voice, facial expressions, eye movements, manual signs, and gestures. Because of this ability, they could make and use tools.

(3) Language and the rise of human culture: mythic culture. Cognitive level of *Homo sapiens*, which transmitted hunting and gathering skills over the generations by using language.

(4) External symbolic storage: theoretic culture. Cognitive level for building symbolic memorials or visuographic records to store meanings externally. This corresponds to all Upper Paleolithic, Mesolithic, and Neolithic societies.

According to Stanford (1999), (2) corresponds to the differentiation of the earliest human beings from chimpanzees, and (3) corresponds to the stage of ambushing large animals. Stage (4) corresponds to the period that came after hunter-gatherer societies. The term human beings should thus include *Sahelanthropus* and *Homo erectus*, and their long history preceded the era of *Homo sapiens*. The history after stage (2) was approximately seven million years.

During these seven million years, humans have had countless experiences, and started by changing their forms of our productive organization and then, our social relationships,[50] such as cooperative hunting, the equal distribution of game, and war among tribes. In fact, the earliest human beings could hunt small animals only through cooperation because of their low productivity, but *Homo erectus* and *Homo sapiens* could hunt large animals such as buffalo, reindeer, red deer, horses, or ibex. This type of large-scale cooperative hunting was conducted by the Blackfoot people by the 19th century in the Great Plains. Blackfoot people skillfully drove large herds of buffalo over the cliffs.

When early *Homo sapiens* hunted mammoths, they also did so cooperatively. Although one man could not hunt a mammoth alone, a team formed to spear the beast collectively could hunt one down for hours at a time. If a member of the team was killed, their family would have been cared for by the other members. Most crucially, the game was probably distributed equally among the hunting party, because it was attained collectively. Therefore, people in such societies were considered equal.

Also in Chapter one in this book, I had mentioned my experience in the Iban village, which was also an equal society. Approximately thirty families were living in a single longhouse in a village, and the chief lived at the center of the house. However, the size of his room was same as that of the other members, and I paid cookies and not money for their

---

[50] As mentioned in note 1 in Chapter 1, strictly speaking, human beings not only produce and consume goods but also reproduce by forming special social relationships. This is seen through anthropoid behavior, such as the male domination of females or enclosure, male-on-male battles, and female control of these male battles. Without these activities, families cannot be formed and private property cannot be protected. These problems are too substantial and lengthy to be discussed here but I hope to address them in the future.

welcome dance, because money was not so important for them. Therefore, they had only a little money, and cookies that I gave them were distributed equally for each family. It was really an equal society.

Human beings have been organized in social classes for ten thousand years at most, which is under one percent of human history. Before class societies, humans were poor, hungry, had no concept of human rights, and lived short lives. However, they also had no social class, that is, they were equal. Human beings have progressed since the advent of social classes, but they have been retained the communist ideal that all should be equal and wealthy, and all should live long lives with sufficient human rights. However, it remained in question whether such a society is possible or not. The aim of Marxism is to answer this question scientifically.

Addendum I
The Fundamental Marxian Theorem with Joint Production and Fixed Capital

The Fundamental Marxian Theorem (FMT) greatly shocked mainstream economics. Here we analyze two of the many reactions it received. The first was Ian Steedman's criticism of Okishio's proof, providing a counterexample in which the above proof cannot hold under joint production; that is, when the same labor produces plural products jointly (Steedman, 1975). However, if we formulate joint production case based on our two-sector model that is strictly divided into two sectors: sector of the means of production and sector of the means of consumption. This can be shown as follows.

First, we rewrite price inequalities as

$$\left.\begin{array}{l} p_1 + b_1 p_2 > a_1 p_1 + \tau_1 w \\ p_2 + b_2 p_1 > a_2 p_1 + \tau_2 w \end{array}\right\} \tag{A.1}$$

where the first sector jointly produces b_1 units of consumption goods per unit of production of the means of production, and the second sector also jointly produces b_2 units of the means of production per unit of the production of consumption goods. These price inequalities can be transformed into

$$1 - R\left\{\frac{(a_2 - b_2)\tau_1 + (1 - a_1)\tau_2}{(1 - a_1) + b_1(a_2 - b_2)}\right\} > 0 \tag{A.2}$$

Here, we assume $1 - a_1 > 0$, $a_2 - b_2 > 0$ because net production must be positive in the first sector and the net input must be positive in the second sector by its nature. On the other hand, the rewritten value equations

$$\left.\begin{array}{l} t_1 + b_1 t_2 = a_1 t_1 + \tau_1 \\ t_2 + b_2 t_1 = a_2 t_1 + \tau_2 \end{array}\right\} \tag{A.3}$$

can be transformed into

$$\left.\begin{array}{l} t_1 = \dfrac{\tau_1 - b_1 \tau_2}{1 - a_1 + b_1(a_2 - b_2)} \\[2mm] t_2 = \dfrac{(a_2 - b_2)\tau_1 - (1 - a_1)\tau_2}{1 - a_1 + b_1(a_2 - b_2)} \end{array}\right\} \tag{A.4}$$

Thus, the above inequalities can be rewritten as

$$1 - Rt_2 > 0 \tag{A.5}$$

which is identical to Okishio's FMT. That is, the joint production has no problem for FMT. Okishio (1977), Chapter 3, Section 5 also discussed this point by providing a countercriticism against Steedman.

However, Steedman's counterexample is a little different from the above formula (A.1) and (A.3), because his counterexample formulated the following inequalities:

$$\left.\begin{array}{l} p_1 + b_1 p_2 > a_{11} p_1 + a_{12} p_2 + \tau_1 w \\ p_2 + b_2 p_1 > a_{21} p_1 + a_{22} p_2 + \tau_2 w \end{array}\right\} \tag{A.6}$$

and value equations

$$\left.\begin{array}{l} t_1 + b_1 t_2 = a_{11} t_1 + a_{12} t_2 + \tau_1 \\ t_2 + b_2 t_1 = a_{21} t_1 + a_{22} t_2 + \tau_2 \end{array}\right\} \tag{A.7}$$

The point of this assumption is that both types of goods can be used as the means of production. For example, pencils can be used both at work and at home. Input-output tables show this structure, and this is the reality. However, even if both pencils at work and at home can be used in the same input structure, we can express this structure by dividing it into two inequalities and two equations of the pencil that is actually used for consumption and the pencil that is used as the means of production. Theoretical models are formed in this abstract way. Our modelling in which the means of production and the means of consumption are clearly identified is reasonable.[1]

The second criticism of the FMT was provided by Morishima (1973), Chapter 14, saying that this theorem will fail under the condition of fixed capital. However, Okishio (1977), Chapter 3, Section 3 solves this problem by introducing the method for treating fixed capital used by John von Neuman, Piero Sraffa, and Michio Morishima himself. Their formulation is an interpretation of the fact that fixed capital that has been used can be used again in the following year, which we understand as the fact that this year's productive activity produces one-year-grown fixed capital as a by-product. This can also be envisaged as a problem of joint production, and express the conditions for the

---

[1] If we stick to the view that the means of consumption are also input in the production process, we need very complicated assumptions to maintain FMT. For example, Roemer (1980), Matsuo (1994), and Yoshihara (2008) aimed to refute Petri's (1980) counter-examples to the FMT. However, again, all these arguments are based on the strange assumption that the means of consumption are also input in the production process.

existence of profit, as in the subsection 3.1.2. That is, the condition for the existence of profit for

$$\left.\begin{array}{l} \text{the first sector in period 0: } p_1^1 + \sigma_1 p_1^0 > p_1^0 + Rp_2\tau_1 \\ \text{the first sector in period 1: } p_1^2 + \sigma_1 p_1^0 > p_1^1 + Rp_2\tau_1 \\ \cdots \\ \text{the first sector in period } n-1: \sigma_1 p_1^0 > p_1^{n-1} + Rp_2\tau_1 \\ \text{the second sector in period 0: } p_1^1 + \sigma_2 p_2^0 > p_1^0 + Rp_2\tau_2 \\ \text{the second sector in period 1: } p_1^2 + \sigma_2 p_2^0 > p_1^1 + Rp_2\tau_2 \\ \cdots \\ \text{the second sector in period } n-1: \sigma_2 p_2^0 > p_1^{n-1} + Rp_2\tau_2 \end{array}\right\} \quad (A.8)$$

Here, one unit of fixed capital is used for all durability periods (n) from zero to $n-1$, but it ages and its price falls every year. The aging is shown by the subscripts in the top right corner of p_1 (the price of the means of production). However, we assume that p_2 (the price of the means of consumption), R (real wage rate), and τ_1 and τ_2 (both sectors' direct labor input) are constant. Furthermore, both sectors are assumed to produce σ_1 and σ_2 units of both goods, respectively, by one unit of the means of production in every period. Therefore, for example, the right side of the first inequality shows that one unit of a new means of production and τ_1 units of labor power are inputs; and the left side shows that σ_1 units of a new means of production and one unit of one-year-old means of production are produced. This inequality between both sides means that there can be a certain profit. p_1 disappears in the nth period because this fixed capital is used up completely in this period. The consumption goods sector has the same structure.

Now, let us solve these inequalities. First, we sum up both sides of the first n inequalities and the second n inequalities in (A.8), and have

$$\left.\begin{array}{l} np_1^0 \sigma_1 > p_1^0 + nRp_2\tau_1 \\ np_2^0 \sigma_2 > p_1^0 + nRp_2\tau_2 \end{array}\right\} \quad (A.9)$$

and as in the above cases, we can combine these two inequalities into

$$1 - \left\{\frac{(n\sigma_1 - 1)\tau_2 + \tau_1}{\sigma_2(n\sigma_1 - 1)}\right\} R > 0^2 \quad (A.10)$$

---

[2] Here, we apply the condition $n\sigma_1 - 1 > 0$, which will be proved shortly.

Therefore, our problem can become a question of whether $\dfrac{(n\sigma_1 - 1)\tau_2 + \tau_1}{\sigma_2(n\sigma_1 - 1)}$ is t_2 or not. In order to answer this question, then we also introduce the following equations:

$$\left.\begin{aligned}
&\text{output = input in the first sector in period } 0: \quad t_1^1 + \sigma_1 t_1^0 = t_1^0 + \tau_1 \\
&\text{output = input in the first sector in period } 1: \quad t_1^2 + \sigma_1 t_1^0 = t_1^1 + \tau_1 \\
&\quad \ldots \\
&\text{output = input in the first sector in period } n-1: \quad \sigma_1 t_1^0 = t_1^{n-1} + \tau_1 \\
&\text{output = input in the second sector in period } 0: \quad t_1^1 + \sigma_2 t_2^0 = t_1^0 + \tau_2 \\
&\text{output = input in the second sector in period } 1: \quad t_1^2 + \sigma_2 t_2^0 = t_1^1 + \tau_2 \\
&\quad \ldots \\
&\text{output = input in the second sector in period } n-1: \quad \sigma_2 t_2^0 = t_1^{n-1} + \tau_2
\end{aligned}\right\} \quad (A.11)$$

Here, the values of one unit of the means of production are shown by t_1 whose subscripts in the top right corner express their aging like the cases of profit existence conditions and the value of one unit of the means of consumption is t_2. Again, as with the inequalities above, summing up both sides of the first n equations and the second n equations yields

$$\left.\begin{aligned} nt_1^0 \sigma_1 &= t_1^0 + n\tau_1 \\ nt_2^0 \sigma_2 &= t_1^0 + n\tau_2 \end{aligned}\right\} \quad (A.12)$$

These two equations can be solved as

$$\left.\begin{aligned} t_1 &= \dfrac{n\tau_1}{n\sigma_1 - 1} \\ t_2 &= \dfrac{(n\sigma_1 - 1)\tau_2 + \tau_1}{\sigma_2(n\sigma_1 - 1)} \end{aligned}\right\} \quad (A.13)$$

The first result shows that $(n\sigma_1 - 1)$ should be positive if t_1 and t_2 are meaningful in the sense of economy, and if we substitute the second result into (A.10), we can take $1 - Rt_2 > 0$, which is the condition of labor exploitation. Therefore, we can confirm here that the FMT can hold in the case with fixed capital.

Addendum II
Decentralized Market Model of the Marxian Optimal Growth Theory

The Marxian optimal growth model is generally expressed as the social planner model, as if the total labor of the entire society could be freely manipulated by a specific planner. Of course, this does not happen in real life, but there is a more important reason than the ease of calculation for solving such a model. The reason is that these calculations reveal the socially optimal state. Only by using this framework can we accurately show that it is socially optimal to raise the capital–labor ratio via capital accumulation after the industrial revolution, and that it is socially optimal to stop capital accumulation after reaching the optimal capital–labor ratio. In other words, this is the proof that capitalism was necessary after the industrial revolution, and that this capitalism will eventually come to an end.

However, things are different in the real world. This is because in reality there is no social planner, and no one is driven by thinking about society as a whole. Households act only to maximize their own utility and corporations act only to maximize their own profit. For this reason, modern economics has modeled this structure as a form of decentralized market model. Therefore, in this addendum, we rewrite the Marxian optimal growth model in this form.

In modern economics, the first theorem of welfare economics has proved that the consequences of both models are the same when perfect competition is assumed with (1) no externalities, (2) perfect information, (3) no transaction costs, and (4) the rationality of economic agents as an ideal type of market. Therefore, even if the Marxian optimal growth model in this book is rewritten as a decentralized market model, the solution must be identical under conditions (1) to (4) above. We explain this as follows.

First of all, we assume that the two sectors of production are formed by one firm each in the Marxian optimal growth model and name the firm in the means of production sector as Firm 1 and the firm in the means of consumption sector as Firm 2. In order to avoid unnecessary complications in the transformation, we make a simplification to set the prices of the means of consumption and the means of production to 1 and p, which were set to p_2 and p_1, respectively, in Chapters 3 and 4 of this book. This simplification means that the means of consumption becomes a numeraire. Wages in firm 1 and in firm 2 are

MARXIAN ECONOMICS

set as w_1 and w_2, respectively. Then, there are four markets and four types of prices in each market in this model. That is:

(1) The means of consumption market (the price of means of consumption as the numeraire, is set to 1).
(2) The labor power market (wage in the means of production sector w_1, wage in the means of consumption production sector w_2).
(3) The asset market (price of the means of production as goods are purchased as assets by households p).
(4) The capital market (price that firms borrow the means of production as factors of production from households r).

First, note that all these prices are functions of time, and the amount of labor input to each of firms 1 and 2 is named L_1 and L_2 ($L_1 + L_2 = L$). The reason why L_1 and L_2 are expressed as $(1 - s)L$ and sL in the text is that a social planner could decide s, but in a decentralized market there is no social planner and individual firms determine their labor employment in a decentralized manner. Thus, we do not use that share s in this stage. Only when we compare the result of calculation with that of social planner model later in this addendum do we use the expressions $(1 - s)L$ and sL.

PROFIT MAXIMIZING BEHAVIOR OF FIRMS

The profits of firms 1 and 2 after paying capital rent and wages become

$$\left.\begin{aligned} \pi_1 &= pBL_1 - w_1 L_1 \\ \pi_2 &= AK^\alpha L_2^\beta - rK - w_2 L_2 \end{aligned}\right\} \quad (A.14)$$

where r is the rental price per unit of capital. At this point, the profit maximizing conditions can be obtained by the partial differentiation of both equations by L_1, L_2, and K. Therefore,

$$\left.\begin{aligned} pB &= w_1 \\ \beta AK^\alpha L_2^{\beta-1} &= w_2 \\ \alpha AK^{\alpha-1} L_2^\beta &= r \end{aligned}\right\} \quad (A.15)$$

UTILITY MAXIMIZATION BEHAVIOR OF HOUSEHOLDS

Next, we consider the optimization problem of a typical household. A household uses labor income and interest income to purchase the means of consumption and the means of production. If we ignore the purchase of the means of production for the moment and assume that the household allocates its income to the formation of asset a and consumption

ADDENDUM II

Y, the budget constraint equation for flow is now as follows (note that a is the total asset owned by the total household, not the asset per capita):

$$\dot{a} = \tilde{r}a + w_2 L_2 + w_1 L_1 - Y \tag{A.16}$$

where $\tilde{r}$ is the interest rate, and all the variables are measured in the monetary base; that is, measured by the means of consumption because here p_2 is set to one. In this case, the capital gains from rising asset prices are assumed to be included in asset income. Households are subject to this budget constraint and the non-Ponzi game condition (the condition that borrowing does not expand indefinitely in the manner of a Ponzi scheme):

$$\lim_{t \to \infty} a(t) \cdot e^{-\int_0^t \tilde{r}(\tau) d\tau} \geq 0 \tag{A.17}$$

Under these conditions, households maximize the following objective function:

$$\int_0^\infty e^{-\rho t} \log Y(t) dt \tag{A.18}$$

by choosing the best paths of Y, a, and s. In practice, however, instead of assuming that the household makes a choice of s, the optimization problem can be solved by substituting the equilibrium condition of the labor market, $w_1 = w_2 = w$, into the budget constraint equation of the household (the equation for $\dot{a}$). This is because the household's optimal choice is that $w_1 = w_2 = w$, if we assume that both goods are produced, that is, that $L_1 > 0$ and $L_2 > 0$ are always satisfied. In the following, we will treat it as such. In this case, the Hamiltonian can be written as follows:

$$H = e^{-\rho t} \log Y + v(\tilde{r}a + wL - Y) \tag{A.19}$$

and its first-order conditions and transversality condition become

$$\left. \begin{array}{l} \dfrac{\partial H}{\partial Y} = \dfrac{1}{Y} e^{-\rho t} - v = 0 \iff v = \dfrac{1}{Y} e^{-\rho t} \\[2mm] \dfrac{\partial H}{\partial a} = v\tilde{r} = -\dot{v} \iff \tilde{r} = -\dfrac{\dot{v}}{v} \\[2mm] \lim_{t \to \infty} a(t) v(t) = 0 \end{array} \right\} \tag{A.20}$$

By solving this system, following two equations can be taken showing the optimal path:

$$\left. \begin{array}{l} \dfrac{\dot{Y}}{Y} = \tilde{r} - \rho^{1} \\[2mm] \lim_{t \to \infty} a(t) \cdot e^{-\int_0^t \tilde{r}(\tau) d\tau} = 0 \end{array} \right\} \tag{A.21}$$

---

[1] This equation can be obtained by $\dot{v} = \left(\dfrac{e^{-\rho t}}{Y} \right)' = \dfrac{(e^{-\rho t})'Y - e^{-\rho t}\dot{Y}}{Y^2} = -\dfrac{e^{-\rho t}}{Y}\left(\rho + \dfrac{\dot{Y}}{Y} \right)$ and the first two conditions in (A. 20).

where the first equation is the Euler equation and the second is the transversality condition. This is the system for expressing the optimal path of this model.

However, note that the second condition has been changed to an equation from the non-negative condition in the above formula. This is because this condition must cover two similar but different conditions. The first is the non-Ponzi game condition that one cannot continue to incur debt, and the second is the transversality condition that one does not leave any assets behind. In this model, utility is derived only from goods for consumption and not from simply leaving assets per se.

CONDITIONS FOR MARKET EQUILIBRIUM

Since the Marxian optimal growth model is a dynamic general equilibrium model, it assumes an equilibrium in all markets. Thus, if we assume a supply and demand equilibrium in three of the four markets, one of the remaining markets also comes into equilibrium by the Walrasian law. All the conditions for a supply and demand equilibrium in each market can be explained as follows.

(1) Means of consumption market

In order to assume an equilibrium in the means of consumption market, first we need to know how much demand for the means of consumption can be generated in this economy, and if we assume the product exhaustion theorem explained in subsection 4.3.3, the equilibrium condition for supply and demand of the means of consumption becomes

$$Y = AK^\alpha L_2^\beta = rK + w_2 L_2 \qquad (A.22)$$

This is equivalent to the assumption that the rental price of capital, r, and the wage, w_2, are determined by the marginal productivities of each factor of production and a constant return to scale, that is, $\alpha + \beta = 1$. Under this assumption, by substituting this equation into the budget constraint equation for households (A.16) the household level of supply and demand is seen to match the means of consumption market.

(2) Labor power market

Since the equilibrium conditions for the labor power market are given by $w_1 = w_2 = w$, substituting (A.15) with this condition $w_1 = w_2 = w$ gives

$$w = \beta AK^\alpha L_2^{\beta-1} = pB \qquad (A.23)$$

(3) Asset market (Equilibrium between the monetary valuation of the accumulated capital and the prices of the means of production)

In the asset market the interest rate on financial assets and the rate of return on capital goods (which, from the perspective of the firm, is the rental price of capital) are equal. If we denote the rate of interest on financial assets as $\tilde{r}$, this condition can be expressed as

$$\tilde{r} = \frac{r}{p} - \delta + \frac{\dot{p}}{p} \qquad (A.24)$$

Both sides of this equation are expressed in monetary term (because here we set the price of the consumption good $p_2 = 1$, this unit is the means of consumption). If we call one unit of consumption goods one yen for simplicity, the left side of this equation is the amount of interest per yen made by lending the financial assets. On the other hand, the first term on the right side of this equation represents the amount of income per yen of capital goods, the second term represents the amount of depreciation per yen of capital goods, and the final term represents the amount of capital gain per yen of capital goods.

Under the above conditions, the asset market must be balanced at:

$$a = pK \qquad (A.25)$$

This indicates that the total asset value a of a household must be the same as the monetary valuation of the total amount of physical capital K in existence.

(4) Capital market (consumers' lending the means of production to firm 2)

Let us show that, assuming equilibrium in the three markets described so far, the capital market automatically becomes an equilibrium, yielding the capital accumulation equation (A.30) = (A.28) of the original social planner model. First, substituting (A.22) and (A.24) into the budget constraint equation for households (A.15), we can obtain

$$\dot{a} = \tilde{r}a + wL - Y = pBL_1 - \delta pK + \dot{p}K \qquad (A.26)$$

On the other hand, since nominal term of asset $a = pK$ can be dynamized to $\dot{a} = \dot{p}K + p\dot{K}$, above budget constraint equation for households can be transformed into

$$p\dot{K} = pBL_1 - \delta pK \qquad (A.27)$$

Dividing both sides by p, we obtain

$$\dot{K} = BL_1 - \delta K \qquad (A.28)$$

This is identical to the capital accumulation function (the production function of the means of production) of the original Marxian optimal growth model. In other words, the above market demand–supply equilibrium equation determines the growth path of the model, and this is consistent with the results of the Marxian optimal growth model solved as the social planner model.

Moreover, behind the equation (A.28), the rental price of capital, r, is determined. If we substitute the third equation of (A.15) into the Euler equation and the transversality condition, we can derive a dynamic equation as the consequence of the equilibrium of the asset market. As the results, the market equilibrium (and equilibrized growth path of the accumulation and consumption) can be expressed by the following five equations. From here on, L_1 and L_2 will be expressed in the form of $(1-s)L$ and sL as in the main model.

$$Y = AK^\alpha (sL)^\beta \tag{A.29}$$

$$\dot{K} = B(1-s)L - \delta K \tag{A.30}$$

$$\frac{\dot{Y}}{Y} = \tilde{r} - \rho = \left(\frac{r}{p} - \delta + \frac{\dot{p}}{p}\right) - \rho = \frac{\alpha A}{p}\left(\frac{sL}{K}\right)^\beta + \frac{\dot{p}}{p} - \delta - \rho \tag{A.31}$$

$$\beta A\left(\frac{sL}{K}\right)^{-\alpha} = pB \tag{A.32}$$

$$\lim_{t\to\infty} a(t) \cdot e^{-\int_0^t \tilde{r}(\tau)d\tau} = 0 \tag{A.33}$$

Here, we assumed $\alpha + \beta = 1$ to derive equations (A.31) and (A.32).

FROM MARKET EQUILIBRIUM TO THE DYNAMIC EQUATIONS OF THE SOCIAL PLANNER MODEL

Then, the last remaining question is whether the dynamic equation of s (4.37) in subsection 4.3.2 can be derived from the market equilibrium, because the dynamic equation (A.30) of K is already identical to equation (4.3) in the basic model in subsection 4.1.1. Therefore, first, let us remember the dynamic equation of s (4.37) derived in section 4.3, as

$$\dot{s} = s\left\{\frac{BL}{K} \cdot \frac{\alpha}{\beta} s - (\rho + \delta)\right\} \tag{A.34}$$

We now derive this dynamic equation in this decentralized market model. The first step is to differentiate the logarithmized (A.29) and (A.32) to

$$\frac{\dot{Y}}{Y} = \alpha \frac{\dot{K}}{K} + \beta \frac{\dot{s}}{s} \tag{A.35}$$

$$\alpha\left(\frac{\dot{K}}{K} - \frac{\dot{s}}{s}\right) = \frac{\dot{p}}{p} \tag{A.36}$$

By substituting equation (A.31) for the left side of equation (A.35), and again substituting the result for the right side of the latter equation (A.36), we obtain

$$\frac{\alpha A}{p}\left(\frac{sL}{K}\right)^\beta + \alpha\left(\frac{\dot{K}}{K} - \frac{\dot{s}}{s}\right) - (\delta + \rho) = \alpha\frac{\dot{K}}{K} + \beta\frac{\dot{s}}{s} \tag{A.37}$$

If we assume $\alpha + \beta = 1$ again, it can be rewritten as

$$\frac{\alpha A}{p}\left(\frac{sL}{K}\right)^{\beta} - (\delta + \rho) = \frac{\dot{s}}{s} \tag{A.38}$$

and substituting $\dfrac{1}{p} = \dfrac{B}{\beta A}\left(\dfrac{sL}{K}\right)^{\alpha}$ which is derived from equation (A.32) for the left side of equation (A.38), we obtain

$$\frac{B}{\beta A}\left(\frac{sL}{K}\right)^{\alpha} \alpha A\left(\frac{sL}{K}\right)^{\beta} - (\delta + \rho) = \frac{\dot{s}}{s} \tag{A.39}$$

This can be rewritten as

$$\frac{B\alpha}{\beta}\left(\frac{L}{K}s\right)^{\alpha+\beta} - (\delta + \rho) = \frac{\dot{s}}{s} \tag{A.40}$$

If we assume $\alpha + \beta = 1$, this equation becomes

$$\dot{s} = s\left\{\frac{BL}{K} \cdot \frac{\alpha}{\beta}s - (\rho + \delta)\right\} \tag{A.41}$$

which is what we wanted to obtain. Therefore, under this assumption, this dynamic path of capital accumulation is identical with that of the basic social planner model.

ON THE RELATIONSHIP BETWEEN THE MARKET MODEL AND THE SOCIAL PLANNER MODEL

These calculations have allowed us to introduce the mechanism of perfectly competitive markets into the Marxian optimal growth model under the condition of a constant return to scale. In addition, the first theorem of welfare economics in modern economics is shown to hold true, and the consequences of perfectly competitive markets and those of the social planner model are identical. However, there may be another way to handle the market. For example, it may be possible to change the setting, such that firm 2 pays interest not on the value of capital pK but on the physical quantity K.

With this in mind, let us consider how s is determined in the market model and in the original social planner model. The path of s in the social planner model was calculated on the assumption that a social planner, who has control over all labor, the essential factor of production, can allocate that labor directly to both sectors of the means of consumption and production and the means of production in order to maximize the intertemporal utility of the entire population. However, s in the market model is determined by the wage w in the labor market and the relative price p of both goods, which reflects the supply and demand conditions of the means of consumption and means of production. Let us consider this relation in more detail.

(1) Households determine the purchase ratio of the means of consumption and the means of production that maximizes their utility under the budget constraint by referring to the relative prices p of both goods and rental prices r of the means of production. This is expressed in the Euler equation, which is a decentralized relationship in which the households' decisions also affect p and r through the asset and capital markets, while conversely being regulated by these two variables.

(2) Next, profit-maximizing firms facing market prices p and r procure labor for two sectors in which they produce the means of consumption and the means of production.

(3) Households supply labor to both sectors, but if the supply ratio $s: 1 - s$ to the two sectors is not well matched, wages w_1 and w_2 in the two sectors will differ. This is irrational because different prices of labor (wages) are charged for the same amount labor expense. Therefore, s is determined so that w_1 and w_2 are equalized.

(4) To summarize these processes, the demand behavior of households is the first step, which determines the labor demand structure of firms through market prices p, r, and w, which in turn determines the labor supply behavior of households. In fact, since demand, supply, and prices are determined simultaneously in the market in the model, there is no temporal back-and-forth or logical precedence relationship, but as the objective function of this system is to maximize household's utility, the final determination of s in 3) is also for that purpose. The only difference is that it is not determined directly as a control variable in the social planner model, but rather as the demand-supply matching behavior in each market through the three types of prices, p, r, and w in the market model. The social planner model does not have any prices, and is formulated as a kind of physical allocation model of labor and capital goods. On the other hand, the market model allows each entity to act decentrally based on price information."

The above discussion shows that if the market is perfectly competitive, the social planner model is more convenient to lead the dynamic path of the various variables. Therefore, it is necessary to solve the decentralized market model when analyzing the diachronic fluctuations of prices r, w, and p, which are generally not included in the original model, or when introducing factors such as government spending, taxes, externalities in production (ex. knowledge spillover by the transfer of ideas among individuals and firms), and asset inequality among households (class division) to examine their effects. In fact, using such a decentralized market model, Liu (2008) found that corporate taxation proportional to the output of the sector of the means of consumption and the resulting distribution of tax revenue to consumers do not lead to distortions in the allocation of resources.

However, instead of discussing the deviation from the optimal solution (the social planner model) in the framework of modern economics, we mention another way to derive this deviation as a result of the interrelationship between politics and economy, or the interrelationship between different interest groups (such as classes). Since politics is not a means to realize the interests of all members of a society evenly, it causes deviations in the realization of interests among interest groups. I have recently discussed this issue systematically and mathematically in Onishi (2021b). One of the subjects of Marxian economics is widespread in this area on politics.

Addendum III
Price Term of Reproduction Scheme

EQUALIZED RETURN ON INVESTED CAPITAL

Although we equalized the profit rate to calculate the price of production in Section 5.3, how to define the profit rate is an issue. This is because Marx's profit rate in Volume 3 of *Capital* is not the ratio of profit to the cost price but the ratio of profit to the total capital advanced, and strictly speaking, it may be close to the concept of "return on invested capital" in a modern economy. In modern Japan, capital invested or Marx's "advanced capital" does not include wages. Regular monthly wages are paid by private Japanese companies on the 25th, close to the end of the month. Thus, unlike in the past, when capitalists (capital providers) were also the managers, a suitable formulation for profit rate equalization in the present conditions should be the equalization of m/c, where money capitalists and functioning capitalists are separate. Here, the income of the functioning capitalist (hired managers) appears as a part of v (which actually constitutes part of m), as with other workers, so that all the profit created appears to have been created by the contribution of constant capital. If so, we call this m/c the "return on invested capital" and change the law of profit rate equalization into the law of the equalization of return on invested capital. The first step of the transformation becomes

$$W_1' = c_1(1+r_i^0) + v_1 \text{ and } W_2' = c_2(1+r_i^0) + v_2 \tag{A.42}$$

where r_i is the equalized return on invested capital. However, this step still does not include the change in the prices of c_1 and c_2, as in the original transformation problem explained in subsection 5.3.1. Therefore,

$$W_1'' = c_1 \frac{W_1'}{W_1}(1+r_i^1) + v_1 \frac{W_2'}{W_2} \text{ and } W_2'' = c_2 \frac{W_1'}{W_1}(1+r_i^1) + v_2 \frac{W_2'}{W_2} \tag{A.43}$$

However, because this equation is also incomplete, we must take the next step. That is,

$$W_1''' = c_1 \frac{W_1''}{W_1}(1+r_i^2) + v_1 \frac{W_2''}{W_2} \text{ and } W_2''' = c_2 \frac{W_1''}{W_1}(1+r_i^2) + v_2 \frac{W_2''}{W_2} \tag{A.44}$$

MARXIAN ECONOMICS

Table A.1 Reproduction scheme of the price of production under the equalized return on invested capital

	c^i	v^i	m^i	Total
Sector 1	$c_1 \dfrac{W_1^*}{W_1}$	$v_1 \dfrac{W_2^*}{W_2}$	$c_1 \dfrac{W_1^*}{W_1} r_i^*$	W_1^*
Sector 2	$c_2 \dfrac{W_1^*}{W_1}$	$v_2 \dfrac{W_2^*}{W_2}$	$c_2 \dfrac{W_1^*}{W_1} r_i^*$	W_2^*
Whole society	$(c_1+c_2)\dfrac{W_1^*}{W_1}$	$(v_1+v_2)\dfrac{W_2^*}{W_2}$	$\left\{(c_1+c_2)\dfrac{W_1^*}{W_1}\right\} r_i^*$	$W_1^* + W_2^* = W_1 + W_2$

Finally, just as before, the following equilibrium returns on invested capital r_i^* and the selling prices W_1^* and W_2^* of each sector are obtained:

$$W_1^* = c_1 \frac{W_1^*}{W_1}(1+r_i^*) + v_1 \frac{W_2^*}{W_2} \quad \text{and} \quad W_2^* = c_2 \frac{W_1^*}{W_1}(1+r_i^*) + v_2 \frac{W_2^*}{W_2} \tag{A.45}$$

The reproduction scheme in the level of price of production is shown in **Table A.1**. In this table, c^p, v^p, and m^p in Table 5.2 in subsection 5.3.1 are converted into c^i, v^i, and m^i, respectively, which might be the price system closest to the real world.

REPRODUCTION SCHEME IN TERMS OF PRICE IN THE MARXIAN OPTIMAL GROWTH MODEL

This shows that the reproduction scheme of the Marxian optimal growth model, which we examined in Chapter 4, can also be transformed at the price level. We see in value terms how Table 4.2 in subsection 4.4.1 can be transformed in terms of price.[1] **Table A.2** shows the results of solving the Marxian optimal growth model at the price level presented in Addendum II. We see (1) that δK and L are re-evaluated (weighted) at the price of p_k and w: (2) similarly, $\dot{K} + \delta K$ and Y are re-evaluated at the price of p_k and p_c (p_k and p_c are the prices of the means of production and the means of consumption in terms of utility, respectively; in Addendum II, it is deemed that $p_k = p$, $p_c = 1$). (3) Since the second sector's m^p represents

---

[1] Kanae (2011) was the first to carry out a comparative analysis of the reproduction scheme at the value and price levels as understood in the Marxian optimal growth model.

ADDENDUM III

Table A.2 Reproduction scheme of the Marxian optimal growth model (simplified version) at the price level

	c^p	v^p	m^p	Total
First sector	0	$w(1-s)L$	0	$w(1-s)L = p_k(\dot{K}+\delta K)$
Second sector	$p_k\delta K$	wsL	$r_c p_c K - p_k \delta K = p_k K\left(\rho - \dfrac{\dot{p}_k}{p_k}\right)$	$r_c p_c K + wsL = p_c Y$
Society as a whole	$p_k \delta K$	wL	$r_c p_c K - p_k \delta K = p_k K\left(\rho - \dfrac{\dot{p}_k}{p_k}\right)$	$r_c p_c K + wL$ $= p_k(\dot{K}+\delta K) + p_c Y$

the profit ($r_c K$) obtained from the capital stock invested minus depreciation, it is naturally different from m at the value level.[2] These three characteristics are largely the same as those found in Table 5.2.[3]

---

[2] The condition under which total surplus value and total profit in tables 4.2 and A.2 are equal is

$$(1-\beta)sL = \left(r + \dfrac{\beta\delta}{B} - p_k\delta\right)K$$

However, this does not generally hold.

[3] Because Table A.2 is based on a simplified version of the model, capital goods are not input in the first sector, which means its profit is set at nil. Below, we show the case in which both sectors input capital goods. This shows the result of calculating the model presented in note 8, subsection 4.1.2. Note that it has basically the same characteristics as Table A.2.

Table A.3 Reproduction scheme of the Marxian optimal growth model (complete version with capital input in both sectors) at the price level

	c^p	v^p	m^p	Total
First sector	$p_k \delta K_k$	wL_k	$\dfrac{K_k}{K}(r_c p_c K - p_k \delta K) = p_k K_k \left(\rho - \dfrac{\dot{p}_k}{p_k}\right)$	$r_c p_c K_k + wL_k = p_k(\dot{K}+\delta K)$
Second sector	$p_k \delta K_c$	wL_c	$\dfrac{K_c}{K}(r_c p_c K - p_k \delta K) = p_k K_c \left(\rho - \dfrac{\dot{p}_k}{p_k}\right)$	$r_c p_c K_c + wL_c = p_c Y$
Society as a whole	$p_k \delta K$	wL	$(r_c p_c K - p_k \delta K) = p_k K\left(\rho - \dfrac{\dot{p}_k}{p_k}\right)$	$r_c p_c K + wL = p_k(\dot{K}+\delta K) + p_c Y$

MARXIAN ECONOMICS

Note that the implications of $p_k K\left(\rho - \dfrac{\dot{p}_k}{p_k}\right)$, which we obtained using the equation (4.42) presented in subsection 4.3.3 for mp in the second sector and, substituting the shadow price of the means of production, μ, with the price of the means of production, p_k.

(1) $\dfrac{-\dot{p}_k}{p_k}(>0)$ shows the degree of the social contribution of activities by functioning capitalists. This means that the fall in the price of the means of production in terms of utility, p_k, represents the social achievements of business activities through capital accumulation, not the achievement of money capitalists who merely lend money. In other words, this represents an authentic achievement by the functioning capitalists involved, and also represents the socio-historical role of exploitation. However, this gradually decreases as capital accumulates, and when it reaches the optimal capital-labor ratio, it becomes zero. Table 4.2 confirms this at value level, and now we also reconfirms that relation at price level. Incidentally, when late Professor Okishio argued that various phenomena on the level of profit are simply a reflection of movements of surplus value, he makes the same point.

(2) Nevertheless, while it may fall, as long as ρ is an exogenous variable, this portion can only reach a constant level; it can never become zero. In other words, in capitalism, while there is a tendency of a falling rate of profit, it will never reach zero.[4] This is a new finding that has not been obtained in Table 4.2.

---

[4] However, a falling rate of profit is inevitable. In both tables A.1 and A.2, if we ignore the difference in price between both sectors and express the price as ρ and the rental price of capital as r, the profit ratio becomes

$$\dfrac{(r-\delta)pK}{wL + \delta pK}$$

Furthermore, using equation (4.42) in subsection 4.3.3, this can be transformed into

$$\dfrac{(r-\delta)pK}{(\beta AK^{\alpha-1}L^\beta)pK + \delta pK} = \dfrac{r-\delta}{\beta AK^{\alpha-1}L^\beta + \delta} = \dfrac{r-\delta}{\dfrac{\beta}{\alpha}(\alpha AK^{\alpha-1}L^\beta) + \delta} = \dfrac{r-\delta}{\dfrac{\beta}{\alpha}r + \delta}$$

Differentiating the rightmost term yields

$$\dfrac{\partial}{\partial r}\left(\dfrac{r-\delta}{\dfrac{\beta}{\alpha}r + \delta}\right) = \dfrac{\left(1+\dfrac{\beta}{\alpha}\right)\delta}{\left(\dfrac{\beta}{\alpha}r + \delta\right)^2} > 0$$

This means that the tendency of the rental price of capital (r) to fall decreases the profit rate. This reconfirms the causal relation in the neoclassical phenomenon of the decline in the rental price of capital as a result of capital accumulation results in Marx's falling rate of profit.

(3) On the other hand, if surplus value continues to exist even in a stable state, then it suggests that all the surplus value = all the profit is used for consumption at that time. This is the case of simple reproduction discussed in subsection 4.2.1 and is also confirmed in Table 4.3. However, the working class can also receive this part of the profit through banks or other financial institutions as interest, and if all members of the society have the same amount of assets, then there is no problem even if exploitation continues to exist in the sense outlined above. This may be why Piketty does not discuss exploitation itself but discusses only the disparity of assets among the members of a society.

(4) Finally, let us examine the implication of the fact that time preference, ρ forms profits even under zero growth. In subsection 4.3.3, we emphasize that this does not alter the fundamental nature of exploitation proved by FMT (the difference between what the workers produce and what they receive). However, since the argument here is that price, as distinct from value, is determined by valid claims—such as profit rate equalization—made by various actors, and capitalists' attitudes towards demanding profit—such as that the future returns must be greater than the present investment—has become important. However, the problem is found at the level of prices, just as in the case of profit rate equalization, and it constitutes a significant issue at this level.[5]

---

[5] According to Professor Sousuke Morimoto at St. Andrew's University, Marx was planning to discuss this problem in Chapter 4 of Volume 3 of *Capital*. However, he died before he could do so, and this section was written by Engels.

Addendum IV
Three-Sector Reproduction Scheme Including the Commercial Sector

Just as we examined the redistribution of surplus value as a result of realizing fairness among industrial sectors in subsection 5.3.1 we can examine another form of equity among industrial sectors that produce value and non-industrial sectors that are not involved in production. The subsection 5.2.1 discussed how commercial profit is determined and the conditions under which this kind of non-productive sector can exist and make a profit. We saw in the former explanation that the rate of profit of the commercial sector should be equal to that of the industrial sector, and thus we must now equalize the rate of profit of the commercial sector again with the equalized rate of profit between two industrial sectors in this section. In other words, we now need to investigate the equalization process of the rates of profit among three sectors: two industrial and one commercial sector.

We first set up the original value structures of these three sectors:

$$\left. \begin{array}{l} W_1 = c_1 + v_1 + m_1 \\ W_2 = c_2 + v_2 + m_2 \\ W_c = c_c + v_c + m_c \end{array} \right\} \quad (A.46)$$

Here, suffixes c indicates the commercial sector, which does not produce any value and c_c must cover the total value of $W_1 + W_2$ divided by the annual number of turnovers of all enterprises because the commercial sector deal in all commodities produced in the society. We assume such a c_c. Under this assumption, we should remember that W_c is zero because the commercial sector does not create any value, but now we can confirm this fact by examining intersectoral transactions among three sectors. That is, if we assume simple reproduction,

$$\left. \begin{array}{l} \text{Intersectoral supply = demand of the means of production} \quad v_1 + m_1 = c_2 + c_c \\ \text{Intersectoral supply = demand of the means of consumption} \quad c_2 = v_1 + m_1 + v_c + m_c \end{array} \right\}$$

$$(A.47)$$

Substituting the second equation for c_2 in the right side of the first equation leads to

$$W_c = c_c + v_c + m_c = 0 \quad (A.48)$$

MARXIAN ECONOMICS

This is the same in the case of extended reproduction, where m_1, m_2, and m_c are divided into $m(c) + m(v) + m(k)$, respectively. Of course, the commercial sector also actually spends $c_c + v_c > 0$, but this spending is not a productive activity and therefore it does not transfer or formulate any value. Therefore, the surplus value in this sector is negative:

$$m_c = -(c_c + v_c) < 0 \tag{A.49}$$

This is because these costs are spent in the non-productive sector. This is reconfirmed by the calculation result of (A.48).

However, as we saw in the last section, the commercial sector can receive a portion of surplus value from the productive sector by shortening the time of circulation. Therefore, after equalizing the rates of profit among the three sectors, the rate of profit in the commercial sector has to become positive, or strictly speaking, it has to become equal to that of the other sectors. Thus, Marx proposed the theory of commercial profit soon after the theory of the equalization of the general rate of profit in *Capital*, and therefore, we now go on to explain the equalization of the rate of profit among the above three equations. For this purpose, we recall that the intersectoral ratio between the productive sectors and the commercial sector calculated in subsection 5.2.1 was

$$\frac{c_c + v_c}{c_p + v_p} = \frac{\Delta c - \Delta c'}{\Delta p} \cdot \frac{r}{1+r} \tag{A.50}$$

If we express the $\frac{\Delta c - \Delta c'}{\Delta p}$ part on the right side of (A.50) as "z," we have

$$c_c + v_c = z \frac{r}{1+r}(c_1 + c_2 + v_1 + v_2) \tag{A.51}$$

and

$$m'_c = z \frac{r}{1+r}(m'_1 + m'_2) \tag{A.52}$$

after the equalization of the rate of profit with the average rate of profit of the productive sectors. Therefore,

$$\left. \begin{aligned} W'_c &= z \frac{r}{1+r}(c_1 + c_2 + v_1 + v_2) + z \frac{r}{1+r}(m'_1 + m'_2) \\ &= z \frac{r}{1+r}(c_1 + c_2 + v_1 + v_2 + m'_1 + m'_2) \\ &= z \frac{r}{1+r}(W'_1 + W'_2) \end{aligned} \right\} \tag{A.53}$$

ADDENDUM IV

However, because still there is a difference in the rate of profits between the two productive sectors, we should start the transformation process again from the following three equations, like subsection 5.3.1. That is, the starting situation is

$$\left.\begin{aligned} W_1' &= (c_1 + v_1)(1 + r^0) \\ W_2' &= (c_2 + v_2)(1 + r^0) \\ W_c' &= z\frac{r^0}{1+r^0}(c_1 + c_2 + v_1 + v_2)(1+r^0) = z(c_1 + c_2 + v_1 + v_2)r^0 \end{aligned}\right\} \quad (A.54)$$

where the rate of profit is specified as that on the first stage of transformation process.

The second stage becomes

$$\left.\begin{aligned} W_1'' &= \left(c_1 \frac{W_1'}{W_1} + v_1 \frac{W_2'}{W_2}\right)(1+r^1) \\ W_2'' &= \left(c_2 \frac{W_1'}{W_1} + v_2 \frac{W_2'}{W_2}\right)(1+r^1) \\ W_c'' &= z\left(c_1 \frac{W_1'}{W_1} + c_2 \frac{W_1'}{W_1} + v_1 \frac{W_2'}{W_2} + v_2 \frac{W_2'}{W_2}\right)r^1 \end{aligned}\right\} \quad (A.55)$$

Therefore, the final and equalized stage must be

$$\left.\begin{aligned} W_1^* &= \left(c_1 \frac{W_1^*}{W_1} + v_1 \frac{W_2^*}{W_2}\right)(1+r^*) \\ W_2^* &= \left(c_2 \frac{W_1^*}{W_1} + v_2 \frac{W_2^*}{W_2}\right)(1+r^*) \\ W_c^* &= z\left(c_1 \frac{W_1^*}{W_1} + c_2 \frac{W_1^*}{W_1} + v_1 \frac{W_2^*}{W_2} + v_2 \frac{W_2^*}{W_2}\right)r^* \end{aligned}\right\} \quad (A.56)$$

This result shows several important points.

(1) The total "production" of the commercial sector W_c^* can be rewritten as

$$W_c^* = z\frac{r^*}{1+r^*}(W_1^* + W_2^*) \quad (A.57)$$

which shows that it is $z\frac{r^*}{1+r^*}$ times larger than the total production of the industrial sectors.

(2) Because z is "1—shortening ratio of the circulation period," this part is technologically determined outside of this system of equations.
(3) Because there are four unknown variables, W_1^*, W_2^*, W_c^* and r^*, one more constraint condition must be introduced into this system: the total price = total value constraint, expressed as $W_1^* + W_2^* + W_c^* = W_1 + W_2 + W_c$. In this case, it can be expressed as $W_1^* + W_2^* + W_c^* = W_1 + W_2$, because W_c is zero, as discussed above. This formula shows the origin of W_c^* much more clearly. This is the result of the Marxian formulation in which surplus value can be created only in the industrial sector not in the commercial sector.[1]

---

[1] This formulation still does not fully take into account the difference in turnover periods between the industrial and commercial sectors. See Onishi (2020c) for a discussion of the changes in the intersectoral ratios resulting from this consideration.

Addendum V
Incorporating Class Dynamics in the Marxian Optimal Growth Model

DYNAMITIZATION OF THE CONCEPT OF EXPLOITATION IN ANALYTICAL MARXISM

It is not that there is no class perspective in the Marxian optimal growth theory. The contradiction between c, v, and m can expressed in the Marxian reproduction scheme with two-sector and three-value components. Furthermore, there have been some attempts to incorporate the concept of exploitation of Analytical Marxism (the second definition of exploitation defined in this book) into the model. Analytical Marxism, as we saw in subsection 3.1.3, uses a static framework but it can be transformed into a dynamic form in the Marxian optimal growth model. In this addendum, we explain this framework. For this purpose, we first reconfigure the model as

$$\left. \begin{array}{l} Y = AK^\alpha L^\beta \\ \dot{K} = BL \end{array} \right\} \quad (A.58)$$

At this stage, it should be understood that this is not a macro-level model, but represents a specific industry in a specific region. This is a situation in which in a certain region, capitalists and workers (as defined by the amount of capital goods they own), are lending and borrowing capital goods within a certain industry in the situation in which they can only reallocate the total capital goods they themselves own. And as we explain later in this addendum, capital depreciation is omitted from the second equation, and the first equation is expressed as $f(K, L)$ to avoid unnecessary complications. The classes of capitalists and workers in the Analytical Marxist framework are then introduced as shown in Table 3.1 in subsection 3.1.3, with the subscripts 1 for the capitalist variables and 2 for the worker variables. Then, since we assume in that subsection that all additional production resulting from lending capital is acquired by capitalists, the gain (G_1) by this class can be expressed as follows:

$$G_1 = f(K_1, L_1) + \left[f\left(\frac{K_0 + K_1}{2}, L_0\right) + f\left(\frac{K_0 + K_1}{2}, L_1\right) - f(K_0, L_0) - f(K_1, L_1) \right] \quad (A.59)$$

MARXIAN ECONOMICS

The first term represents the output of capitalists in the absence of lending capital, and the next term in parentheses is the increase in aggregate output resulting from lending capital. The above can be summarized as

$$G_1 = f\left(\frac{K_0 + K_1}{2}, L_0\right) + f\left(\frac{K_0 + K_1}{2}, L_1\right) - f(K_0, L_0) \quad (A.60)$$

This indicates the dynamic path of the capital accumulation of the two classes. This addendum can thus be understood as a trial to dynamitize the concept of exploitation in Analytical Marxism.

Based on the above settings, the capitalist's additional gain from the increase in K_1 can be calculated as follows:

$$\left.\begin{aligned}\frac{\partial G_1}{\partial K_1} &= \frac{\partial\left(f\left(\frac{K_0 + K_1}{2}, L_0\right) + f\left(\frac{K_0 + K_1}{2}, L_1\right) - f(K_0, L_0)\right)}{\partial K_1} \\ &= \frac{f_K\left(\frac{K_0 + K_1}{2}, L_0\right) + f_K\left(\frac{K_0 + K_1}{2}, L_1\right)}{2} \\ &= f_K\left(\frac{K_0 + K_1}{2}, L\right)\end{aligned}\right\} \quad (A.61)$$

Here, f_K denotes the partial differentiation of the function f with respect to K, and the last equation does not distinguish between L_0 and L_1, and is written only as L, because we assume that their labor power holdings are identical (as we did in Table 3.1 in subsection 3.1.3). On the other hand, the additional gain from directly inputting labor in the production of consumption goods becomes

$$\left.\begin{aligned}\frac{\partial G_1}{\partial L_1} &= \frac{\partial\left(f\left(\frac{K_0 + K_1}{2}, L_0\right) + f\left(\frac{K_0 + K_1}{2}, L_1\right) - f(K_0, L_0)\right)}{\partial L_1} \\ &= f_L\left(\frac{K_0 + K_1}{2}, L_1\right)\end{aligned}\right\} \quad (A.62)$$

Here, the last f_L denotes the partial differentiation of the function f with respect to L.

As mentioned in subsection 4.1.2, where we explained how to calculate the target value of capital accumulation, the condition for this class to maintain a constant capital–labor ratio is that the additional gain from the additional input of L_1 is equalized both in the direct production of the means of consumption and by an indirect route via machine production. This problem can be addressed more rigorously by taking into account the existence of a time preference rate ρ (which is assumed not to differ across classes) and the use of machines over time, and therefore, we can lead to the next equilibrium condition:

$$\frac{\partial G_1}{\partial K_1} \frac{dK_1}{dL_1} \frac{1}{\rho} = \frac{\partial G_1}{\partial L_1} \tag{A.63}$$

Therefore, by substituting the result of the calculations in (A.61) and (A.62) into this equation (A.63) and considering that the original production function of means of consumption is $Y = AK^\alpha L^\beta$, we obtain

$$A\alpha L_1^\beta \left(\frac{K_0 + K_1}{2}\right)^{\alpha-1} B\left(\frac{1}{\rho}\right) = A\beta L_1^{\beta-1}\left(\frac{K_0 + K_1}{2}\right)^\alpha \tag{A.64}$$

and this equation can be solved as

$$\frac{K_0 + K_1}{L} = \frac{2\alpha}{\beta} \frac{B}{\rho} \tag{A.65}$$

This result is interesting because it is twice the original target value of capital accumulation, without taking into account the two classes and capital depreciation (δK) shown in subsection 4.1.2. It means if the target value of capital accumulation of one class is the same as the original target value ($\alpha B/\beta\rho$), the target value of the other class should automatically be the same. If we express this as an equation, it becomes

$$\left(\frac{K_0}{L_0}\right)^* = \left(\frac{K_1}{L_1}\right)^* = \frac{\alpha}{\beta} \frac{B}{\rho} \tag{A.66}$$

In fact, if we focus the accumulation behavior of the workers, we find that this class does not benefit at all from the increase in production due to capital lending, and therefore, based on its original production function, the worker's equilibrium condition becomes

$$\frac{\partial Y}{\partial K} \frac{dK}{dL} \frac{1}{\rho} = \frac{\partial Y}{\partial L} \tag{A.67}$$

Substituting the results of differentiating Y by K, K by L, and Y by L for $\partial Y/\partial K$, dK/dL, and $\partial Y/\partial L$ in (A. 67), respectively, using first two productions in (A.58), this equilibrium condition leads the following target of capital accumulation = optimal capital–labor ratio:

$$\left(\frac{K_0}{L_0}\right)^* = \frac{\alpha}{\beta} \frac{B}{\rho} \tag{A.68}$$

This means that the target value $(K/L)^*$ of the two classes attains the same level. In other words, if workers can accumulate, the result is that the levels of capital holdings of both classes are equalized, and capital lending and exploitation as the second definition disappear.

Furthermore, besides the above case, we can calculate three other cases in which the total increase in production due to capital lending is acquired both by capitalists and workers. These are:

(1) When the workers acquire all the increase in production. In this case, the gain (G_0) of the worker becomes

$$G_0 = f(K_0, L_0) + \left[f\left(\frac{K_0 + K_1}{2}, L_0\right) + f\left(\frac{K_0 + K_1}{2}, L_1\right) - f(K_0, L_0) - f(K_1, L_1) \right] \quad (A.69)$$

and the gain of the capitalist (G_1) becomes

$$G_1 = Y_1 \quad (A.70)$$

(2) When capitalists and workers share the increase in production in the ratio $\mu : 1 - \mu$ ($0 \leq \mu \leq 1$). In this case, the capitalist (G_1) gains

$$G_1 = f(K_1, L_1) + \mu \left[f\left(\frac{K_0 + K_1}{2}, L_0\right) + f\left(\frac{K_0 + K_1}{2}, L_1\right) - f(K_0, L_0) - f(K_1, L_1) \right] \quad (A.71)$$

and the worker (G_0) gains

$$G_0 = f(K_0, L_0) + (1 - \mu) \left[f\left(\frac{K_0 + K_1}{2}, L_0\right) + f\left(\frac{K_0 + K_1}{2}, L_1\right) - f(K_0, L_0) - f(K_1, L_1) \right]$$

$$(A.72)$$

(3) When, in addition to the assumption in (2) above, the population ratio of capitalists and workers is $c: w$ ($c + w = 1$) and the means of production are used in that ratio. In this case, capitalist (G_1) gains

$$G_1 = f(K_1, L_1) + \mu [f(wK_0 + wK_1, L_0) + f(cK_0 + cK_1, L_1) - f(K_0, L_0) - f(K_1, L_1)] \quad (A.73)$$

and worker (G_0) gains

$$G_0 = f(K_0, L_0) + (1 - \mu)[f(wK_0 + wK_1, L_0) + f(cK_0 + cK_1, L_1) - f(K_0, L_0) - f(K_1, L_1)]$$

$$(A.74)$$

In all these cases, Onishi and Fujiyama (2003) confirm that workers accumulate the above target values.

EFFECT OF THE EXISTENCE OF WORKERS' RIGHTS TO DECIDE ON CAPITAL ACCUMULATION

If the whole increase in production through lending capital is acquired by capitalists, who have much stronger bargaining power against workers, capital lending benefits the

ADDENDUM V

capitalists from the beginning. Therefore, because they want this situation to continue, they try to maintain the disparity in capital holdings, which is the condition for capital lending, if possible. And to do so, the capitalists must prevent workers from accumulating capital, which they can in fact do under certain conditions, such as when workers are so poor that their income is below subsistence level, and they are deprived of the right to accumulate capital in exchange for a guarantee from the capitalists that their wage will not fall below that level. This is because workers earning below the subsistence level have no choice but to accept such offers from capitalists.

In fact, if this assumption seems a little strange as a theoretical model, it can actually be said to an assumption used by Marx, except for the assumption that at the initial stage workers own a very small amount of capital. This is because workers' income in this case is completely guaranteed by capitalists as a subsistence wage that makes subsequent capital accumulation impossible, as was the reality of working-class conditions at the time of Marx. Strictly speaking, even though Marx admitted some exceptions, he also stated in the beginning of Chapter 31 of *Capital*, Volume 1, that workers can also rise to be capitalists.

This can be expressed in a mathematical model. Yamashita (2005) models the cases in which both capitalists and workers can choose the optimal path of capital accumulation and cases in which the workers cannot accumulate capital as described above. Specifically:

(1) The case in which both capitalists and workers can choose the optimal path for capital accumulation

First, we set the subscripts for the various symbols as before; namely, 1 for the capitalists and 2 for the workers. Since all the capital held by the workers is used by the workers, but the capital held by the capitalists is divided and used by both classes, the ratio used by the capitalists is set as v, and the ratio used by the workers is $1 - v$. Furthermore, in the condition where both classes hold same amount of labor power L, the ratio of labor used in the production of consumption goods and capital goods is set as s and $1 - s$ for the capitalists and u and $1 - u$ for the workers. In this case, the objective function of workers becomes

$$\max \int_0^\infty e^{-\rho t} \log \left\{ A K_{0t}^\alpha (u_t L)^\beta \right\} dt \tag{A.75}$$

This is the same as the original objective function of workers in the case where class relations are not included, because the workers receive almost no benefit as a result of capital lending (as all the fruits of increased production are acquired by the capitalists). We now calculate the capitalist's accumulation path under the condition where the workers choose the accumulation path shown in (A.75), and in this case, the objective function of the capitalist becomes

$$\max \int_0^\infty e^{-\rho t} \log[A(v_t K_{1t})^\alpha (s_t L)^\beta + A\{(1-v_t)K_{1t} + K_{0t}\}^\alpha (\bar{u}_t L)^\beta - A K_{0t}^\alpha (\bar{u}_t L)^\beta] dt \tag{A.76}$$

Here, the reason why we draw an overline on the top of "u" is to make it clear that this is predetermined by the workers before the capitalist's decisions, and therefore, not an instrumental variable for the capitalist.

(2) Cases where the workers cannot accumulate

In this case, under the assumption explained at the beginning of this subsection, the worker's income is guaranteed by the capitalists, and they receive only a substance wage that does not allow for capital accumulation, so they have no room to choose at all. However, since the capitalist also controls the substance wage, the latter's objective function seems to be

$$\max \int_0^\infty e^{-\rho t} \log[A(v_t K_{1t})^\alpha (s_t L)^\beta + A\{(1-v_t)K_{1t} + K_{0t}\}^\alpha (u_t L)^\beta - AK_{0t}^\alpha (u_t L)^\beta]dt \quad (A.77)$$

However, as we saw earlier, since the capitalists want to lend maximum capital to maximize exploitation (second definition of exploitation), they want to minimize capital accumulation on the workers' side. Therefore, the optimal choice on the capitalist side of u_t becomes $u_t = 1$ (the lack of accumulation by workers constitutes the capitalists' interest!) and the objective function of the capitalists can be simplified as

$$\max \int_0^\infty e^{-\rho t} \log[A(v_t K_{1t})^\alpha (s_t L)^\beta + A\{(1-v_t)K_{1t} + K_{0t}\}^\alpha (L)^\beta - AK_{0t}^\alpha (L)^\beta]dt \quad (A.78)$$

These equations can be solved as a dynamic optimization problem, and it was solved by Yamashita (2005) in a discrete form.[1] And for each case, the following results are derived.

(i) Assuming that the production technology and the time preference are the same, both the workers and the capitalists eventually accumulate the same level of capital, at which point capital lending and exploitation (second definition) disappear.
(ii) While the workers' capital does not accumulate, the capitalists' capital does, which widens the gap in capital holdings and thus increase exploitation (second definition). In this process, the worker's income does not increase, but the capitalist's income increases further due to the increase in both capital accumulation and exploitation (second definition). In other words, the gap between assets and income widens. This case closely resembles the social situation envisioned by Marx.

---

[1] In fact, as Yamashita himself says, there is a calculation error in Yamashita (2005), and he fails to draw the specific conclusion that the existence of asset inequality does not lead to capital lending when the instantaneous utility function is linearized in the form $U = Y$. However, when the instantaneous utility function is diminishing, asset inequality almost certainly leads to capital lending.

In reality, the development of capitalism also increases the real income of the working class and this is not the case in (ii). Workers also hold financial assets, which leads to their ownership of capital via banks and securities firms. This was different in Marx's day.

However, this disparity has again become a major problem in developed economies after the end of high economic growth. For this reason, the re-emergence of the disparity problem must be explained using a logic and causality that differ from the context discussed above. For example, there is a suspicion that the assumption in (i) that technology and time preference are identical may not hold. In particular, the issue of time preference needs to be examined. This issue was discussed in the subsections 1.3.5, 4.4.5, and 4.5.3 of this book, but we note that this question will be examined again in future.

Addendum VI
Converting the Analytical Marxist Model to a Labor Hire Model to Express Historical Trends of Disparities in Firm Size

CONVERTING THE ANALYTICAL MARXIST MODEL TO A LABOR HIRE MODEL

The Marxian optimal growth model that has been developed in this book shows the historical trends of income inequality among nations by introducing early-starter nations and latecomer nations, as seen in subsection 4.4.5. In addition, we saw in Addendum V that the static class model of Analytical Marxism has a possibility to be incorporated into our dynamic model. In this addendum we use these frameworks and extend them to a model that explains the historical trend of disparities in firm size.

For this purpose, we first convert capital lending, as formulated by Analytical Marxism, into labor hire. This is because the total production of the whole society can be maximized by transferring labor power from the workers' own homes to the workplaces owned by the capitalists, instead of by lending capital to the workers by the capitalists, as proposed by Analytical Marxists. At the beginning of Chapter 3, we discussed how command over capital comes into play in the workplace, but the point here is that capital lending can be rewritten as hired labor. Consequently, we now show Table A.4, which has been translated from **Table 3.1** in subsection 3.1.3 to change direction from an analysis focused on capital lending to one focused on labor hire. In Table 3.1, the four units of capital held by the capitalist were lent to the workers, but **Table A.4** assumes that a portion of capital in the capitalist's factory is used by the workers for the use of 2/3 of their labor.

Table A.4 clearly shows that different amounts of labor are used in the two workplaces, which expresses the variation in the size of the two firms. When measured in terms of the ratio of used labor, the disparity in the firm scale has increased to 1.67 : 0.33 from 1 : 1 before employed labor emerged. This means that hiring labor makes some firms larger, and this disparity in firm size is due to the uneven development of capital accumulation throughout history, especially in the agricultural sector. Although small-scale farming is the mainstream of agriculture in developed countries today, large-scale agriculture with a large labor power existed in the past, such as the latifundium in Europe and slave-based agriculture in the U.S.A.

MARXIAN ECONOMICS

Table 3.1 Capital lending and exploitation in Analytical Marxism (reprint from Chapter 3)

	Capitalist			Worker			Total		
	Machine	Labor	Production	Machine	Labor	Production	Machine	Labor	Production
Initial holding	10	1	3	2	1	1	12	2	4
After capital lending	6	1	2.5	6	1	2.5	12	2	5

Table A.4 Conversion of capital lending in Table 3.1 to labor hire

	Capitalist			Worker			Total		
	Machine	Labor	Production	Machine	Labor	Production	Machine	Labor	Production
Initial holding	10	1	3	2	1	1	12	2	4
After labor hiring	10	1.67	4.2	2	0.33	0.8	12	2	5

However, the shift to small-scale farming has again become common worldwide, and the scale of farming has once more leveled off. In other words, disparities in the scale of agriculture move in a Kuznets curve: initially lacking disparity, but then expanding, and finally contracting again. Historical materialism can explain these historical trends. We start with a general explanation in Table A.4, because this change in scale disparity is also a change from self-employment to hired labor, but now change to self-employment again. In this addendum, we present a model that determines the optimal amount of hired labor by generalizing the initial capital holdings of capitalists and workers as a and b ($a > b$ is naturally assumed), relying on Yoshii (2018), and by setting the production function appropriately.[1] For this purpose, Table A.4 can be rewritten as **Table A.5**.

In Table A.5 the labor power moves as hired labor to the firm with more capital, and the amount of this movement is denoted by λ, resulting in a change in total output. The size of this change is $[a^\alpha(1+\lambda)^\beta + b^\alpha(1-\lambda)^\beta] - (a^\alpha + b^\beta)$ which is clearly positive in the case that $0 < \lambda \leq 1$, and $a > b$, and shows the total output is increasing. And since this increase is acquired by the

---

[1] As can be seen, this is a production function of the Cobb-Douglas type, in which the total factor productivity is set to 1 by setting the units of labor and capital inputs appropriately. This setting is also in accordance with Yoshii (2018).

ADDENDUM VI

Table A.5 Generalized case of Table A.4

	Capitalist			Worker			Total		
	Machine	Labor	Production	Machine	Labor	Production	Machine	Labor	Production
Initial holding	a	1	a^α	b	1	b^α	$a+b$	2	$a^\alpha + b^\beta$
After labor hire	a	$1+\lambda$	$a^\alpha(1+\lambda)^\beta$	b	$1-\lambda$	$b^\alpha(1-\lambda)^\beta$	$a+b$	2	$a^\alpha(1+\lambda)^\beta +$ $b^\alpha(1-\lambda)^\beta$

capitalist, who is stronger than the worker (the same assumption as in the first half of Addendum V), in essence, the capitalist seeks to maximize this increase, that is, to maximize their total output. In this case, how much labor power is employed by the capitalist as hired labor? This magnitude can be derived by finding the λ that maximizes the total output. Therefore:

$$\frac{\partial}{\partial \lambda}\{a^\alpha(1+\lambda)^\beta + b^\alpha(1-\lambda)^\beta\} = a^\alpha \beta(1+\lambda)^{\beta-1} + b^\alpha \beta(1-\lambda)^{\beta-1} = 0 \tag{A.79}$$

Solving this equation, we get the optimal λ for capitalists: $\lambda^* = \dfrac{a^{\frac{\alpha}{1-\beta}} - b^{\frac{\alpha}{1-\beta}}}{a^{\frac{\alpha}{1-\beta}} + b^{\frac{\alpha}{1-\beta}}}$, which is clearly $0 < \lambda^* < 1$ from the assumption that $a > b > 0$. This means that when $b > 0$, the entire labor of the worker is not transferred to the capitalist as hired labor, and there is still a corresponding amount of labor on the worker's side as long as b amounts of capital exist. However, if workers have no capital at all ($b = 0$), they have no choice but to sell all their labor to the capitalist as hired labor. The results of this calculation show that if β, which represents the contribution of labor input to production, is 1, in other words, if the production function is linear, then the increase in total output becomes

$$[a^\alpha(1+\lambda)^\beta + b^\alpha(1-\lambda)^\beta] - (a^\alpha + b^\alpha) = (a^\alpha - b^\alpha)\lambda + (a^\alpha + b^\alpha) \tag{A.80}$$

This result shows that the production function is monotonically increasing under the assumption that $a > b$, and as a result, it reaches a maximum at $\lambda = 1$ when all labor is hired by the capitalist. Since this is a matter of the shape of the production function, it means that production is fully specialized in the case of a linear production function, but not in the case of the Cobb-Douglas type of production function where substitution between factors of production is assumed.

HISTORICAL TREND OF DISPARITIES IN FIRM SIZE

In this way, we have shown that the disparity of the capital holding gives rise to a disparity in labor usage in terms of labor employment, which in turn gives rise to a disparity at the level of production. Next, we show this disparity in terms of output and examine how it might change historically. For this purpose, we first calculate the initial disparity in production between the two business units as $(a/b)^\alpha$ before the labor hire, and its disparity after hiring labor becomes

$$\frac{a^\alpha(1+\lambda)^\beta}{b^\alpha(1-\lambda)^\beta} = \left(\frac{a}{b}\right)^\alpha \left(\frac{1+\lambda}{1-\lambda}\right)^\beta = \left(\frac{a}{b}\right)^\alpha \left(\frac{a}{b}\right)^{\frac{\alpha}{1-\beta}} = \left(\frac{a}{b}\right)^{\frac{\alpha(2-\beta)}{1-\beta}} \quad \text{[2]} \tag{A.81}$$

In this case, $\frac{\alpha(2-\beta)}{1-\beta}$ is always greater than α in the definition range of $0 < \beta < 1$, which indicates that the disparity has widened. In other words, as long as a/b is greater than 1 (as long as $a > b$, i.e., as long as the capitalist owns more capital than the worker), this disparity will always increase due to the movement of labor power by employing labor.

If this is so, then, the disparity in firm size depends entirely on the relationship in size between a and b. For this we need only the historical tendencies of a and b. For example, in the agricultural sector, as mentioned above, the historical disparity in Chinese farmers is very typical. It is because the use of ox-drawn plows, which finally appeared in the Spring and Autumn and Warring States Period, spread from a very few farmers in the beginning to all farmers later. As a result, there were three stages (1) small disparity before the appearance of ox-drawn plows, (2) widened disparity when some farmers used ox-drawn plows, and (3) narrowed disparity when all farmers used ox-drawn plows. This is understood to mean that the presence or absence of oxen and plows, which are "capital" for farmers, caused fluctuations in a/b among farmers, which in turn led to the disparity in production.

This disparity can be simulated by a numerical example in which only a among a and b rises to a certain point, but then stops rising, and then b starts to catch up. The results are shown in **Table A.6**. Here, we assume that $\alpha = \beta = 0.5$.

Table A.6 illustrates the historical changes in the agricultural sector: starting with the homogenization of small farmers => class differentiation among farmers => homogenization of small farmers. This is evidenced by the fact that every farmer in Japan now owns some agricultural machinery (and, incidentally, every small-scale self-employed fisherman in Japan who goes fishing in coastal and offshore waters owns their own fishing boat). In subsection 4.4.5 of this book, we show the widening and then narrowing trends of the income disparity between countries. If there is an optimal amount of capital accumulation for farmers (and fishermen), we can draw a Kuznets curve in which the income disparity between countries first widens and then gets smaller.

---

[2] In this transformation, we used the result of calculation of λ^*.

Table A.6 Images of historical changes in disparity expansion and contraction

Period	1	2	3	4	5	6	7	8	9
Capital holding of the preceding firm (a)	1	2	3	4	5	5	5	5	5
Capital holding of the following firm (b)	1	1	1	1	1	2	3	4	5
Capital disparity (a/b)	1	2	3	4	5	2.5	1.67	1.25	1
Output disparity	1	2.83	5.20	8.00	11.18	3.95	2.15	1.40	1.00

Mathematical Appendix
How to Solve the Dynamic Optimization Problem

Marxian optimal growth theory in this book sets up and solves various models using the assumption that an economic agent behaves optimally from the present into the distant future. This method, which uses Hamiltonian, is generally established as dynamic optimization problems. Since understanding Hamiltonian is crucially important to understand this book, we give an intuitive explanation in this appendix. Concretely speaking, we explain the Lagrangian as a discrete model and the Hamiltonian as two continuous models. Although main body in this book only uses a continuous model, we start by introducing the discrete model, since it's first-order conditions are simple and intuitively understandable.

LAGRANGIAN TO SOLVE DISCRETE MODEL

To begin with, let take an example a discrete type of example. This seeks to maximize $\sum_{t=0}^{\infty} \beta^t \log Y_t$ under the following three constraints:

$$\max \sum_{t=0}^{\infty} \beta^t \log Y_t$$
$$s.t. \begin{cases} Y_t = AK_t^\alpha (s_t L)^{1-\alpha} \\ K_{t+1} - K_t = B(1-s_t)L - \delta K_t \\ 0 \le s_t \le 1 \end{cases} \quad (M.1)$$

Here, Y_t and K_t and s_t, respectively, denote the means of production, the means of production, and the allocations of total labor L for the consumption goods sector. $0 < \beta < 1$ denotes the discount rate by time, and $0 < \delta \le 1$ denotes the depreciation rate. Y_t, K_t, s_t may vary with time t. ρ, δ are constant variables and s_t is a control variable. The economic entity chooses the optimal value s_t at any time. $\sum_{t=0}^{\infty} \beta^t \log Y_t$ is the objective function and must be maximized. $Y_t = AK_t^\alpha (s_t L)^{1-\alpha}$ is a production function of the means of consumption and is a linear homogeneous Cobb-Douglas type of function. $K_{t+1} - K_t = B(1-s_t)L - \delta K_t$ is a capital accumulation equation. To solve this optimization problem, transpose all terms to the right-hand side from the left-hand side in the capital accumulation equation:

$$B(1-s_t)L + (1-\delta)K_t - K_{t+1} = 0 \quad (M.2)$$

MARXIAN ECONOMICS

Multiplying (M.2) by the Lagrange multiplier μ_t and adding the instantaneous utility function, we obtain the Lagrangian ($\mathcal{L}$). The Lagrange multiplier μ_t is the price of a unit of the means of production measured in utility from an economics perspective:

$$\mathcal{L} = \sum_{t=0}^{\infty} \beta^t [\log Y_t + \mu_t \{B(1-s_t)L + (1-\delta)K_t - K_{t+1}\}] \tag{M.3}$$

Using the Lagrangian, we can transform a complicated problem with constraints into a simple problem without constraints. That is to say, our solutions can be obtained as the first-order conditions of the Lagrangian ($\mathcal{L}$):

$$\begin{cases} \dfrac{\partial \mathcal{L}}{\partial s_t} = 0 \\ \dfrac{\partial \mathcal{L}}{\partial K_t} = 0 \\ \dfrac{\partial \mathcal{L}}{\partial \mu_t} = 0 \end{cases} \tag{M.4}$$

Technically speaking, there are certain mathematically difficult problems, but this is formally the same as the normal way of using the Lagrange multiplier. This Lagrangian ($\mathcal{L}$) is transformed into the following:

$$\mathcal{L} = \cdots + \beta^{t-1}[\log Y_{t-1} + \mu_{t-1}\{B(1-s_{t-1})L + (1-\delta)K_{t-1} - K_t\}] \\ + \beta^t [\log Y_t + \mu_t \{B(1-s_t)L + (1-\delta)K_t - K_{t+1}\}] + \cdots \tag{M.5}$$

Therefore, its first-order conditions become as follows.

Since $\dfrac{\partial \mathcal{L}}{\partial s_t} = 0$, $\dfrac{1}{Y_t}\dfrac{\partial Y_t}{\partial s_t} - \mu_t BL = 0$, then

$$\dfrac{1-\alpha}{s_t} = \mu_t BL \tag{M.6}$$

is obtained. Since $\dfrac{\partial \mathcal{L}}{\partial K_t} = 0$, $-\beta^{t-1}\mu_{t-1} + \beta^t \left\{ \dfrac{1}{Y_t}\dfrac{\partial Y_t}{\partial K_t} + \mu_t(1-\delta) \right\} = 0$, then

$$\mu_{t-1} = \beta \left\{ \dfrac{\alpha}{K_t} + \mu_t(1-\delta) \right\} \tag{M.7}$$

is obtained. Since $\dfrac{\partial \mathcal{L}}{\partial \mu_t} = 0$, $B(1-s_t)L + (1-\delta)K_t - K_{t+1} = 0$, then

$$K_{t+1} - K_t = B(1-s_t)L - \delta K_t \tag{M.8}$$

is obtained.

MATHEMATICAL APPENDIX

Because the last equation is a capital accumulation equation, it turns out that the Lagrangian ($\mathcal{L}$) contains the original constraints. In other words, using Lagrangian can transform an optimization problem with constraints into another with no constraints.

PRESENT VALUE HAMILTONIAN TO SOLVE CONTINUOUS MODEL

Next, we set up the optimization problem of continuous model corresponding to the above discrete type of model:

$$\max \int_0^\infty e^{-\rho t} \log Y \, dt \\ s.t. \begin{cases} Y = AK^\alpha (sL)^{1-\alpha} \\ \dot{K} = B(1-s)L - \delta K \\ 0 \leq s \leq 1 \end{cases} \tag{M.9}$$

The symbols in this equation are identical with those in the discrete model. For the sake of simplicity, we omit time t. The objective function is the integral function and the capital accumulation equation is differential equation. Thus, you can think of $K_{t+1} - K_t$ in the discrete model as $\dot{K}$ in the continuous one. However, take note that the control variable s is special, because it can be jumped from 0 to 1 at a certain point (ex bang-bang control).

Although Y, the amount of consumption at time t, is evaluated as $\log Y$ instantaneous at time t from the viewpoint of utility, Y must be devaluated by multiplying $e^{-\rho t}(<1)$ if evaluated at time 0 (refer subsection 4.3.2). The sum of the instantaneous utility at any time evaluated at time 0 becomes $\int_0^\infty e^{-\rho t} \log Y \, dt$, and named intertemporal utility.

Here, we measure the utility at time t from time 0. Producing the means of production is not directly useful for consumption in the present term, but it is indirectly useful to produce the means of consumption in the future. Although instantaneous utility obtained from the means of production is $\log Y$, we do not know how much producing the means of production contributes to total utility using this information alone. However, since the means of production is used to produce the means of production in the next term or later, the price of the means of production must exist. Let λ be the price of the means of production measured in utility, and we then obtain an indicator that measures the contribution to total utility of the means of consumption and the means of production in the present. This is called the present value Hamiltonian, defined as follows:

$$H = e^{-\rho t} \log Y + \lambda \dot{K} = e^{-\rho t} \log Y + \lambda \{B(1-s)L - \delta K\} \tag{M.10}$$

As you see, this is the sum of the present value utility taken by consumption and the newly accumulated capital stock measured in utility ($\lambda \dot{K}$). In this way, here λ is the transformer of the measurement from the physical term into utility. If we take one period as one year, H means the net national income measured in utility. The first-order conditions of Hamiltonian are as follows:

$$\begin{cases} \dfrac{\partial H}{\partial s} = 0 \\ \dfrac{\partial H}{\partial K} = -\dot{\lambda} \\ \dfrac{\partial H}{\partial \lambda} = \dot{K} \end{cases} \quad \text{(M.11)}$$

The Hamiltonian has a significant characteristic. It is optimal on the whole to choose control variables to maximize the Hamiltonian in each period. This characteristic is called Pontryagin's maximum principle. If the solution is not the extreme point ($s \neq 0,1$), but the interior point ($0 < s < 1$), this condition means $\dfrac{\partial H}{\partial s} = 0$.

In other words, $\dfrac{\partial H}{\partial K} + \dot{\lambda} = 0$ (transformed from $\dfrac{\partial H}{\partial K} = -\dot{\lambda}$) can be interpreted as follows. On the left side, $\dfrac{\partial H}{\partial K}$ is the marginal income (income gain) obtained per unit of capital. $\dot{\lambda}$ is the increase in the price of the means of production (capital gain). Summing up these two (total return) forms the total revenue from operating the means of production, that is, the marginal revenue obtained by operating one unit of the means of production. The point where total revenue is maximized is zero if the means of production are used efficiently. If it is negative, we must reduce our use of the means of production, and if it is positive, we must increase it. As a result, and at this point, its marginal value becomes zero.

The formula (income gain + capital gain = total return) itself is easy to understand if we take the case of the stock market. Share dividends are income gains and gains from price increases are capital gains. Their total is the total return. Both are generally positive, but share dividends in some special, rapidly growing companies are set at zero so that all the profit can be invested, such as Apple, Google, and Amazon until recently. This is because it is better to raise funds for dividends by turning all profits into investments than to pay dividends in these companies. In this case, the share price (i.e., corporate value) has increased by that amount, and so shareholders support such a policy.

However, this formula is different from the stock market in that that $\dfrac{\partial H}{\partial K}$ can also be negative. There can be zero dividends, but there is no negative dividend in the stock market. Also, since asset prices may fall, there can be negative $\dot{\lambda}$. This is a capital loss.

The third equation in (M.11) $\dfrac{\partial H}{\partial K} = \dot{K}$, leads $\dot{K} = B(1-s)L - \delta K$. This is the capital accumulation equation. In this way, the Hamiltonian contains the original constraints, hence an intertemporal optimization problem can be transformed into an optimization problem in one term. Furthermore, the following condition besides these three first-order conditions must be satisfied. This equation is called the transversality condition.

$$\lim_{t \to \infty} \lambda K = 0 \quad \text{(M.12)}$$

MATHEMATICAL APPENDIX

This condition shows that the price of the means of production (λ) converges to 0 if measured at time 0. If it converges to a positive value, the means of production is inefficiently produced.

CURRENT VALUE HAMILTONIAN TO SOLVE CONTINUOUS MODEL

Although above method measured utility at time 0, there is another method to measure at time t. When the price of a unit of the means of production is measured at time t be μ, we can set a different type of Hamiltonian as:

$$H_c = \log Y + \mu \dot{K} = \log Y + \mu\{B(1-s)L - \delta K\} \quad \text{(M.13)}$$

It is called current value Hamiltonian (H_c), and the subscript c indicates that this is measured at time t. Therefore, there are the following relations between λ and μ and between H and H_c:

$$\lambda = e^{-\rho t}\mu, \quad H = e^{-\rho t}H_c \quad \text{(M.14)}$$

Under these circumstances, how different are the first-order conditions of the present value Hamiltonian and the current value Hamiltonian? To find out, we need the first-order conditions with respect to K. For this purpose, the relation $\lambda = e^{-\rho t}\mu$ can be differentiated by t into

$$\dot{\lambda} = (e^{-\rho t}\mu)' = (e^{-\rho t})'\mu + e^{-\rho t}\dot{\mu} = -\rho e^{-\rho t}\mu + e^{-\rho t}\dot{\mu} = -e^{-\rho t}(\rho\mu - \dot{\mu}) \quad \text{(M.15)}$$

Then, using this relation, the following can be obtained:

$$\frac{\partial H_c}{\partial K} = \frac{\partial(e^{\rho t}H)}{\partial K} = e^{\rho t}\frac{\partial H}{\partial K} = -e^{-\rho t}\dot{\lambda} = \rho\mu - \dot{\mu} \quad \text{(M.16)}$$

Since partial derivatives with respect to s or μ do not influence $e^{-\rho t}$, after all, the first-order conditions of the current value Hamiltonian become as follows:

$$\left.\begin{aligned}\frac{\partial H_c}{\partial s} &= 0 \\ \frac{\partial H_c}{\partial K} &= \rho\mu - \dot{\mu} \\ \frac{\partial H_c}{\partial \mu} &= \dot{K}\end{aligned}\right\} \quad \text{(M.17)}$$

They differ from the first-order conditions of the present value Hamiltonian only at $\rho\mu$. We should memorize these equations as formulas.

We considered the economic interpretation only of the first-order conditions of the present value Hamiltonian, but we can do the same in the case of the current value Hamiltonian. Transforming the second equation, the following is obtained:

MARXIAN ECONOMICS

$$\frac{\partial H_c}{\partial K} + \dot{\mu} = \rho\mu \tag{M.18}$$

Both sides in this equation are measured in utility. Suppose that you have an asset μ, and you invest it. The left side means the benefit you obtain from purchasing and using the means of production. $\frac{\partial H_c}{\partial K}$ is the direct revenue obtained from direct productive activity, and $\dot{\mu}$ is capital gain (or loss). The right side indicates the return taken by investing μ at ρ, which is he interest rate measured in utility. Therefore, the first-order conditions of this current value Hamiltonian express that economic entities are carrying out their activities comparing their rate of return with ρ. In addition, the transversality condition is:

$$\lim_{t \to \infty} e^{-\rho t} \mu K = 0 \tag{M.19}$$

Here μK is the price of the means of production at time t measured in utility, and multiplying $e^{-\rho t}$, it can be transformed into its current value measured at time t as $e^{-\rho t} \mu K$. Hence this equation shows that the price of the means of production converges to 0 over time.

HAMILTONIAN TO SOLVE MARKET MODEL

We now consider how this explanation on the Hamiltonian in the social planner's model is related to the decentralized market model (in Addendum II). To examine this question. we need to revise the continuous model in the previous subsection as follows:

$$\begin{aligned} & \max_{K_2, L_1, L_2} \int_0^\infty e^{-\rho t} \log Y \, dt \\ & s.t. \begin{cases} Y = AK_2^\alpha L_2^{1-\alpha} \\ \dot{K} = BL_1 - \delta K \\ 0 \leq K_2 \leq K \\ L_1, L_2 \geq 0, \ L_1 + L_2 \leq L \end{cases} \end{aligned} \tag{M.20}$$

Here, K_2, L_2 and L_1 are the amount of the means of production, the labor allocated to produce the means of consumption, and the labor allocated to produce the means of production, respectively. Subscripts 1 and 2 represent the first and second sectors, respectively. K_2, L_1, L_2 are not negative values, and by these definitions, $K = K_1 + K_2$ and $L = L_1 + L_2$.

We set the current value Hamiltonian $\bar{H}_c$ with constraints as follows. Here we set $U(Y) = \log Y$ to facilitate visualization:

$$\begin{aligned} \bar{H}_c &= U(Y) + \mu \dot{K} + R(K - K_2) + W(L - L_1 - L_2) \\ &= \log Y + \mu(BL_1 - \delta K) + R(K - K_2) + W(L - L_1 - L_2) \end{aligned} \tag{M.21}$$

MATHEMATICAL APPENDIX

$$\begin{cases} \dfrac{\partial \bar{H}_c}{\partial K_2} = 0 & \Leftrightarrow R = \dfrac{\partial \log Y}{\partial K_2} \quad \Leftrightarrow R = \dfrac{\partial \log Y}{\partial Y}\dfrac{\partial Y}{\partial K_2} \\[2pt] \dfrac{\partial \bar{H}_c}{\partial L_1} = 0, \dfrac{\partial \bar{H}_c}{\partial L_2} = 0 \Leftrightarrow W = \mu \dfrac{\partial \dot{K}}{\partial L_1} = \dfrac{\partial \log Y}{\partial L_2} \Leftrightarrow W = \mu \dfrac{\partial \dot{K}}{\partial L_1} = \dfrac{\partial \log Y}{\partial Y}\dfrac{\partial Y}{\partial L_2} \\[2pt] \dfrac{\partial \bar{H}_c}{\partial K} = \rho\mu - \dot{\mu} & \Leftrightarrow R - \mu\delta + \dot{\mu} = \rho\mu \\[2pt] \dfrac{\partial \bar{H}_c}{\partial R} = 0 & \Leftrightarrow K_2 = K \\[2pt] \dfrac{\partial \bar{H}_c}{\partial W} = 0 & \Leftrightarrow L_1 + L_2 = L \\[2pt] \dfrac{\partial \bar{H}_c}{\partial \mu} = \dot{K} & \Leftrightarrow \dot{K} = BL_1 - \delta K \end{cases} \quad (\text{M.22})$$

In the first equation in (M.22), $\dfrac{\partial \log Y}{\partial Y} = \dfrac{\partial U(Y)}{\partial Y}(=p_2)$ means the price of consumption goods measured in utility. $\dfrac{\partial Y}{\partial K_2}$ is the marginal productivity of the means of production to produce the means of consumption. Hence R means the rental price of the means of production measured in utility. In the second equation in (M.22), $\dfrac{\partial \dot{K}}{\partial L_2}$ is the labor productivity to produce the means of production and μ is the price of the means of production measured in utility, hence W is equal to the price of marginal product measured in utility in both sectors of the means of production and consumption. As a result, W is interpreted as the wage rate measured in utility.

The third equation means there is no arbitration condition in the asset market. For example, suppose that we invest asset μ measured in utility. $R - \mu\delta + \dot{\mu}$ indicates the revenue measured in utility in the case that we purchase α unit of the means of production. R is rental price measured in utility, $\mu\delta$ is the depreciation of a unit of the means of production measured in utility, and $\dot{\mu}$ is the capital gain (or loss) measured in utility. On the other hand, $\rho\mu$ indicates the revenue obtained in a bank deposit with the interest rate ρ, measured in utility. This equation means that the rate of return is the same in all markets.

The last three equations in (M.22) show that this current value Hamiltonian contains the original constraints. Capital accumulation equation $\dot{K} = BL_1 - \delta K$ comes from the $\mu(BL_1 - \delta K)$ part of the Hamiltonian, and other two constraints $K_2 = K$ and $L_1 + L_2 = L$ come from the $R(K - K_2) + W(L - L_1 - L_2)$ part of Hamiltonian. Here, because the part $R(K - K_2) + W(L - L_1 - L_2)$ can be understood as the unused factors of production measured in utility, these first-order conditions means that unused factors should be zero if $(\bar{H}_c)$ is maximized. Hence, the current value Hamiltonian H_c can be understood as the real net national income, and the current value Hamiltonian with constraints $(\bar{H}_c)$ can be understood as the largest possible national net income (potential net national income).

MARXIAN ECONOMICS

What we want to clarify on the comparison of the dynamic optimization problems is the difference of the measurement. That is, as we explained here, all these variables μ, R and W are measured in utility, and p_2 is the price of the means of consumption measured in utility because it is derived as $\frac{\partial U(Y)}{\partial Y}$. Therefore, p, r, and w in the market model explained in Addendum II can be led by dividing μ, R, and W by p_2, respectively. That is,

$$p = \frac{\mu}{p_2}, \; r = \frac{R}{p_2}, \; w = \frac{W}{p_2} \tag{M.23}$$

Finally, we introduce the (real) interest rate $\tilde{r}$ which was shown by equation $\frac{\dot{Y}}{Y} = \tilde{r} - \rho$ in the Addendum II. However, because this left side formula $\left(\frac{\dot{Y}}{Y}\right)$ was deduced only from a special instantaneous utility function $U(Y) = \log Y$, here we introduce $\tilde{r}$ in the general formulation. For this purpose, first, picking up the equation $R - \mu\delta + \dot{\mu} = \rho\mu$ from seven first-order conditions of the current value Hamiltonian $\bar{H}_c$, and dividing both sides by μ, we then subtract $\frac{\dot{p}_2}{p_2}$ from both sides, and obtain

$$\frac{R}{\mu} - \delta + \frac{\dot{\mu}}{\mu} - \frac{\dot{p}_2}{p_2} = \rho - \frac{\dot{p}_2}{p_2} \tag{M.24}$$

Then, because $\frac{R}{\mu} = \frac{\frac{R}{p_2}}{\frac{\mu}{p_2}} = \frac{r}{p}$ and $\frac{\dot{p}}{p} = \frac{\dot{\mu}}{\mu} - \frac{\dot{p}_2}{p_2}$ ($\because \log p = \log \mu - \log p_2$), the equation (M.24) becomes:

$$\frac{r}{p} - \delta + \frac{\dot{p}}{p} = \rho - \frac{\dot{p}_2}{p_2} \tag{M.25}$$

Now, considering the interest rate equalization condition in the asset market in Addendum II: $\tilde{r} = \frac{r}{p} - \delta + \frac{\dot{p}}{p}$, it becomes

$$\tilde{r} = \rho - \frac{\dot{p}_2}{p_2} \tag{M.26}$$

This is a special case in which instantaneous utility function is set as a logarithm form. Therefore, now we go to generalize this formula. First, because by definition the differentiation of the instantaneous utility by consumption measured in utility is the price of the means of production measured in utility, if we express $\frac{\partial U(Y)}{\partial Y}$ as $U_Y(Y)$, and $U_Y(Y) = p_2$, we can transform the equation (M.26) into

MATHEMATICAL APPENDIX

$$\tilde{r} = \rho \frac{\frac{dU_Y(Y)}{U_Y(Y)}}{\frac{dY}{Y}} \cdot \frac{\dot{Y}}{Y} \quad \left(or \quad \tilde{r} = \rho - \frac{d\log U_Y(Y)}{d\log Y} \frac{\dot{Y}}{Y} \right) \tag{M.27}$$

where we used the differential formulas for composite functions, and $\frac{\dot{Y}}{Y}$ part and the $-\frac{\frac{dU_Y(Y)}{U_Y(Y)}}{\frac{dY}{Y}}$ part mean the growth rate of consumption and the elasticity of the marginal utility, respectively. In this case, because $-\frac{\frac{dU_Y(Y)}{U_Y(Y)}}{\frac{dY}{Y}} = -\frac{\frac{dp_2}{p_2}}{\frac{dY}{Y}}$, $-\frac{\frac{dU_Y(Y)}{U_Y(Y)}}{\frac{dY}{Y}}$ can be understood as the elasticity of price measured in utility to consumption. In other words, it expresses the rising rate of the price of the means of production when consumption increases by 1 percent.[1]

From the above, we can see that the second term on the right side of the equation (M.27) is the rate of return (benefit) obtained from consumption, and adding the time preference rate ρ to it equals the real interest rate $\tilde{r}$ on the left side. Consumption takes place in the present, and interest is the price paid in the future for present consumption to endure and postponing it to the future. The time preference rate indicates the degree to which the future is discounted to evaluate the present value. Thus, this equation can be understood as the equation of equalization of present and future rates of return and is called the Euler equation.

In addition, because the instantaneous utility function was set to $U(Y) = \log Y$ in Addendum II, price of the means of consumption measured in utility becomes $p_2 = U_Y(Y) = \frac{1}{Y}$ and therefore $\frac{\dot{p}_2}{p_2} = -\frac{\dot{Y}}{Y}$. In this case, the Euler equation becomes

$$\frac{\dot{Y}}{Y} = \tilde{r} - \rho \tag{A.109}$$

This is not a detailed explanation on how the first-order conditions are deduced from the Hamiltonian. If you wish to study Hamiltonians in more detail, see Mathematical Appendix A.3 in Barro and Sala-i-Martin (2004).

(This appendix is written by Ryo Kanae.)

---

[1] Because usually $\frac{dU_Y(Y)}{dY} < 0$, $-\frac{\frac{dU_Y(Y)}{U_Y(Y)}}{\frac{dY}{Y}} > 0$.

References

Aoyagi, Kazumi. 2010. *Feminism and Political Economy*, 2nd ed. [In Japanese]. Tokyo: Ochanomizu Shobo.

Ardrey, Robert. 1961. *African Genesis: A Personal Investigation into the Animal Origins and Nature of Man*. New York: Atheneum Books.

Ardrey, Robert. 1976. *The Hunting Hypothesis*. Glasgow: William Collins Sons.

Barro, Robert J. and Xavier I. Sala-i-Martin. 2004. *Economic Growth*, 2nd version. Cambridge, MA: MIT Press.

Binmore, Ken. 2007. *Game Theory: A Very Short Introduction*, Oxford: Oxford University Press.

Böhm-Bawerk, Eugen von. 1890. *Capital and Interest, The History and Critique of Interest Theories*. Translated by William Smart, M.A. London: Macmillan.

Bortkiewicz, Ladislaus. 1906. "Wertrechnung und Preisrechnung im Marxschen system" [In German]. *Archiv für Sozialwissenschaft und Sozialpolitik* 23 (1): 1–50.

Boserup, Ester. 1965. *The Conditions of Agricultural Growth: The Economics of Agrarian Change Under Population Pressure*. London: George Allen & Unwin.

Bowles, Samuel. 2004. *Microeconomics*. Princeton, NJ: Princeton University Press.

Boyd, Robert and Joan B. Silk. 2009. *How Humans Evolved*, 5th ed. New York: W. W. Norton.

Buchannan, James M. and Richard E. Wagner. 1977. *Democracy in Deficit: The Political Legacy of Lord Keynes*. New York: Academic Press.

Dart, Raymond. 1953. "The Predatory Transition from Ape to Man." *International Anthropological and Linguistic Review* 1 (4): 201–17.

De Vries, J. 1975. "Peasant Demand Patterns and Economic Development: Friesland 1550–1750." In *European Peasants and Their Markets: Essays in Agrarian Economic History*, edited by W. N. Parker and E.L. Jones, 205–66. Princeton, NJ: Princeton University Press.

Diamond, Jared. 1997. *Guns, Germs, and Steel: The Fates of Human Societies*. New York: W.W. Norton.

Doi, Masaoki. 1966. *Jesus Christ* [In Japanese]. Tokyo: San'ichi Shobo.

Donald, Merlin. 1991. *Origins of the Modern Mind: Three Stages in the Evolution of Culture and Cognition*. Cambridge, MA: Harvard University Press.

Engels, Frederik. 1880. "Socialism: Utopian and Scientific." In *Marx/Engels Selected Works*, Volume 3, 95–151.

Engels, Frederik. 1902. *The Origin of Family, Private Property and State*. Translated by Ernest Untermann. Chicago: C.H. Kerr.

Engels, Frederik. 1954. *Dialectics of Nature*. Translated by C. Dutt. Moscow: Foreign Languages Publishing House.

Fagan, Brian. 2004. *The Long Summer: How Climate Changed Civilization*. New York: Basic Books.

Fang, Xing. 2000. "From Serfs to Middle Class Farmers in the Qing Dynasty" [In Chinese]. *Chinese Academy* 2: 44–61.

Fukutomi, Masami. 1972. "The Asian Mode of Production and the Concept of 'State Feudalism'." *Historical Journal (Rekishi Hyoron)* [In Japanese]. 262: 18–30.
Gao, Wende. 1980. *Study of the Mongolian Slavery System*. Hohhot: Inner-Mongolian Peoples' Publisher [In Chinese].
Graeber, David. 2011. *Debt: The First 5,000 Years*. Brooklyn, NY: Melville House.
Hamasaki, Reizo. 2012. *History of the Sea People and the Archipelago: Activities Spread to Fishing, Salt Production, Trade, etc.* [In Japanese]. Kanazawa: Hokuto Shobo.
Hara, George. 2017. *"Public Interest" Capitalism: The End of the US–British Type of Capitalism* [In Japanese]. Tokyo: Bungeishunju.
Harvey, David. 2014. "Afterthoughts on Piketty's Capital", available from http://davidharvey.org/2014/05/afterthoughts-pikettys-capital/ (Accessed: 3 November 2022).
Hayashi, Toshio. 2007. *Scythae and Hsiung-Nu Huns: Nomad Culture* [In Japanese]. Tokyo: Kodansha.
Hayashi, Toshio. 2008. *Birth of the Nomad Nation* [In Japanese]. Tokyo: Yamakawa-Shuppan.
Hicks, John Richard. 1969. *A Theory of Economic History*. Oxford: Oxford University Press.
Inoue, Yuichi and Yuuho Yamashita. 2013. "Industrialization and Social Capital Accumulation of Developing Country" [In Japanese]. *Dokkyo Economic Journal* 92: 41–9.
Institute for Fundamental Economic Science. 1995. *Japanese Corporate Society and Women* [In Japanese]. Tokyo: Aoki Shoten.
Institute for Fundamental Economic Science. 2011. *The Global Economic Crisis and Marxian Economics* [In Japanese]. Tokyo: Otsuki Shoten.
Ito, Koichi. 1995. *Thinking Mongolian Economic History* [In Japanese]. Kyoto: Horitsubunkasha.
Izumi, Hiroshi and Jie Li. 2005. "Total Factor Productivity and Total Labor Productivity" [In Japanese]. *Statistics* 89: 18–34.
Izumi, Hiroshi. 2014. *Calculation of Embodied Labor and Basic Economic Indicators* [In Japanese]. Tokyo: Otsuki Shoten.
Kakiuchi, Keiko. 2015. *Introduction of the Cheng-Zhu School* [In Japanese]. Kyoto: Minerva Shobo.
Kamiya, Kunihiro. 1967. "Fishermen's Stratification and Ruling Structure (2): Through the Actual State of Election Behavior in a Typical *Amimoto* Fishing Village" [In Japanese]. *Sociology* 13 (3): 67–98.
Kanae, Ryo. 2008. "The Reality of the 'Marxian Optimal Growth Model' and Its Value and Pricing Issues" [In Japanese]. *The Economic Review* 182 (5–6): 615–26.
Kanae, Ryo. 2011. "Marxian Economics and Macroeconomic Dynamics" [In Japanese]. *Letters of Economic Science* 126: 108–11.
Kanae, Ryo. 2013. *Marxian Optimal Growth Theory* [In Japanese]. Kyoto: Kyoto University Press.
Kawaoka, Takeharu. 1987. *People of the Sea: History and Folklore of Fishing Villages* [In Japanese]. Tokyo: Heibonsha.
Kikunami, Hiroshi. 2018. *How Might Be Discussed by 200 Years Old Marx?* [In Japanese]. Kyoto: Kamogawa Shuppan.
Kitamura, Yasuhiro. 2015. *Large Land Management and Society in Ancient Japan* [In Japanese]. Kyoto: Doseisha.
Kuruma, Samezo. 1957. *Theory of Form of Value and Theory of Exchange* [In Japanese]. Tokyo: Iwanami Shotten.

REFERENCES

Kusano, Yasushi. 1970. "The Reclamation of the Swamps and the Evolution of the Custom of Double Landownership in the Sung and Yuan China (Cont.)" [In Japanese]. *Reports of the Oriental Society* 53 (1): 42–77.
Lee, James and Feng Wang. 1999. *One Quarter of Humanity: Malthusian Mythology and Chinese Realities 1700–2000*. Cambridge, MA: Harvard University Press.
Lenin, Vladimir Ilyich. 1933. *Imperialism, the Highest Stage of Capitalism: A Popular Outline*. London: M. Lawrence.
Lenin, Vladimir Ilyich. 1977. "On the So-called Market Question." In *Lenin Collected Works*, Volume 1, 75–125. Moscow: Progress Publishers.
Liu, Yang. 2008. "Government in the Marxist Optimal Growth Model" [In Japanese]. *The Economic Review* 182 (4): 469–81.
Malthus, Thomas Robert. 1798. *An Essay on the Principle of Population*. London: J. Johnson.
Mao, Sanliang. 2003. "Trend of Regional Disparity and Regional Policy." In *Quantitative Analysis of the Chinese Economy* [In Japanese]. Edited by Hiroshi Onishi and Go Yano, 166–188. Kyoto: Sekaishisosya.
Marcos, and Yvon Le Bot. 1997. *El sueño zapatista*. Barcelona: Plaza & Janes.
Marx, Karl. 1857. *Pre-Capitalist Economic Formations*. Available from https://www.marxists.org/archive/marx/works/1857/precapitalist/index.htm.
Marx, Karl. 1871. *The Civil War in France*, 3rd ed. London: Edward Truelove.
Marx, Karl. 1887. *Capital: A Critique of Political Economy, Volume I*. First English edition, Translated by Samuel Moore and Edward Aveling, edited by Frederick Engels, 4th German edition changes included as indicated with some modernization of spelling by Progress Publishers, Moscow. Available from www.marxists.org/archive/marx/works/download/pdf/Capital-Volume-I.pdf (Accessed: 11 May 2021).
Marx, Karl. 1894. *Capital: A Critique of Political Economy, Volume II*. First English edition, 1907. Edited by Frederick Engels. Moscow: Progress Publishers. Available from www.marxists.org/archive/marx/works/download/pdf/Capital-Volume-II.pdf (Accessed: 11 May 2021).
Marx, Karl. 1907. *Capital: A Critique of Political Economy, Volume III*. First German edition, 1894. Edited by Frederick Engels, translated and published by International Publishers, New York. Available from www.marxists.org/archive/marx/works/download/pdf/Capital-Volume-III.pdf (Accessed: 11 May 2021).
Marx, Karl. 2022. *Critique of the Gotha Program*. Oakland, CA: PM Press.
Marx, Karl and Frederik Engels. 1994. *Marx Engels Collected Works*, Volume 34. New York: International Publishers.
Marx, Karl and Frederik Engels. 1910. *Manifesto of the Communist Party*. Chicago: Charles H. Company.
Mason, Paul. 2015. *Postcapitalism*. London: Exarcheia Ltd.
Matsuo, Tadasu. 1994. "On the Aporia of Joint Production" [In Japanese]. *The Bulletin of Japan Society of Political Economy* 31: 83–99.
Matsuo, Tadasu. 2007. "Labor Exploitation Theory as a Normative Theory" [In Japanese]. *Political Economy Quarterly* 43 (3): 55–67.
Mayer, Tom. 1994. *Analytical Marxism*. London: Sage.
Minami, Ryoshin. 1990. *Economic Development of China* [In Japanese]. Tokyo: Toyokeizai Shimposha.

Miyamoto, Kazuo. 2005. *A History of China 01: From Mythology to History* [In Japanese]. Tokyo: Kodansha.

Morimoto, Sousuke. 2011. "Labor Theory of Value and Time: Böhm-Bawerk's Critique of Marx" [In Japanese]. *The Economic Review* 185 (5): 47–62.

Morimoto, Sousuke. 2014. "New Interpretation as an Interpretation of *Capital*" [In Japanese]. *Political Economy Quarterly* 51 (3): 54–64.

Morioka, Koji. 2000. *Fraudulent Accounts* [In Japanese]. Tokyo: Iwanami-shoten.

Muto, Masayoshi. 2015. "Social Dilemma and Environmental Problem." In *Reading Society Using Mathematics* [In Japanese]. Edited by Kazuo Seiyama, 41–80. Tokyo: Yuhikaku.

Nagata, Takahiro. 2020. "Existential Condition of Mediator: Focusing on the Number of Negotiations" [In Japanese]. *Political Economy Quarterly* 57 (2): 104–11.

Nagaura, Kenji. 1985. "Steedman's Grasp of Labour Theory of Value—An Aspect of the Value Controversy" [In Japanese]. *Hitotsubashi Ronso* 93 (2): 153–68.

Nakamura, Satoru, ed. 1993. *The Despotic State and its Socio-economy in East Asia* [In Japanese]. Tokyo: Aoki Shoten.

Nakamura, Satoru. 1977. *Theory of Slavery and Serfdom System* [In Japanese]. Tokyo: University of Tokyo Press.

Nakamura, Satoru. 2013. "Theoretical Problems of the Chinese Despotic State" [In Japanese]. *For New Historical Science* 282: 68–79.

Negishi, Takashi. 1985. *Economic Theories in a Non-Walrasian Tradition*. Cambridge: Cambridge University Press.

Neumann, Erich. 1963. *The Great Mother: An Analysis of the Archetype,* translated from German by Ralph Manheim. Princeton, NJ: Princeton University Press.

Niida, Noboru. 1962. *Study of Chinese History of Law* [In Japanese]. Tokyo: University of Tokyo Press.

Nishida, Masanori. 1986. *The Settlement Revolution in Human History* [In Japanese]. Tokyo: Shinyosha.

Oakley, Kenneth P. 1959. *Man the Tool-Maker*. Chicago: University of Chicago Press.

Obata, Hiroki. 2016. *Jomon People the Seed-Sower: Latest Science Reverses Origin of Agriculture* [In Japanese]. Tokyo: Yoshikawakobunkan.

Obata, Michiaki. 2009. *Principle of Economics* [In Japanese]. Tokyo: University of Tokyo Press.

Oguri, Takanori. 2005. "Livedoor vs. Fuji Television Case and Japanese Capitalism" [In Japanese]. *Economy*, August, 143–51.

Ohnishi, Hiroshi. 2010. "Uneven Development of the World Economy: From Krugman to Lenin." *World Review of Political Economy* 1 (1): 51–69.

Okishio, Nobuo. 1957. "On Aggregate Supply Function" [In Japanese]. *Kobe University Economic Review* 7: 165–77.

Okishio, Nobuo. 1961. "Technical Change and the Rate of Profit" [In Japanese]. *Kobe University Economic Review* 4: 85–99.

Okishio, Nobuo. 1967. *Accumulation Theory* [In Japanese]. Tokyo: Chikuma Shobo.

Okishio, Nobuo. 1977. *Marxian Economics: Theory of Value and Price* [In Japanese]. Tokyo: Chikuma Shobo.

Okishio, Nobuo. 1978. *Development of Contemporary Economics* [In Japanese]. Tokyo: Toyokeizai Shinpo.

REFERENCES

Okishio, Nobuo. 1997. "Surplus Value and Technological Progress" [In Japanese]. *Economy*, October.
Okishio, Nobuo. 2001. "Competition and Production Prices." *Cambridge Journal of Economics*, 25 (4): 493–501.
Okoshio, Nobuo. 1965. *Basic Theory of Capitalist Economy (enlarged edition)* [In Japanese]. Tokyo: Sobunsha.
Okishio, Nobuo and Masanori Nozawa, eds. 1982. *Democratic Reform of the Japanese Economy and Vision of Socialism* [In Japanese]. Toyko: Otsuki Shoten.
Okishio, Nobuo and Masanori Nozawa, eds. 1983. *Quantitative Analyses of the Japanese Economy* [In Japanese]. Tokyo: Otsuki shoten.
Onishi, Hiroshi. 1990. "Learning from the Reality of Capitalism and Socialism." In *Where Are Socialism and Capitalism Heading?* [In Japanese]. Edited by Yamaguchi, Masayuki, Koji Morioka and Hiroshi Onishi, 69–98. Kyoto: Kamogawa Shuppan.
Onishi, Hiroshi. 1991. "On the Historical Nature of Productivity" [In Japanese]. *Bulletin of Japan Society of Political Economy* 28: 85–99.
Onishi, Hiroshi. 1992. *Socialism as Pre-capitalism and Post-capitalism* [In Japanese]. Tokyo: Otsuki Shoten.
Onishi, Hiroshi. 1998a. "Changes in the Capital-Labor Ratios of National Currency Units and Macroeconomic Yields" [In Japanese]. *The Economic Review* 161 (1): 93–107.
Onishi, Hiroshi. 1998b. *Rise and Fall of the Pacific Rim Countries: Structure and Simulations of the Kyoto University Pacific Rim Model* [In Japanese]. Kyoto: Kyoto University Press.
Onishi, Hiroshi. 2001. "Marxian Economics in the 21st Century and the Task of the New Century" [In Japanese]. *Letters of Economic Science* 95: 85–90.
Onishi, Hiroshi. 2003a. *From Globalization to Military Imperialism* [In Japanese]. Tokyo: Otsuki Shoten.
Onishi, Hiroshi. 2003b. "Achievements of American Indian Studies of the East Part of North-America and Engels' Origin (1)" [In Japanese]. *The Economic Review* 172 (4): 1–19.
Onishi, Hiroshi. 2003c. "Achievements of American Indian Studies of the East Part of North-America and Engels' Origin (2)" [In Japanese]. *The Economic Review* 172 (5–6): 1–13.
Onishi, Hiroshi. 2004. "Implications of Native American Studies of the East Part of North-American on Historical Materialism" [In Japanese]. *Journal of Japanese Scientists* 39 (10): 546–51.
Onishi, Hiroshi. 2005. "Revision of Engels' 'Origin' Based on the New Achievements of Native American Studies" [In Japanese]. *Materialism and the Present*, 36: 29–43.
Onishi, Hiroshi. 2007. "Forming Kuznets Curve among Chinese Provinces." *Kyoto Economic Review* 76 (2): 155–63.
Onishi, Hiroshi. 2008. *Just What is the Tibet Issue?* [In Japanese]. Kyoto: Kamogawa Shuppan.
Onishi, Hiroshi. 2011. "Beijing Consensus: The Best Mix of State and Market in High Growth Period" [In Japanese]. *Political Economy Quarterly* 48 (3): 18–31.
Onishi, Hiroshi, ed. 2012. *Minority Problems and the Gap with the Han Chinese in China* [In Japanese]. Kyoto: Kyoto University Press.
Onishi, Hiroshi. 2016. "Middle Income Trap of China." *Proceedings of International Conference: Implications of a Possible PRC Growth Slowdown for Asia*, held by the Asian Development Bank Institute, Tokyo, November 25–26, 2015.
Onishi, Hiroshi. 2018. "Conditions Where the Ruled Class Unites for the Revolution: Applicability of a Game Theory on Social Dilemmas" [In Japanese]. *Political Economy Quarterly* 55 (2): 53–7.

Onishi, Hiroshi. 2019. "A Proof of Labor Theory of Value Based on Marginalist Principles." *World Review of Political Economy* 10 (1): 85–94.
Onishi, Hiroshi. 2020a. "Marxist-Leninist Models of Uneven Development of the World Capitalism, Imperialist War and Hegemon Change" [In Japanese]. *Political Economy Quarterly* 56, (4): 7–16.
Onishi, Hiroshi. 2020b. "Extensions of Onishi (2018) Social Movement Model into Majority Politics Model" [In Japanese]. *Political Economy Quarterly* 56 (2): 70–6.
Onishi, Hiroshi. 2020c. "Reproduction Scheme Including Commercial Sector and Optimal Weight of Commercial Sector Considering Turn-over of Capital" [In Japanese]. *Study on Politics and Economy*, 115, 27–35.
Onishi, Hiroshi. 2021a. "A Generalized Marxian Differential Rent Theory Based on Cobb-Douglas Production Function—Referring to Engels' Way of Calculation" [In Japanese]. *Mita Journal of Economics* 114 (1): 83–99.
Onishi, Hiroshi. 2021b. *Marxian Formal Economics on Politics* [In Japanese]. Tokyo: Keio University Press.
Onishi, Hiroshi and Atsushi Tazoe. 2011. "Organic Composition of Capital, Falling Rate of Profit and 'Preferential Growth of the First Sector' in the Marxian Optimal Growth Model." *Marxism* 21, 237–59.
Onishi, Hiroshi and Hideki Fujiyama. 2003. "Exploitation of Capital by Labor in the Marxian Optimal Growth Theory" [In Japanese]. *Working Papers of Graduate School of Economics of Kyoto University*, J-33, 1–13.
Onishi, Hiroshi and Ryo Kanae. 2015. "'Age of Large-Population Countries' and the Marxian Optimal Growth Theory" [In Japanese]. *Mita Journal of Economics* 107 (3): 139–55.
Otani, Teinosuke. 2011. *Marx's Theory of Association* [In Japanese]. Tokyo: Sakurai Shoten.
Ozaki, Yoshiharu. 1990. *Economics and Historical Change* [In Japanese]. Tokyo: Aoki Shoten.
Pasinetti, Luigi, L. 1977. *Lectures on the Theory of Production*. New York: Columbia University Press.
Petri, Fabio. 1980. "Positive Profits Without Exploitation: A Note on the Generalized Fundamental Marxian Theorem." *Econometrica* 48 (2): 531–33.
Piketty, Thomas. 2013. *Le capital au 21e siècle* [In French]. Paris: Editions du Seuil.
Policy Research Institute. 2001. *Report of the Study Group of Independence of the Local Economy and Public Investment* [In Japanese]. Ministry of Finance.
Pomeranz, Kenneth. 2000. *The Great Divergence: China, Europe, and the Making of the Modern World Economy*. Princeton: Princeton University Press.
Renfrew, Colin. 2007. *Prehistory*. London: Weidenfeld & Nicolson.
Robinson, Warren and Wayne Schutjer. 1984. "Agricultural Development and Demographic Change: A Generalization of the Boserup Model." *Economic Development and Cultural Change*, 32, 355–66.
Roemer, John. 1980. "A General Equilibrium Approach to Marxian Economics." *Econometrica*, 48 (2): 505–30.
Roemer, John. 1982. *A General Theory of Exploitation and Class*. Cambridge, MA: Harvard University Press.
Rosenstein-Rodan, P.N. 1961. "Notes on the Big Push." In *Economic Development for Latin America*, edited by H.S. Ellis, 57–81. Cambridge, MA: MIT Press.

REFERENCES

Rostow, Walt Whitman. 1960. *The Stages of Economic Growth*. Cambridge: Cambridge University Press.

Roxiangul, Wuful and Ryo Kanae. 2009. "Three Sector 'Marxian Optimal Growth Models' and Strong Accumulation Period" [In Japanese]. *The Economic Review* 183 (1): 79–87.

Sawada, Isao. 1996. *Hsiung-Nu Huns: Rise and Fall of the Ancient Nomad Nation* [In Japanese]. Tokyo: Toho Shoten.

Schumpeter, Joseph Alois. 1942. *Capitalism, Socialism and Democracy*. New York: Harper and Brothers.

Scott, James C. 2017. *Against the Grain: A Deep History of the Earliest States*. London: Yale University Press.

Sekine, Jun'ichi. 2017. "Mathematical Considerations of Large-scale Industry in Marx's *Capital*" [In Japanese]. *Mita Journal of Economics* 110 (2): 71–92.

Service, Elman R. 1962. *Primitive Social Organization: An Evolutionary Perspective*. New York: Random House.

Shaikh, Anwar M. and E. Ahmet Tonak. 1994. *Measuring the Wealth of Nations*. Cambridge: Cambridge University Press.

Shareholder Ombudsman. 2002. *Companies Can Be Changed* [In Japanese]. Tokyo: Iwanami Shoten.

Shibata, Kei. 1935. *Theoretical Economics:* Volume 1 [In Japanese]. Tokyo: Koubundou.

Shiozawa, Kimio and Tetsuo, Kondo. 1989. *Introduction to Economic History* (new edition) [In Japanese]. Tokyo: Yuhikaku.

Stanford, Craig Britton. 1999. *The Hunting Apes: Meat Eating and the Origins of Human Behavior*. Princeton, NJ: Princeton University Press.

Steedman, Ian. 1975. "Positive Profit with Negative Surplus Value." *Economic Journal* 85 (337): 114–23.

Steedman, Ian. 1977. *Marx After Sraffa*. London: NLB.

Stout, Dietrich. 2016. "Tales of a Stone Age Neuroscientist." *Scientific American,* April, 2016.

Tabata, Minoru. 2015. *Marx and Association*. [In Japanese]. Tokyo: Shinsensha.

Taguchi, Satsuki. 2018. "Institution of Coastal Fisheries in Japan and Democratization of Fisheries" [In Japanese]. *Norin-Kinyu*, April 2018.

Takahashi, Miki. 1995. *A Study of the Social History of Fisheries in the Early Modern Period: The Development and Rise of Fishery Policy in the Early Modern Period.* [In Japanese] Osaka: Seibundo Shuppai.

Takeda, Nobuteru. 1983. *Money and Form of Value* [In Japanese]. Matsudo: Azusa Shuppan.

Takeda, Nobuteru. 1984. "Theory of Form of Value and Theory of Exchange—Controversies on the Inevitability of money." In *Capital as a System 2* [In Japanese]. Edited by Ryozo Tomizuka et al., 359–79. Tokyo: Yuhikaku.

Tazoe, Atsushi and Huan Liu. 2011. "Production Growth and Increasing Intensity Under Human Pressure" [In Japanese]. *Political Economy Quarterly* 49 (2): 55–64.

Tazoe, Atsushi. 2011. "Exploitation Vanishing on the Balanced Growth Path" [In Japanese]. *The Economic Review* 185 (2): 73–81.

Tazoe, Atsushi. 2015. "The Effectiveness of Capital Accumulation in the Japanese Economy" [In Japanese]. *Statistics* 109: 1–11.

Tazoe, Atsushi. 2016. "From Marxian Optimal Growth Theory to Theory of Matured Society" [In Japanese]. *Letters of Economic Science* 139: 61-6.
Terada, Hiroaki. 1983. "The Legal Characteristics of Tianmian Tiandi (田面田底) Custom: A Conceptual Analysis" [In Japanese]. *Journal of Institute for Advanced Studies on Asia*, 93: 33-131.
Terasawa, Kaoru. 2000. *Birth of Kingdom* [In Japanese]. Tokyo: Kodansya.
Tomoyori, Hidetaka. 2019. *AI and Capitalism* [In Japanese]. Tokyo: Hon'noizumisha.
Toner, Jerry. 2014. *How to Manage Your Slaves by Marcus Shidonius Falx*. London: Profile Books.
Tsutsumi, Mika. 2008. *A Big But Poor Country: USA* [In Japanese]. Tokyo: Iwanami Shoten.
Warlas, Leon. 1926. *Éléments d'économie politique pure; ou, Théorie de la richesse sociale*. Paris: R. Pichon et R. Durand-auzias.
Weber, Max. 1924. *Gesammelte Aufsätze zur Sozial- und Wirtschaftsgeschichte*, Tübingen, J.C.B. Mohr.
Wittfogel, Karl August. 1957. *Oriental Despotism—A Comparative Study of Total Power*. New Haven, CT: Yale University Press.
Wuritaoketaohu. 2006. *Mongolian's Nomad Economy and its Transformation* [In Chinese]. Beijing: Minzu University of China Press, in Chinese.
Yamaguchi, Toru. 2007. *History of Coastal Fisheries* [In Japanese]. Tokyo: Seizando Shoten.
Yamamoto, Shichihei. 1971. *Spirit of Japanese Capitalism* [In Japanese]. Tokyo: Kobunsha.
Yamashita, Yuuho and Hiroshi Onishi. 2002. "Reconstructing Marxism as a Neoclassical Optimal Growth Model" [In Japanese]. *Study on Politics and Economy (Seikeikenkyu)* 78: 25-33.
Yamashita, Yuuho. 2005. "Roemer's Exploitation in the Neo-classical 'Marxist Model' of Growth" [In Japanese]. *Political Economy Quarterly* 42 (3): 76-84.
Yasuda, Yoshinori. 2004. *History of Civilization and Climate Change* [In Japanese]. Tokyo: NTT Press.
Yasuda, Yoshinori. 2009. *Rice Farming and Fishing Civilization: from the Yangtze River Civilization to the Yayoi Culture* [In Japanese]. Tokyo: Yusankaku.
Yoneda, Kenjiro. 1968. "The Establishment of the System of 240 Bu 步 equaling 1 Mu 畝: One Aspect of the Reforms of ShangYang" [In Japanese]. *Study of Oriental History* 26 (4): 417-50.
Yoshihara, Naoki. 2008. *Toward a Welfare Theory of Labor Exploitation* [In Japanese]. Tokyo: Iwanami Shoten.
Yoshii, Shunya. 2018. "The Historical Transition Model of Agricultural Management Scale Disparity—Generalization of Onishi (2012)'s Addendum 3" [In Japanese]. *Studies of Politics and the Economy (Seikeikenkyu)* 110: 101-11.
Yoshii, Shunya. 2020. "A Marxian Optimal Growth Model Incorporating Ox-Drawn Plows as a Means of Production" [In Japanese]. Mimeo.
Yukijima, Koichi. 2008. *Scythae; History and Archaeology of Horse-Riding Nomad Nation* [In Japanese]. Tokyo: Yuzankaku.

Index

fig refers to a figure; *n* to a note; *t* to a table

Afghan war 141
Africa, agriculture in 204–6
agricultural revolution 167, 199*n*, 201–4
agriculture 19–20, 23, 41, 161*fig*, 161*n*,
    199–213, 214*fig*
  and symbolic memorials 214–5
  use of ox-drawn plows in 210, 212–3, 266
altruism 42, 43*fig*
Analytical Marxism 77–83, 78*n*, 79*n*, 263–7,
    264*t*, 265*t*
Aoyagi, Kazumi 23
apprenticeship system 84, 96, 99
Ardrey, Robert 226
artificial intelligence (AI) 145*n*
artisans *see* craft economy
Asia, agriculture in 203–4
Association of Southeast Asian Nations
    (ASEAN) 148
Aztec civilization 211

Balkan Bulgarian Airlines 154
banknotes 65–6
banks and banking 50*n*, 162, 185, 188, 188*n*
barter economy 61, 63
Binmore, Ken 31*n*
Blackfoot people 228
Böhm-Bawerk, Eugen von 122–3, 124
Borneo
  Chinese farmers in 200–1
  headhunting in 22, 200
Boserup, Ester 204–5
bourgeois economics 25–6, 39
Bowles, Samuel 42–4, 44*n*, 80, 146
Boyd, Robert and Joan B. Silk 226*n*
Brahmanism 18*n*
British colonialism 162, 163

Buchanan, James M. and Richard E. Wagner 163
Buddhism 21, 26*n*, 29
Bush, George W. 39

Cambridge capital controversy 130*n*
capital 71–2
  circulation of 172–5, 173*fig*
  commercial capital 172–3, 174
  constant and variable 90, 187*n*, 245
  self-valorization of 66–7, 70–7, 82–5
capital accumulation 86, 106, 117–8, 121–2,
    121*fig*, 139, 140*fig*
  by workers 258–9
  see also primitive accumulation of capital
capital turnover 175–6, 176*t*, 176*n*
capital/labor relations 88, 94, 144–5, 144*t*
capitalism 25–6, 40, 82, 86–9, 89*fig*
  growth and death of 105–6, 139
  legitimacy of 132–3
  stages of 166*t*
  violence of 164
  see also monopoly capitalism; post-capitalist;
    pre-capitalist; state capitalism
capitalist class 256, 29–30, 31, 223
  relations with working class xii, 134, 144–5
capitalists 82, 116
  personality of 40–2
child labor 92, 98, 163
China 48, 148*n*, 168–70
  agriculture in 41*n*
  Cultural Revolution in 168
  election systems in 22–3
  ethnic conflict in 23
  Xinjiang Uygur Autonomous Region 21, 23
  for other topics relating to China, *see* the
    topic, e.g. serfdom

Chinese Marxism 25n, 26, 26n
Christianity 20–1, 29
Chu-tzu doctrine 20
class xi–xii, 23–4, 29
　advocates for 28–9, 28fig
　see also capitalist class; ruled and ruling class; working class
class dictatorship 30n
class division 16, 216, 242
class struggle 23–5, 91–2, 144
class-exploitation correspondence principle 77–8, 78t
clerical work 96
climate change 202
Clinton, Bill 141
Cobb-Douglas function 4, 106–7, 190, 264n, 265
colonialism 161–2
commercial sector 103–4, 251–4
commodity economy 47–50, 53, 60fig, 65
commodity exchange 48, 66–7
commodity-producing economies 53–7, 54fig, 86
Communist Party of China 169–70
Confucianism 20, 29, 40, 84
construction industry 99–102
　carpenters in 99
　trade unions in 99, 101–2t
contestable exchange theory 80, 81fig, 146, 146n
craft economy 39–41, 84–5, 217–8
credit system 50n, 188
Cuba 169n

Dart, Raymond 226
deferred gratification 44–5
democratic regulation theory 152
Deng Xiaoping 169, 170
dialectical method 87–8, 89
division of labor 11, 47, 50, 53, 93–5, 99, 103–4, 224
doctors 102
Doi, Masaoki 29
Donald, Merlin 227–8
Dutch slave trade 161–2, 163

early starter and latecomer countries 147–50, 147fig, 149fig, 150fig, 151fig
East India Company 162
economic bubbles 141, 143–4, 194n, 195–6, 195fig
education 97–8, 104
employment laws 160
enclosures 157–9
Engels, Friedrich 30n, 154, 216
　Origin of the Family, Private Property and the State 2n
　'The Part Played by Labor in the Transition from Ape to Man' 225
　Socialism: Utopian and Scientific xi, 106
exchange value 57–8, 65, 171
exploitation 72, 77, 124–6, 134, 164, 171

factory acts 92, 92n, 97–8
Fagan, Brian 202
family system 98
feudalism 20, 40, 84, 96, 139, 209n
fishing industry 201–2, 222–5, 224n
Franklin, Benjamin 10
French Revolution 35
Fuji Television 154
Fukushima nuclear disaster (2011) 24, 142
Fundamenntal Marxism Theorem (FMT) xii, 73–7, 73fig, 106, 124, 183, 185, 197, 231–4

game of chickens 32–3, 33t
game theory 31–2, 31n, 80–1
Gao, Wende 220, 221
gender 23–4
　see also women
Gladstone, William Ewart 160
Global Financial Crisis (2008) 52
gold standard 65–6
government bond system 162–3
grant economy 57

Hamiltonian method 123, 270–7
Harvey, David 126

headhunting societies 22
health services 89, 102–3
Hegelian dialectics 139n
hegemonic and non-hegemonic countries 37, 37t
Hicks, John Richard 52–3
Hinduism 26n
historical materialism xi–xii, 1, 36, 87, 106, 170
homo economicus hypothesis 42–3, 45
horse-riding 219
hospitals 102–3
hostile takeovers 154–5
Hsiung-Nu Hun people 219–21
Hua Guofeng 170n
human development 225–9
human personalities 41–2, 44–5
humans
    and imagination 9
    and reproduction 2n
    as a species being 1–2
hunter-gatherer societies 22, 43–4, 200, 201
hunting 226–7, 226n

ideologues 26–8, 38
Inca civilization 19
Indonesia, coup (1967) 167–8
industrial capital 116, 160–3, 173–4, 177, 185
Industrial Revolution 15, 30, 52, 99
    and investment 164–5, 165fig
information goods 134n
interest
    agio theory of 122, 124
    financial 124, 185–6
    see also self-interest
intertemporal utility 118
investment 143–4, 165–6, 165fig
    decline in 117, 117fig
Iraq war 39, 141
Islam 26n, 29
Israel, and money 65n
Ito, Koichi 220, 221
Izumi, Hiroshi and Jie Li 72n

Japan
    clothing 21–2
    copper weaponry in 18n, 20, 21
    earthquake (2011) 142
    Edo era 21, 104, 209n
    expressway project 143fig
    Jomon era 222
    post-WWII 41
    Meiji Restoration era 35
    Yayoi people 200–1, 222
    for other topics relating to Japan, *see* the topic, e.g. labor; slavery
Japanese economy 141, 142
Japanese Marxism 86
joint production 231–4
joint stock companies 153–5
justice, category of 38–9

Keynes, John Maynard 130n
Kuznets curve 148, 264

labor 2, 121n
    abstract labor 58n
    alienation of 17–18
    dead labor 11–12, 18
    nerve labor 145, 145n
    productive and non-productive 2, 52, 52n, 76, 76fig
    unpaid labor 16, 184
    see also division of labor; skilled and unskilled labor
labor hire 263–5
labor input 3–8, 128fig, 129t
labor theory of value (LTV) 4, 7–8, 11–12, 58, 124, 191
Lagrangian method 269–70
land 183–93, 189n, 194
    accumulation of 203–8, 205fig, 212–3
    ownership of 157, 190–2, 196–8, 209–10, 211n
    fertility of 189–90
    see also rent

latecomer countries *see* early starter and latercomer countries
Leftism 27, 28
Lenin, V.I. *Imperialism: the Highest Stage of Capitalism* 148
Liberal Democratic Party (Japan) 143
liberation theology 29
Livedoor (internet provider) 154
livestock farming 202, 219–20
luxury goods 93n, 138n

machine-based production 14n, 16 51–2, 83–5, 83*fig*, 87, 95n, 106
  cooperation in 93–5
  division of labor in 93, 94
macro-monetary interpretation 182–4, 183n
Malthus, Thomas *An Essay on the Principles of Population* 206
market equilibrium 238–40
market system xii, 80, 86, 134, 171, 236, 241–2
Marx, Karl and Marxism 6–7, 9, 27–8, 39, 82, 154, 156, 170
  Capital xii, 6–7, 20, 47, 60, 71, 79, 93, 134, 171–2, 191
  The Civil War in France 41–2
  Critique of the Gotha Program 45
  Foundations of the Critique of Political Economy 209n
  Pre-Capitalist Economic Formations 199
  and 'Robinson Crusoe' story 126
  for topics discussed by Marx, *see* the topic, e.g. capital accumulation
  see also Analytical Marxism; Chinese Marxism; Japanese Marxism
Marx, Karl and Friedrich Engels *The Communist Manifesto* 40
  Marxian optimal growth theory xiii, 106, 107*fig*, 113–7, 114*t*, 129*fig*, 133, 138, 187, 195–6, 195*fig*, 235–8
  and class 255–61
Marxist economics 1–2, 25–7, 52, 134–5, 155
Mason, Paul 134n
Matsuo, Tadasu 8n

means of production 11, 15–16, 12*fig*, 15*fig*, 17–18, 90, 225–6
meat-eating 225–6
medical services 102–3
Mexico 168n
Miyamoto, Kazuo 202
modes of production 17, 21, 29, 40, 95, 145–6, 202
monetary expression of labor time (MELT) 182–4
money 60–6
Mongolia 23, 220–1
monopolies 162, 190, 193
monopoly capitalism 193
Morishima, Michio 232

Nakamura, Satoru 208–9
Nash equilibrium 32, 32n
nationalization theory 151–5
nature
  labor input in 2–3, 3*fig*, 7–8
  worship of 18–19
Negishi, Takashi 124
New Interpretation 180–4
Nippon Broadcasting 154
Nishida, Masanori 202, 222, 227n
Nixon, Richard 65
nomadic societies 21, 37, 65n, 218t, 219–20, 221n
North Korea 169n

Oakley, Kenneth P. 10n
Obata, Michiaki 167n, 202
Okishio, Nobuo xii, 77, 91, 106, 136, 232
  see also Fundamental Marxism Theorem
Okishio and Nozawa Study Group 152
Onishi, Hiroshi 14n, 27, 117, 189–90
optimal growth theory *see* Marxian optimal growth theory
'oriental despotism' 204

Pasinetti, Luigi 130n
pastoral economies 217, 219–22

INDEX

peasantry 16, 156–8, 158*n*, 209
Petty, William 7
Piketty, Thomas 125–6, 194*n*
Pomeranz, Kenneth 111
population control 200*n*
population growth 204–8
post-capitalist societies 17, 41, 42, 44, 139, 145–6, 151–3
pre-capitalist societies 16–17, 171, 217
primates 9–10
　hunting by 226–7
primitive accumulation of capital 155–60, 158*fig*, 161*n*, 164, 167, 202
primitive communism 21, 69, 100, 220
Prisoner's Dilemma 31–2, 32*fig*
private property 86–7, 216
production 12–14
　specialization in 50, 53
　see also machine-based production; means of production; modes of production; relations of production
profit 66–9, 77, 185–8, 251–4
　falling rate of 136–8
protectionism 161, 163
Protestantism 20–1
public projects
　opposition to 142–4

racism 27*n*
relations of production 6, 29–31, 87
religions 19–20, 26*n*, 38–9
　as tool of ruling class 29
Renfrew, Colin 215
rent 185, 189–93, 190*n*, 194–8, 194*n*, 197–8
　absolute and differential 189*fig*, 191*fig*, 203, 203*n*
　monopoly rent 190, 190*n*
reproduction scheme 112, 115–6, 122, 177, 180*t*, 246–54, 246*t*, 247*t*
retail trade 103–4
Rightism 27*n*
Robinson, Joan 130–2*n*
Roxiangul, Wuful and Ryo Kanae 204*n*

ruled and ruling class 26, 29–36, 32*t*, 33*t*, 34*t*, 169
　uniting of ruled class 33–4
Russian Revolution 35

Samuelson, Paul 130–2*n*
scavenger hypothesis 226*n*
school system 96–8, 104
Schumpeter, Joseph 139, 140
Scott, James 202
Scythia 218–21
sedentarization 202, 214
Sekine, Jun'ichi 94*n*
self-interest 38–9, 193*4n*
self-sufficient economies 47–8, 53
selfishness 39, 42, 44
serfdom 16, 156*n*, 171–2, 208–11, 220–1
service industries 52–3, 146
shamans 19, 215
Shareholder Ombudsman movement 152*n*
shareholder rights 152–5
Shibata-Okishio theorem 136–8
shipping industry 52
shops 103–4
simultaneous single-system interpretation (SSSI) 183
skilled and unskilled labor 85, 96–8
slavery 16, 69–70, 159, 161–2, 163, 208–11, 209*n*, 220–1
Smith, Adam 4, 50
socialism 139
software-based society 146*n*
soldiers, status of 215
social leaders 19
social planning model 241–3
Soviet Union 17, 86, 151, 164, 168
Stanford, Craig Britton 226, 226*n*, 228
state capitalism 17, 82–3
　transition to private capitalism 166–70, 166*fig*
state serfdom 210–1, 221
state slavery 82*n*, 209–10, 216
state violence 157
Statute of Apprentices (England, 1563) 160

291

Statute of Laborers (England, 1349) 92
steady state 117–8, 120, 121*fig*, 122, 134, 139
Steedman, Ian 133, 231
stock markets 153–4
Stout, Dietrich 10n
supermarkets 103–4
superstructure 18, 22–4, 30–1, 35–8, 200
surplus population 106, 135–6
surplus value xi, 72, 90–4, 116–7, 124, 171–2
  absolute and relative 94
  distribution of 177
  special 193
symbolic systems 213–5

Tabata, Minoru 44n
taxation system 162–3
Tazoe, Atsushi 139n, 140n, 146n
temporal single-system interpretation (TSSI) 124n, 183–4
Terasawa, Kaoru 213–4
textile industry 21
Thailand 201
Tibet
  Buddhism 26n, 29
  serfdom and slavery in 166–7, 209n
time preference 112, 124, 148–9
tools and tool-making 9–11, 84, 217, 226
  iron and bronze 217–8
trade unions 134, 160
Trans-Pacific Partnership (CTTP) 23
Tsutsumi, Mika 159n

uklad 82
unemployment 135, 136
United States 83, 159n
  99 percent movement 142
  borrowing by 162
  dollar and gold standard 66
  economy 141–2
  scalping of Native Americans 162
  Three Strikes Law 159n

Unoist economics xii, 81
unskilled labor *see* skilled and unskilled labor
use value 8, 8n, 57
usury 187–8n
Uzawa, Hirofumi 112

vagrancy 157–60
  punishments for 159–60
Vagrancy Act (England, 1547) 159
value 6–8
  see also exchange value; surplus value; use value
Van, Henry 226n
Vietnam 26, 168, 169
violence 164

wages 91
  regulation of 160
  subsistence wage 85n
Walras, Léon 80
wealth 216
Weber, Max 20
Wittfogel, Karl 204
women, and work 23, 98
working class 23, 24–5, 91
  relations with capitalist class xii, 134, 144–5
working conditions 85
  regulations of 91–2, 160

Yamamoto, Shichihei 21
Yamashita, Yuuho 165n, 259, 260n
Yasuda, Yoshinori 222n
Yoshihara, Naoki 77n
Yoshino River dam 142
Yugoslavia 17

Zen Buddhism 21
zero growth 139–42, 144, 149
Zionism 27n
Zoroastrianism 18n